Ocean Engineering & Oceanography

Volume 3

Series editors

Manhar R. Dhanak, Florida Atlantic University, Boca Raton, USA
Nikolas I. Xiros, New Orleans, USA

More information about this series at http://www.springer.com/series/10524

C.M. Wang · B.T. Wang
Editors

Large Floating Structures

Technological Advances

Springer

Editors
C.M. Wang
Engineering Science Programme, Department of Civil and Environmental Engineering
National University of Singapore
Singapore
Singapore

B.T. Wang
Allens
Sydney, NSW
Australia

ISSN 2194-6396 ISSN 2194-640X (electronic)
ISBN 978-981-10-1342-3 ISBN 978-981-287-137-4 (eBook)
DOI 10.1007/978-981-287-137-4

Springer Singapore Heidelberg New York Dordrecht London

Printed on acid-free paper

Springer is part of Springer Science+Business Media (www.springer.com)

Preface

The recent decades have been marked by an unprecedented acceleration in the growth of the world's population. Now standing at over seven billion people, the world's population is expected to rise in this fashion until most countries become fully developed. The need to accommodate this expanding population through new spaces for habitation, work, infrastructure, recreation, storage, and food production; and the necessities of exploiting land-locked resources have increased pressure on governments to release and rezone near-city land-parcels for urban expansion. Ocean space colonisation is one way in which engineers, architects, and urban planners have been engaging with the challenge of providing more space and energy resources for people.

This book surveys key projects that have seen the construction of large floating structures or have attained detailed conceptual designs. These projects add valuable vision to the existing discourse, and include:

- Floating performance stage at the Marina Bay, Singapore
- Yumemai floating swing arch bridge of Osaka, Japan
- Floating oil storage bases in Kamigoto and Shirashima islands, Japan
- Ujina floating ferry pier and Kan-on breakwater, Japan
- Floating offshore wind turbine in Nagasaki, Japan
- Large marine concrete structures in the North Sea, Norway
- Mega-Float in Tokyo Bay, Japan
- Large spar drilling and production platforms
- OTEC platforms
- Mobile offshore base
- Lilypad—a floating ecopolis

This compilation of key floating structures in a single volume captures the innovative features that mark the technological advances made in this field of engineering, and will provide a useful reference for ideas, analysis, design, and construction of these unique and emerging urban projects to offshore and marine engineers, urban planners, architects, and students.

C.M. Wang
B.T. Wang

Contents

Contributors

Vincent Callebaut Vincent Callebaut Architectures, Paris, France

M. Fujikubo Department of Naval Architecture and Ocean Engineering, Graduate School of Engineering, Osaka University, Osaka, Japan

J. Halkyard John Halkyard and Associates, Offshore Engineering Consultants, Houston, TX, USA

E. Inui Fuyo Ocean Development and Engineering Co Ltd, Taito-Ku, Tokyo, Japan

S. Ishida National Maritime Research Institute, Mitaka, Tokyo, Japan

H.S. Koh Naval Systems Programme Centre, Defence Science and Technology Agency, Singapore, Singapore

Tadasu Kusaka Former Chuden Engineering Consultants Co. Ltd, Higashi-Hiroshima, Japan

Y.B. Lim Land Systems Programme Centre, Defence Science and Technology Agency, Singapore, Singapore

T. Maruyama Komaihaltec Inc., Osaka, Japan

Anders Myhr Dr.techn.Olav Olsen, Lysaker, Norway

Tor Ole Olsen Dr.techn.Olav Olsen, Lysaker, Norway

Paul Palo Naval Facilities Engineering and Expeditionary Warfare Center, Cataumet, MA, USA

I. Sato Toda Corporation, Chuo-Ku, Tokyo, Japan

T. Shiraishi Hitachi Ltd, Hitachi, Ibaragi, Japan

H. Suzuki Department of Ocean Technology, Policy, and Environment, Graduate School of Frontier Sciences, University of Tokyo, Chiba, Japan

H. Tanaka Yoshida-Gumi, Osaka, Japan

S. Ueda IDEA Consultants Inc. and Tottori University, Tokyo and Tottori, Japan

S. Ueda Institute of Environment Informatics, Yokohama, Japan

T. Utsunomiya Department of Ocean Energy Resources, Kyushu University, Fukuoka, Japan

B.T. Wang Allens, Sydney, NSW, Australia

C.M. Wang Engineering Science Programme and Department of Civil and Environmental Engineering, National University of Singapore, Kent Ridge, Singapore

E. Watanabe Osaka Regional Planning Institute and Kyoto University, Toyonaka, Osaka and Kyoto, Japan

Olav Weider Dr.techn.Olav Olsen, Lysaker, Norway

A.A. Yee Precast Design Consultants Pte Ltd, Singapore, Singapore; Yee Precast Design Group Ltd., USA, Honolulu, HI, USA

About the Editors

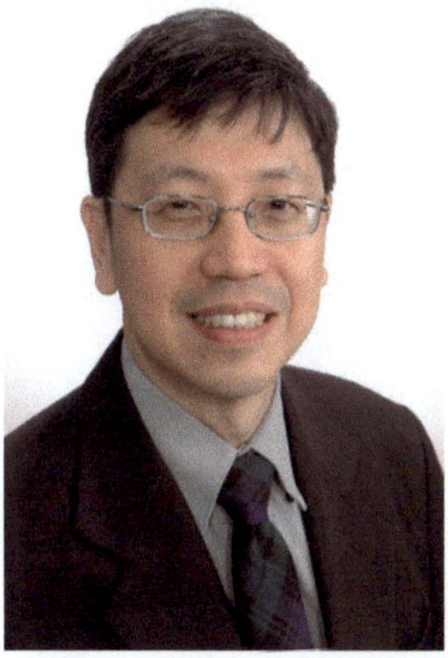

C.M. Wang is a Professor of Civil and Environmental Engineering at the National University of Singapore (NUS). He received his Ph.D. from Monash University in 1982. Currently, he is the Director for the Engineering Science Programme and the Director of the Global Engineering Programme, at NUS. He is also an Adjunct Professor in the Department of Civil Engineering, Monash University, Australia. He is a Chartered Structural Engineer, a Fellow of the Academy of Engineering Singapore, a Fellow of the Institution of Engineers Singapore and a Fellow of the Institution of Structural Engineers. He is presently the Chairman of the IStructE Singapore Regional Group. Professor Wang's research interests are in the areas of structural stability, vibration, optimization, nanostructures and very large floating structures. He has published over 400 technical papers and seven books on the aforementioned areas. His recent awards include the Keith Eaton Award 2014, the Grand Prize in the Next Generation Container Port Challenge 2013, the IES Prestigious Engineering Achievement Award 2013, the IES/IStructE Best Paper Award 2010 and the Lewis Kent Award 2009.

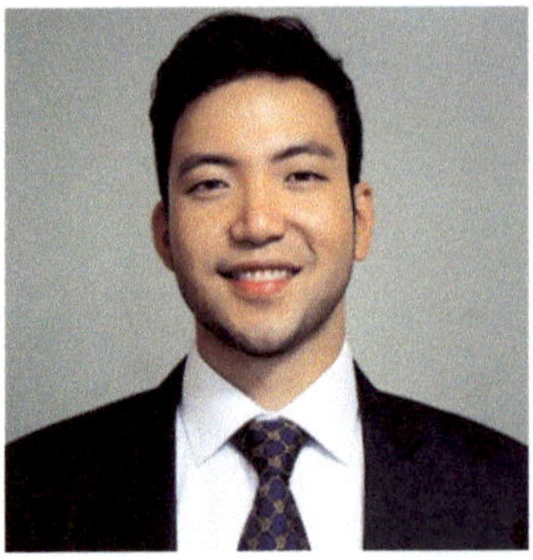

B.T. Wang is a lawyer at Allens, Sydney, Australia. He holds a Master of Laws (Juris Doctor) and a Master of Public Policy and Management from Monash University, as well as a Bachelor of Architecture with Honours from the University of Melbourne. He was a Public Transport Authority CEO Scholar, a finalist in the 2014 Australian Law Awards, and has been a Project Manager with the Infrastructure Delivery team within the State Government of Western Australia. Having worked across a number of sectors including Rail, Defence, Heritage, and Education, his projects have received awards such as the Master Builders Excellence in Construction Award and National Sustainability Award. His research interests lie in the intersection of Public Policy, Law and Infrastructure, and he is drawn to new ways of conceptualizing the city, such as through Ocean Space Colonisation and Technologies.

Great Ideas Float to the Top

C.M. Wang and B.T. Wang

Abstract Coastal land pressures due to urban population growth have seen various cities embrace the use of floating structures to extend their cities into their surrounding ocean space. Simultaneously, the need to go further into deeper waters to extract resources has also driven the development of innovative floating structures. This chapter introduces the developments in the area of predominantly static large scale floating structures. It begins with a review of the advantages of employing floating structures over traditional land reclamation technologies in availing habitable space; and the basic technical elements that underpin the technology. A brief history of the evolution of floating structures is then presented, before a survey of the examples of floating structures in a variety of uses and sites is ventured. Finally, the authors conclude with the ambitious and possible future uses of these large-scale floating structures in meeting our coming urban needs.

C.M. Wang (✉)
Engineering Science Programme and Department of Civil and Environmental Engineering, National University of Singapore, Block EA, #06–10, 9 Engineering Drive 1, Kent Ridge 117575 Singapore
e-mail: ceewcm@nus.edu.sg

B.T. Wang
Allens, Level 28, Deutsche Bank Place, 126 Phillip Street, Sydney, NSW 2000 Australia
e-mail: brydon.wang@allens.com.au

C.M. Wang and B.T. Wang (eds.), *Large Floating Structures*,
Ocean Engineering & Oceanography 3,
DOI 10.1007/978-981-287-137-4_1

1 The Ocean as Our New Home

The increasing global phenomenon of coastal land pressure highlights the ongoing historical attraction to living close to the ocean. Given the unprecedented acceleration in the growth of urban populations, the consequent development towards and along coastlines has produced challenges as urban planners engage with the lack of space and demands on amenities. The chapters in this book offer the 'near shore' and adjacent ocean space[1] as potential frontiers for coastal expansion of cities to alleviate these urban planning demands, as well as document the technological innovations that have permitted deep ocean installations.

The significant advantage of shifting development to the adjacent offshore area is the potential to reduce the impact of urban development on inland resources such as arable land, green belts and other nature reserves. This shifts the current discourse on balancing urban population growth and its accompanying urban encroachment to one which looks at a sensitive colonisation of adjacent offshore areas to meet demands for land, energy and food. Ocean space colonization is one way in which engineers, architects and urban planners have been engaging with the challenge of providing more space, minerals and energy resources for people.

The selection of projects presented explores the use of large-scale floating structures in innovative and environmentally-sensitive ways. These projects have been constructed or have matured to detailed design phases that are available for exploitation. They include:

- Floating performance stage at the Marina Bay, Singapore
- Yumemai floating swing arch bridge of Osaka, Japan
- Floating oil storage bases in Kamigoto and Shirashima islands, Japan
- Ujina floating ferry pier and Kan-on breakwater, Japan
- Floating spar-type wind turbine of the coast of Nagasaki, Japan
- Mega-float at Tokyo Bay, Japan
- Large marine concrete structures in the North Sea
- Large spar drilling and production platforms
- OTEC platform
- Mobile offshore base
- Lilypad—a floating ecopolis.

This compilation of key floating structures in a single volume captures the innovative features that mark the technological advances made in this field of engineering. It aims to be a useful reference for ideas, analysis, design, and construction of these unique and emerging urban projects to offshore and marine engineers, urban planners, architects and students.

In this first chapter, we introduce the developments in the area of predominantly static large-scale floating structures, a category which excludes the equally

[1] The authors in this book use 'ocean space' in a broad manner to encompass the surface of large water bodies adjacent to urban centres.

complex and dynamic field of marine vessels. We begin by highlighting the advantages of utilising such structures over traditional land reclamation technologies in availing habitable space and the basic technical elements that underpin the technology. We subsequently explore the history and evolution of floating structures, surveying examples of floating structures in a variety of uses and sites, before presenting future ambitions of the applications of these large-scale floating structures.

2 Advantages of Floating Structures over Land Reclamation

The use of large-scale floating structures offers significant advantages over land reclamation. These include:

2.1 Rapid Deployment and Ease of Construction

Reclamation works often require a consolidation time of 2–5 years in order for the reclaimed land to settle before construction occurs. The period of consolidation increases by several years if the fill material used is marine clay pumped from the seabed. In contrast, floating structures may be occupied and utilized as soon as they are assembled. Given the modular nature of these structures, they are relatively easy to assemble to the desired built-form (Koekoek 2010) and the construction process can be undertaken in multiple locations to achieve desired construction deadlines. When completed, these individual structural parts can be installed with minimal on-site work, significantly reducing the development period and the return on investment period.

However, the authors observe that technology to build these structures requires a higher skills-base and if employed in a country without the same technical expertise or access to manufacturing facilities as, say, Japan, the cost and construction time for the individual structural components may well make the employment of such floating structures uneconomical. The authors observe that it took less than four months to construct the 1-km long Mega-Float, a floating test runway in the bay of Tokyo (see Fig. 1 and Chap. 8).

2.2 Flexibility in Deployment

Floating structures can easily be relocated and reassembled to their new desired built-form. The modular nature of floating structures also affords the superior advantages of mobility and flexibility that are not inherent in their reclaimed

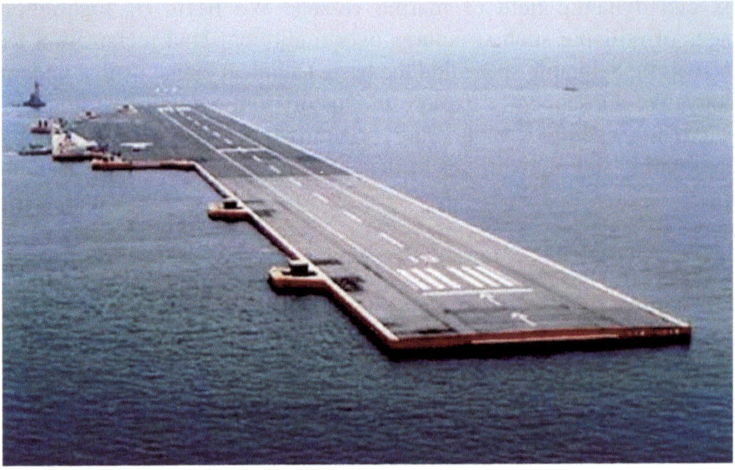

Fig. 1 Floating runway, Japan.

counterparts. Where necessary, floating facilities can be removed if they become obsolete, towed and sunk as artificial reefs, or expanded and grouped with other floating structures as needed. Japan has been a clear leader in demonstrating the effectiveness of this technology in the various iterations of floating dormitories and plant barges that have been built in Japan and subsequently towed and installed in different jurisdictions (Hara et al. 2004; Suzuki 2005).

2.3 Suitable for Deep Water or Soft Sea Bed Conditions

Reclamation works are less cost effective where they are undertaken in deep water or soft sea bed conditions. Such conditions require the use of significant foundation works to transfer loads adequately to the deep hard bed rock. In contrast, floating structures require much less foundation work, depending on a mooring system for anchorage. Free buoyancy force takes care of the gravity loads.

2.4 Relationship to Water Surface

The mooring system of a floating structure permits free vertical movement of the structure to follow the changing sea level and to cater to different payloads. The surface of a floating structure thus remains a constant distance from the water surface. This is particularly desirable in installations such as marinas, docks, and piers where the berthing of boats and ships will be problematic where there are large tidal variations.

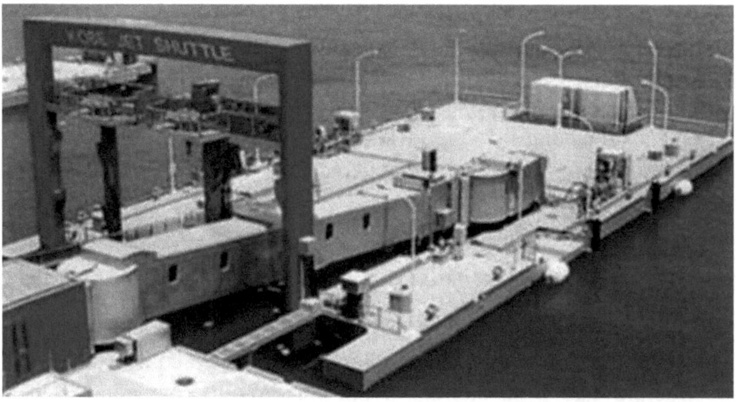

Fig. 2 Floating K-CAT terminal in Kobe, Japan, which connects mega float to shore (*Photo courtesy* Look at the sea. The Floating Structures Association of Japan)

2.5 Earthquake Resistance

Floating structures are protected from seismic shocks since they are inherently base isolated, granting a higher measure of safety to people and structures floating on the surface of the water body. For example, the floating *K-CAT terminal* in Kobe (see Fig. 2), a pier for high-speed shuttle boats linking the *Kansai International Airport* and Kobe Port Island, suffered no damage in the Great Hanshin Earthquake of 1995, which measured 7.2 on the Richter scale. The inherent flexibility of deployment and base isolation, makes floating structures ideal for emergency bases in earthquake conditions where follow-up tremors may affect the disaster site.

2.6 Less Disruptive to Environment

Floating structures are less disruptive to the environment, particularly if they are designed with adequate cut-outs in their footprints to permit sunlight to penetrate below the ocean surface. Reclamation works destroy the immediate marine ecology below its footprint and may disrupt ocean currents or increase sedimentation, creating unintended effects on neighbouring coastal environments.[2]

The environmental impacts of reclamation works can also trigger international tension. In 2002, Malaysia criticised Singapore's land reclamation works, asserting that the doubling of two islands (Pulau Tekong Besar and Pulau Tekong Kecil) in the Johor Straits would constrict current flow in the area, create stronger currents

[2] "Riau sand exports take toll on environment", Haidir Anwar, *The Jakarta Post*, July 3, 2001.

and tidal movements that would increase sedimentation and block river discharge. Malaysia asserted that this would put settlements along the river at risk of flood.[3] Further, it was alleged that the reclamation works were depleting coral reef cover, thereby damaging fish stocks and the livelihoods of local fishermen.[4] There have also been suggestions that the degraded water quality from reclamation works has contributed to deaths in the local endangered dugong populations.[5]

2.7 Modular Units Provide Extra Floor Space

Floating installations are often an assemblage of watertight compartments. Where a floating structure has a substantial depth due to environmental or structural requirements, these watertight compartments can be re-purposed to open up extra floor space. These interior spaces can be re-modelled and utilised as car parks, offices and storage rooms. The floating breakwater in Monaco, also called Monaco seawall, (see Fig. 3) is a double hulled precast structure which houses a 380-lot car park and dry dock for recreational craft.[6] This offers a significant advantage over land reclamation given the ease in ability to extend habitable space below the water line.

3 Science Behind Floating Structures

3.1 Types of Floating Structures

Large-scale floating structures are broadly classified as either pontoons or semi-submersibles. Pontoons are essentially floating slabs that are characterised by their low depth-to-width ratios and are usually deployed in a benign sea state condition such as in waters adjacent to the coast, inside a cove or a lagoon, or where breakwaters and other protective installations can be constructed to protect the structure from large waves and swells. In order to restrain lateral movements, they may be anchored to the sea bed through the use of mooring lines comprising chains, ropes, sinkers, anchors or tethers. Where greater restraint is necessary, the pier/quay wall method or the dolphin-frame guide mooring system may be adopted. Pontoons are generally cost effective with low manufacturing costs and are relatively easy to repair and maintain. Figure 4 shows the main components of

---

[3] "Flood threat from Singapore alleged", *The Straits Times*, March 21, 2002.

[4] "Singapore reclamation works 'hurt Johor fishing", *The Straits Times*, July 9, 2003.

[5] "Local media absorbed by land reclamation issue", *The Straits Times*, Singapore, March 13, 2002.

[6] "Time And Money Are No Object for Harbour Upgrade in Monaco", *Peter Reina,* www.enr.constrcution.com, June 3, 2002.

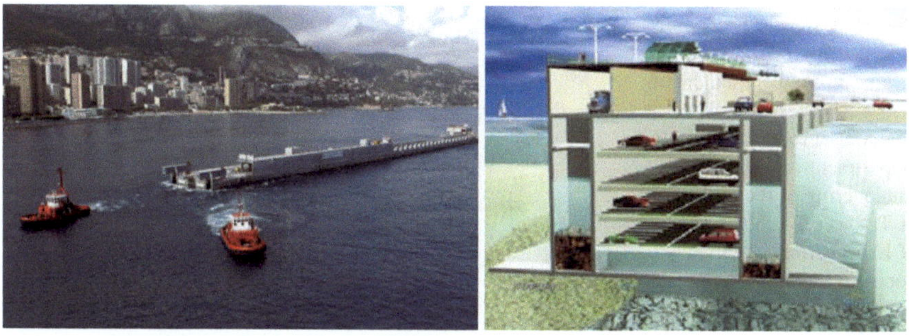

Fig. 3 Floating breakwater, Monaco seawall, in Monaco (*Photo courtesy* FCC construction, http://enr.construction.com/news/environment/archives/020603.asp)

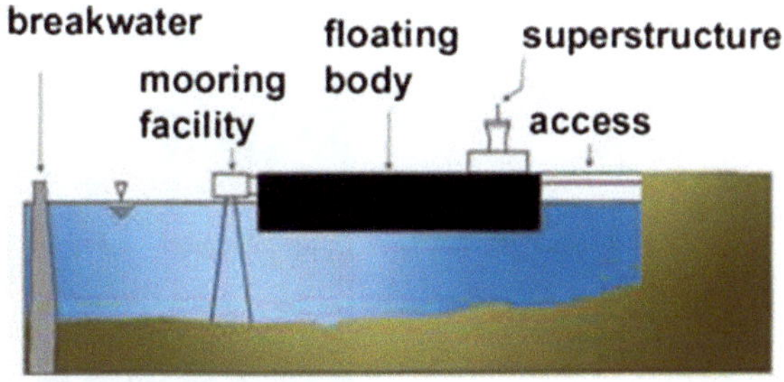

Fig. 4 A pontoon VLFS with associated components

a pontoon: the floating body, the mooring facility, the access bridge/gangway/linkway or floating road and the breakwater.

Semi-submersibles have a structural form that is thicker than their pontoon counterparts and have been deployed in deep sea environments since as early as the 1970s. They are partly raised above the sea level using column tubes or watertight ballast structural compartments at the base to compensate for the effects of larger wave-lengths and -heights. Floating oil drilling platforms and semi-submersible type floating wind farms (see Fig. 5) are typical examples of this category of floating structures.

While semi-submersibles are often towed, some models are hybrids between a pure static floating structure and a marine vessel with their own propulsion system for short-range transport. When deployed to the required location, these floating structures are then attached to the seabed using mooring cables or tethers.

3.2 Movements, Stability and Forces on Floating Structures

A floating structure works on Archimedes' principle of buoyancy. The buoyant force is an upward thrust that opposes the weight of the immersed structure with the magnitude equal to the weight of the fluid displaced by the structure. The draft,

Fig. 5 A semi-submersible type floating offshore wind turbine in Fukushima, Japan (*Photo courtesy* Takeshi Ishihara—Professor in the Department of Civil Engineering at the University of Tokyo's Graduate School of Engineering—"Japan hopes to blow ahead in renewable with floating wind farms", www.japantimes.co.jp)

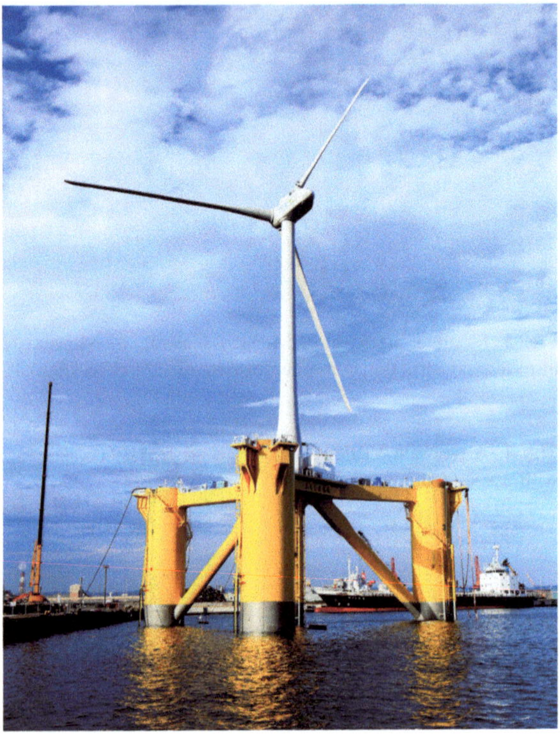

or the depth of the portion of the structure submerged, depends on its weight and geometry. The structure can also be idealized as a rigid beam or plate resting on a continuous elastic Winkler foundation that models the buoyancy force.

Consider a floating structure as shown in Fig. 6. It is susceptible to translational movements and rotations in all the three directions when loads are applied, i.e. it has six degrees of freedom, namely:

1. Surge: translation in the x-direction (forward and backward)
2. Sway: translation in the y-direction (side to side)
3. Heave: translation in the z-direction (up and down)
4. Roll: rotation about longitudinal x-axis
5. Pitch: rotation about transverse y-axis
6. Yaw: rotation about vertical z-axis.

These degrees of freedom may be restricted by appropriate mooring. Where a floating structure is moored, the sway, surge and yaw of the structure can often be effectively neglected. The buoyancy force acting from below the structure stabilises heaving movements and oscillations caused by gravity and dynamic loads. However, eccentric loads and/or moments can cause roll and pitch movements in the floating structure.

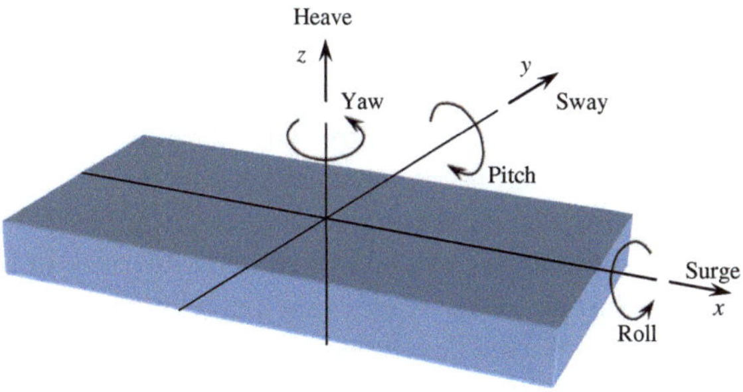

Fig. 6 Types of motions of a floating structure

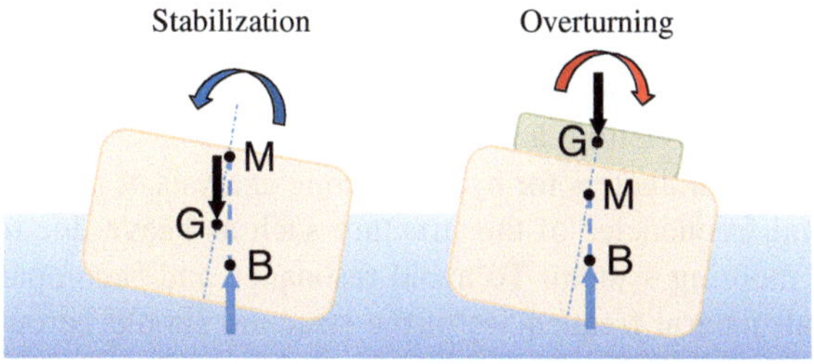

Fig. 7 Stabilization and overturning of floating body. G = centre of gravity, B = centre of buoyancy, M = metacentre

One of the key design criteria is the static stability requirement of a floating structure. Where a structure is subject to eccentric loads and tilts with one of its sides rotating deeper into the water, the counteracting buoyancy force acting to restore the structure to its original position creates a 'bob'. To prevent such movements, a tilt by permanent loads can be avoided by increasing the depth of the structure at the point subject to the load (increasing the draft of the structure) or placing counteracting loads, such as ballast.

The tilt and stability of the structure can be explained in terms of its metacentric height (GM) which is the distance between the centre of gravity (G) and metacentre (M) of the structure as shown in Fig. 7. The metacentre is the point of intersection of the imaginary vertical line passing through centres of gravity and buoyancy and a vertical line passing through the new centre of buoyancy when the structure is tilted.

A small GM increases the risk of capsizing of the structure as the centre of buoyancy is not shifted much and the small upright moment due to buoyancy may not be sufficient to counter the applied load or moment. Therefore an optimum value ensuring safety and serviceability is adopted. For the floating structure to be stable, its centre of gravity should be below the metacentre and the metacentric height should be large. As it can be seen in Fig. 7, a negative metacentric height (i.e., when G is above M) would capsize the floating body. It is thus desirable that G is located as low as possible, which is described as *weight stability*. In ships, ballast tanks at the bottom help in lowering the centre of gravity; thereby ensuring stability. Further, the centre of buoyancy (B), the point where the buoyancy force acts, varies as the structure rotates. The more the centre of buoyancy shifts while the structure tilts, the higher the metacentre will be which makes it more stable. This can alternatively be called as *shape stability*.

In addition to the applied loads, floating structures are subjected to wave forces. Waves cause heaving oscillations, roll and pitch and variances in buoyancy along the length and width. Surface water waves may be generated by wind, tidal bore, earthquakes or landslides. Waves developed in an area may endure even after the wind ceases and may propagate to another area as swell with decaying intensity. Long period swell travels a very long distance as long-crested waves. Wind-generated waves consist of a large number of sinusoids of different heights, periods and directions superimposed on one another. Although regular waves are not found in real seas they can closely model some swell conditions. Regular waves are characterized by the wave period and height. They are commonly used to establish wave conditions for modelling and design.

Other important parameters for hydrodynamic analysis of floating structures are the many natural frequencies of the structure such as heave due to buoyancy, or surge due to a mooring system. To avoid resonance and accompanying response enhancement, all natural frequencies of the structure should be outside the bandwidth of wave excitation. Typically, one can anticipate forces associated with the fluid inertia, its weight, viscous stresses and secondary effects such as surface tension or Coriolis force. It is often not possible to include all forces simultaneously in a mathematical model. In order to determine which pair of forces dominates, it is useful to estimate the orders of magnitude of the inertia, the gravity and the viscous forces and moments, separately, allowing the least effective force to be eliminated. Generally viscous effects are often ignored which simplifies the problem considerably. One uses Newton's law to predict the dynamic behavior of structure.

A very large floating structure can be classified as a rigid structure or a flexible structure under wave action (see Fig. 8). Suzuki et al. (1996a), Suzuki and Yoshida (1996b) provide quantifiable criteria to establish whether there is a need for a hydroelastic analysis of a very large floating structure. They point out that length of the structure must be greater than the wavelength as well as the characteristic length. The characteristic length corresponds to the length of the structure which is deflected due to a static concentrated load.

3.3 Design Considerations

The design considerations for a floating structure may be grouped into: (1) elements that satisfy the structural requirements that address the operating conditions, structural strength, serviceability, durability and safety standards; and (2) socio-political criteria that address the aesthetics, environmental sustainability, budgetary and legal constraints. The calibration of a design response to these considerations

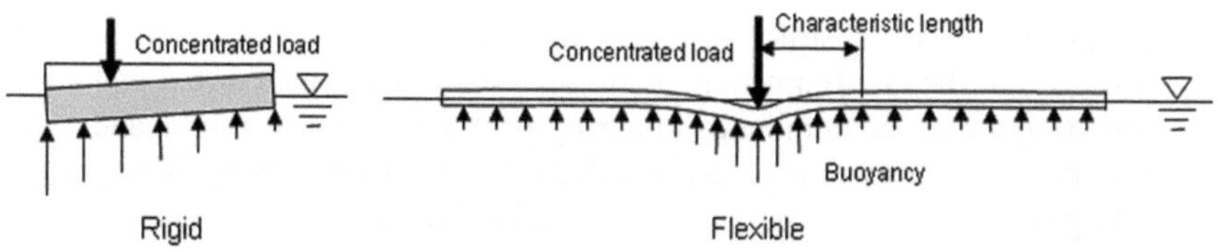

Fig. 8 Rigid and flexible behavior of floating structure

will determine an appropriate design life that caters to the importance of the structure and environmental loads, likely to be between 50 to 100 years, preferably with a low maintenance cost.

The serviceability criterion ensures that the structure fulfils its function as specified by the client. Serviceability requirements include attending to low frequency motion that may cause sea sickness due to continuous heaving, or addressing differential displacement that may render equipment non-operational. Safety requirements ensure that there are no casualties, fatalities, property or environmental damage. Global structural failure, capsizing, sinking, and drifting off-station are failure modes which are taken care of under the safety criteria. Property damage could be acceptable financially, but human loss and environmental damage are permanent effects that are eliminated or minimised under strict design guidelines. Given this need to preserve life, evacuation and rescue are also important design considerations of floating structures.

The analysis and design of floating structures need to have special consideration when compared to land-based structures. These are summarised below (Clauss et al. 1992; Moan 2004):

i. Horizontal forces due to waves are generally several times greater than the (non-seismic) horizontal loads on land-based structures and the effect of such loads depends upon how the structure is connected to the seafloor. A rigid mooring system virtually prevents the horizontal motion while a compliant mooring will allow maximum horizontal motions of a floating structure of the order of the wave amplitude.
ii. In a floating structure, the static selfweight and payloads are carried by the buoyancy force of the water body. As such, there is no need for vertical supporting foundation as opposed to land-based structures. However, the mooring system has to be carefully designed to keep the floating structure in position even if the forces in the mooring system are small. This is due to possible displacement arising from slow-drift wave forces as well as steady current and wind forces. If a floating structure has a compliant mooring system, such as catenary chain mooring lines, the horizontal wave forces are balanced by inertia forces. Where the horizontal size of the structure is larger than the wave length, the resultant horizontal forces will be reduced given that different phases (direction and size) of the wave force will act on various parts of the structure, resulting in smaller forces in the mooring system relative to the total wave force.

iii. Sizing of the floating structure and its mooring system depends on its function and also on the environmental conditions, such as waves, current and wind. The design may be dominated either by peak loading due to permanent and variable loads or by fatigue strength due to cyclic wave loading. Moreover, it is important to consider possible accidental events such as ship impacts and to ensure that the overall safety is not threatened by a possible progressive failure induced by such damage.
iv. Unlike land-based constructions with their associated foundations poured in place, very large floating structures are usually constructed at shore-based building sites remote from the deep water installation area and without extensive preparation of the foundation. Each module must be capable of floating so that they can be floated to the site and assembled in the sea.
v. Owing to the corrosive sea environment, floating structures have to be provided with a good corrosion protection system.
vi. Possible degradation due to corrosion or crack growth (fatigue) requires a proper system for inspection, monitoring, maintenance and repair during use.

3.4 Materials

In the old days, the most successful type of material for floating structures was made of vulcanized rubber. They were lighter, easier to move and could be easily repaired and replaced. For larger structures to carry heavy loads, wood and canvas were adopted. But they have become obsolete and are seldom used today.

In present times, the most popular materials in use are steel, concrete, steel-concrete composite, advanced concrete, and plastics. Since water tightness of concrete is important to avoid or limit corrosion of the reinforcement, either watertight concrete or offshore concrete should be used. High-performance concrete containing fly ash and silica fume is most suitable in corrosive salt water (Hosam et al. 2010). The world's longest floating bridges are made of specialized reinforced concrete. The Evergreen Point floating bridge in Seattle, currently the longest pontoon bridge, used nearly a quarter of a million tonnes of concrete. The effects of creep and shrinkage are considered only when the pontoons are dry, i.e. during their construction and transportation, and hence not considered once they are launched in water. For example, in the Evergreen Point bridge, in order to maintain the structural integrity of the pontoon, the bottom slab of the pontoon was re-heated when its walls were being poured, so that both elements could cool and shrink together, making it 'virtually crack proof' (SR520 program, Washington State Department of Transportation).

The strength, light-weight and flexibility of steel make it a desirable material for floating structures. The steel used must satisfy the appropriate standard specifications (such as the Technological Standard and Commentary of Port and Harbor Facilities 1999). The world's largest floating platform at Marina Bay, Singapore is

120 m long, 83 m wide and 1.2 m depth. It consists of 15 individual steel pontoons, 12 mm thick at top and 8 mm thick at bottom, connected to each other via diamond shaped connectors kept in place by steel pins (Koh and Lim 2008). The pontoon units are easily portable and can be reconfigured into desired shape, which can be attributed to the light weight of steel. Materials like High Density Polyethylene (HDPE) are also promising materials which have emerged recently. HDPE is strong, durable, light and UV resistant, and hence very suitable for long-term use. HDPE is popularly employed in docks, jetties, parking space for private boats and jets, and walkways. The floating wetlands at Punggol, Singapore are made from an assembly of honeycomb shaped HDPE modules (Wong et al. 2013).

3.5 Mooring Systems and Breakwaters

A mooring (or station keeping) system is used to secure a floating structure to a pier or wharf by keeping it in position under wave and other dynamic actions like drift. Mooring prevents horizontal movements and, to a certain extent, vertical motion. The effect of mooring systems on hydroelastic behaviour of floating structures has been frequently analysed. Operating conditions and environmental factors such as waves, wind forces and depth determine the type of mooring system to be chosen. The most common type of mooring methods include dolphin frame-guide, chain/cable, pier wall, mooring pile, tension leg, and jerk-free moorings (see Fig. 9).

Breakwaters are required for protection of floating structures when sea conditions are harsh. A general rule of thumb is to have a breakwater when the significant height of the wave is greater than 4 m. A breakwater can be of a bottom sloping-type (mounded), vertical type, composite-type or wave-energy dissipating blocks. The most basic sloping type is a multilayered rubble mounded breakwater, sometimes with a fine core. The material for the mound is usually reclaimed sand, rubble and concrete blocks. They can also be modified as access roads to a floating structure. These are susceptible to wave impulse and scouring and are preferred only in shallow waters. Vertical type breakwaters are more suitable for deep water (Takahashi 1996).

In order to reduce the wave reflection and the breaking wave force on the vertical wall, concrete blocks are placed in front of it. Caissons are chosen as vertical walls for their low cost and reliability. However, caissons need a hard seabed. Although breakwaters are very effective in mitigating the response of floating structures, their environmental impacts are severe as they destroy the seabed, affect marine life and interrupt ocean currents. Moreover, the dredging of sand and construction technology involved is expensive and requires a lot of resources. To cope with these limitations, floating breakwaters have been proposed. Oscillating Water Column (OWC) may be embedded in floating breakwaters to extract the wave energy, effectively reducing the hydroelastic response. Maeda et al. (2001) found 50 % reduction in response of a large-scale pontoon unit housing three OWCs. Other anti-

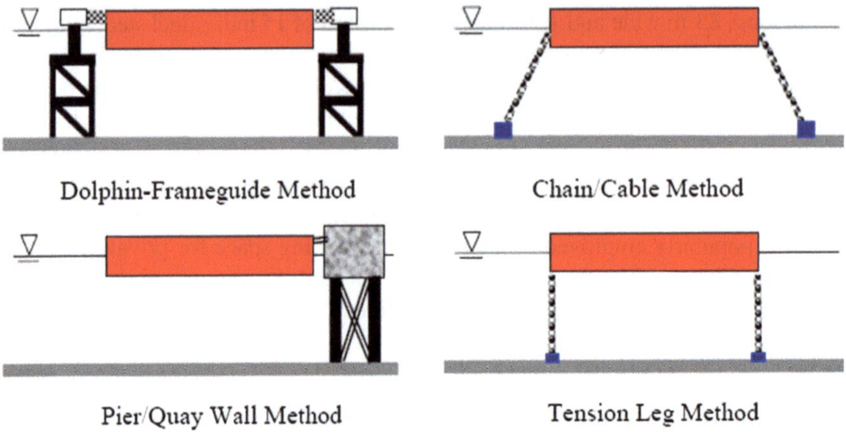

Fig. 9 Mooring systems for pontoon type floating structures

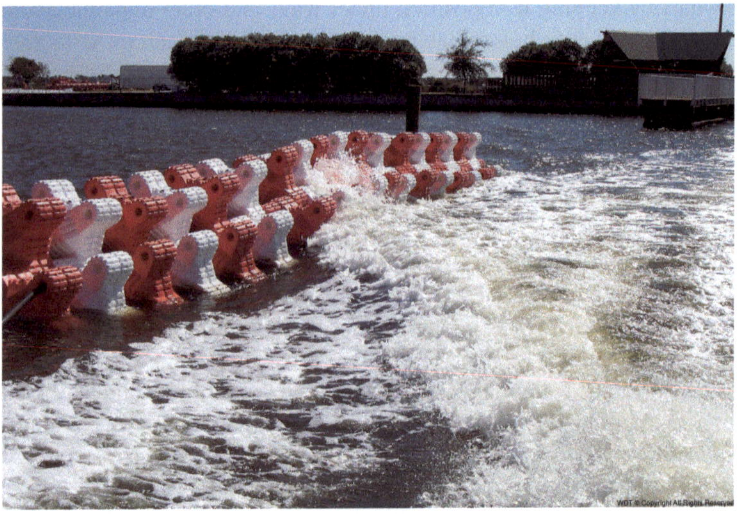

Fig. 10 WhisprWave floating breakwater inspired by Buckminster Fuller (*Photo courtesy* www.whisprwave.com)

motion devices like air cushions not only reduces the motion, but also bending moments and shear forces generated in the pontoon, though it depends greatly on the ratio of length of air cushion unit to wavelength, which could have converse effects if not in the desirable range. Anti-motion devices coupled with OWC can be so effective that breakwaters may be eliminated. Unique modular WhisprWave Floating Breakwater, an advanced version of Buckminster Fuller's floating breakwater that assembles to attenuate wave forces, may permit deployment of large–scale pontoon structures in deeper waters where more permanent breakwater solutions are not feasible (see www.whisprwave.com and Fig. 10).

4 Evolution of Floating Structures

4.1 Floating Villages

Floating human settlements are not a novel concept. Floating villages have thrived in various forms and on a variety of water bodies throughout history. They have appeared in different parts of the world, such as the floating villages in the Cambodian lakes where large communities engage in aquaculture and fishing; or on the Ganvie in Benin, Africa, where the Tofinu tribe adopted the form of a floating village from as early as the 16th Century as a protective measure against slave trading. Beyond these two examples, Vietnam (see Fig. 11), Indonesia, Thailand, China, Peru and Bolivia also have a history of floating villages.

These floating settlements may appear like normal houses and huts or may exist in the form of house boats. The community of Uros resides in the Titicaca Lake, Peru and dwell in houses made from a special kind of reed (totora reed) which grows in that particular lake. Some communities in Aberdeen, Hong Kong also live on the surface of the sea in their boats.

Modern day floating homes can be traced to the 1980s and arose due to scarcity of land and economic concerns. At the time, *International Marine Floatation System Inc.* came up with a new technology of creating real estate on water. This system was based on a core of polystyrene foam and concrete shell, creating a structure that was inherently buoyant and 'unsinkable'. This technology resulted in the formation of large floating clusters of homes in Vancouver (see Fig. 12). A large effort is underway in the Netherlands to embrace floating buildings and houses.

4.2 Floating Bridges

Floating bridges have also enjoyed a prominent position in the history of floating structures. Borne out of necessity, these structural innovations have been employed in cities to connect urban nodes across deep channels and rivers, alleviating traffic pressures on existing crossings. Floating bridges were also built as a necessity of war, permitting the movement of soldiers and equipment between ground bases and marine vessels.

4.2.1 Historical Floating Bridges

Historical floating bridges predominantly consisted of trafficable timber walkways built over an array of boats that were secured together and anchored to the floor of the waterway. These structures were unlikely to have a significant design lifespan and had to be renewed on a regular basis. They were not sturdy nor could they be used for extensive periods in harsh environmental conditions or for heavy loads.

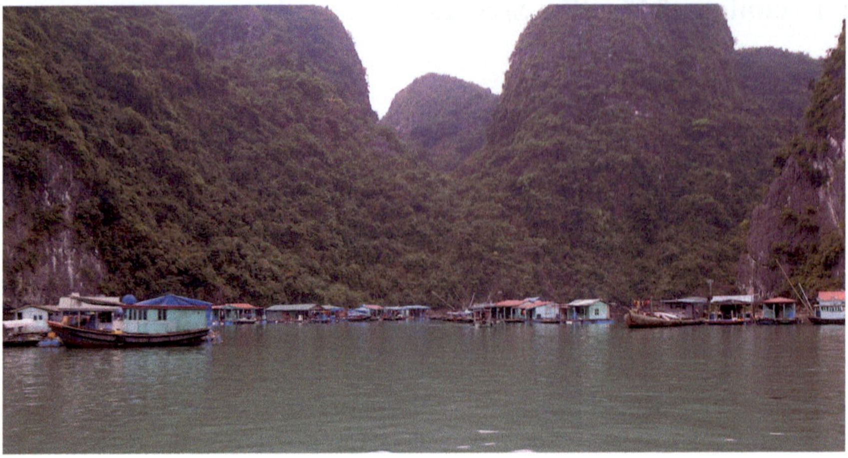

Fig. 11 Floating houses in Ha Long Bay, Vietnam

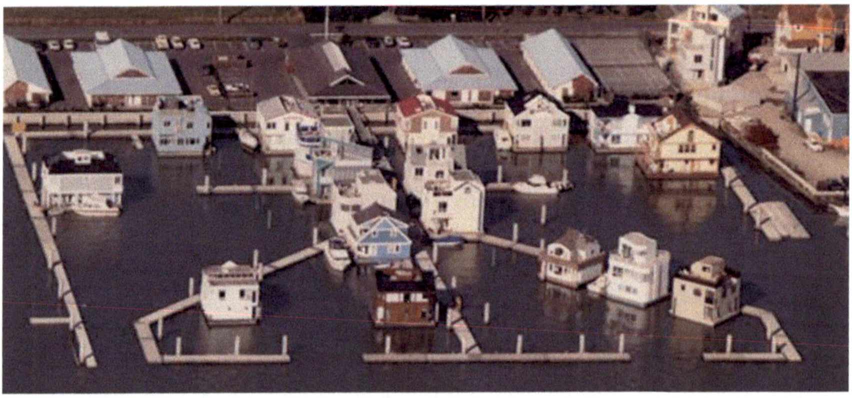

Fig. 12 Floating homes in Canoe Pass Village in Vancouver, Canada (*Photo courtesy* www.floatingstructures.com)

The use of these structures gradually declined as cities grew and the need for more permanent solutions with greater structural capacity became a necessity

One of the earliest floating bridges of note was created in 480 BC during the invasion of Greece by the Persian king Xerxes. The desirability of a maritime assault over an attack via land led to the engineering of a floating bridge that was created over the Dardenelles Strait, Turkey, comprising 300 small ships tied together and anchored at both the ends by large ships (see Fig. 13).

600 years before Xerxes, King Wen of Zhou Dynasty created a pontoon bridge in 11th Century BC to cross rivers. Later, the *Dongjin* Bridge in Eastern China, dating back to 1163–1173 AD was constructed during the Chinese *Song Dynasty*. The 400 m long pontoon bridge was constructed over Zhang and Gong Rivers in

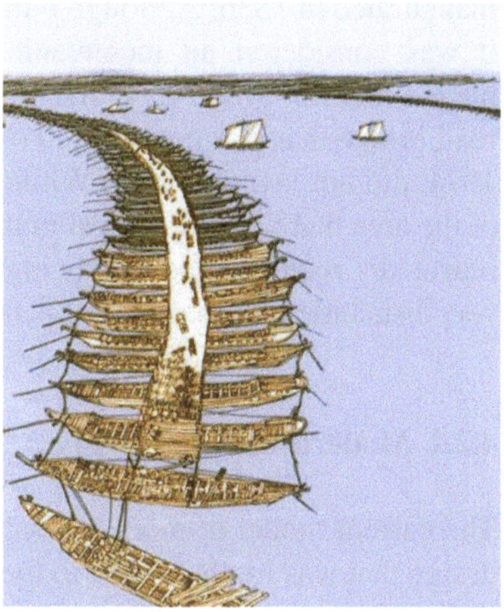

Fig. 13 Floating Xerxes bridge built in 480 B.C

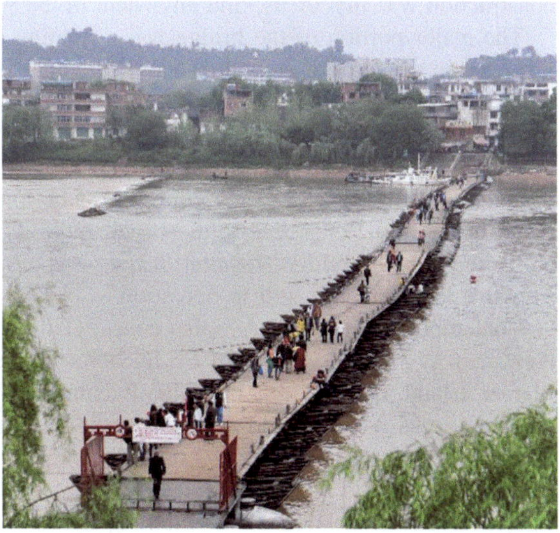

Fig. 14 Floating Dongjin bridge over Gong River, China built in 1163–1173 A.D

the same tradition of wooden planks installed over chain-linked wooden boats. The entire structure is renewed on a regular cycle and remains in operation (see Fig. 14).

A contemporary of the *Dongjin Bridge* was the pontoon bridge of Seville, Spain, built over River Guadalquivir in 1171 AD. It served to connect the cities of Triana and Alcor, Spain and was similarly made of large barges of wood hooked together by iron pins to alleviate the effect of tides. It was repaired and its parts were replaced repeatedly until in 1852 when it was removed.

The ease of deploying boats and setting up timber floating bridges made this structural option popular well into the latter half of the 19th Century, even with the need to continually renew the structure. Some examples persisted well into the early 20th Century. The *Woolston Floating Bridge* over Itchen River in UK was inaugurated in 1836. Although built as a temporary solution to the increased traffic, it was considered an inconvenient traffic route. Despite this, the bridge was maintained for over 140 years. In contrast, the 124-m long floating wooden railroad bridge that was constructed over the Mississippi River in Marquette, Iowa, in 1874, did not last that long. While the structure was integral to the major railyard of the area that was the busiest in the region, the onerous requirements of having to repeatedly rebuild it led to it being finally abandoned in 1961. The pontoon bridge was dismantled and taken to Lacrosse, Wisconsin.

4.2.2 Modern Floating Bridges

The current model of modern floating bridges can be traced to the pontoon bridge design that was implemented in the *Hobart Bridge* in 1943. Connecting the eastern and western parts of Hobart, Australia over the Derwent River, the bridge's unique construction was first of its kind anywhere in the world.

The major portion of the bridge was a floating concrete structure that curved upstream in the form of an arch. This feature enabled it to withstand the forces of wind and current without the use of anchors that had been utilised in traditional forms of floating bridges. The bridge was constructed in 12-pontoon sections that were then towed out into the river and connected to the banks and to each other in the middle. However, it did not enjoy the same longevity as its historic counterparts. In the mid-1950s, due to increased traffic flow, the bridge was dismantled and a new floating bridge, *Tasman Bridge*, was proposed and completed in 1964. This five-lane bridge is still in operation.

Contemporaneously, the US also constructed its first floating concrete bridge, the *Lacey V. Murrow Memorial Bridge* across Lake Washington from Seattle to Mercer Island, Washington in 1940. It comprised a combination of fixed and movable spans which would retract into the fixed spans and give way to waterborne traffic. It was regarded as 'original, distinctive, striking and graceful' and a landmark of Seattle city. It is the second largest floating bridge (at 2020 m) in existence. The *Evergreen Point Floating Bridge* (at 4750 m), and built on the same lake 23 years later, is currently the longest pontoon bridge (see Fig. 15).

Other examples of modern floating bridges include the 931 m long *Bergsøysund Bridge* (see Fig. 16a) and the 1614 m long *Nordhordland Bridge* (see Fig. 16b) in Norway. The *Nordhordland Bridge* is a combined cable stayed and pontoon bridge with large spans and minimal supports. The *Yumemai Bridge* in Japan is the world's largest floating steel arch bridge supported by two large pontoons on either side of its 410 m span (see Fig. 16c and Chap. 3). It connects two reclaimed islands of Japan where high-level bridges or tunnels are not feasible, and is capable of rotating in plan to allow navigation. The floating concrete bridge of Dubai, inaugurated with much fanfare in 2007, is built across the Dubai creek and connects Bur Dubai with the Deira section of Dubai (see Fig. 16d).

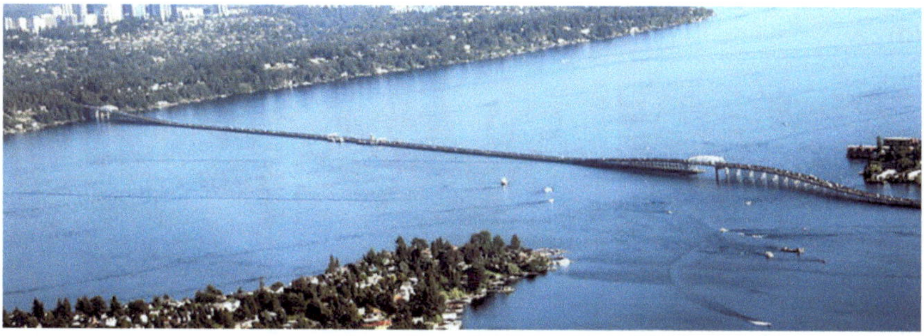

Fig. 15 World's longest floating bridge, Governer Albert D Rosellini bridge (also known as the Evergreen Point Bridge)

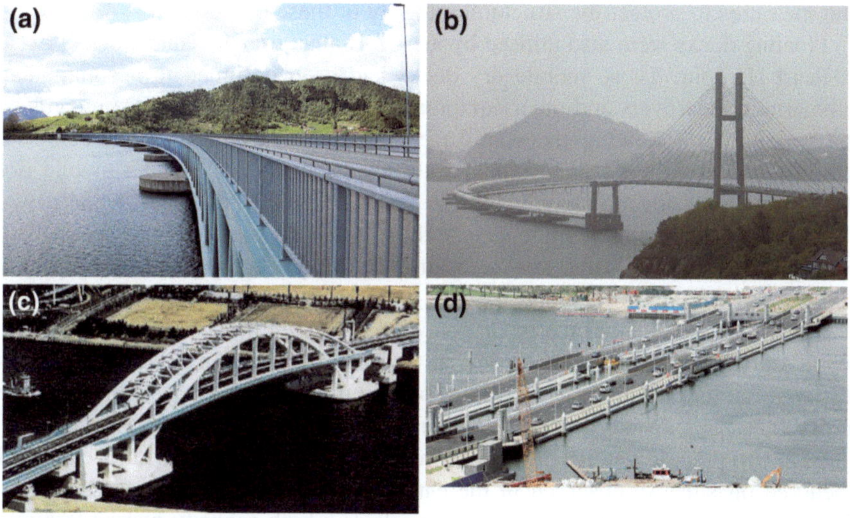

Fig. 16 **a** Bergsøysund bridge (*Photo courtesy* Erik Johan Bjertnæs, http://bridge-info.org/) **b** Nordhordland bridge (*Photo courtesy* Erik Johan Bjertnæs, http://bridge-info.org/) **c** Yumemai Bridge **d** Dubai floating bridge (*Photo courtesy* Shihab, http://www.khaleejtimes.com, July 2, 2013)

4.3 Floating Ports, Docks and Piers

Floating docks came to the fore during World War II. When securing an existing harbour proved impractical, the idea of a movable harbour was conceived. *Mulberry Harbour* was an excellent example of a floating dock employed by the British in World War II (The Hetch Co. 1945). The harbour was designed in three sections: breakwaters, pier head, and walkway from pier to beach (see Fig. 17). Given the need to deploy these floating structures in harsh weather and ocean conditions, there has been much investment in research and development to achieve remarkable

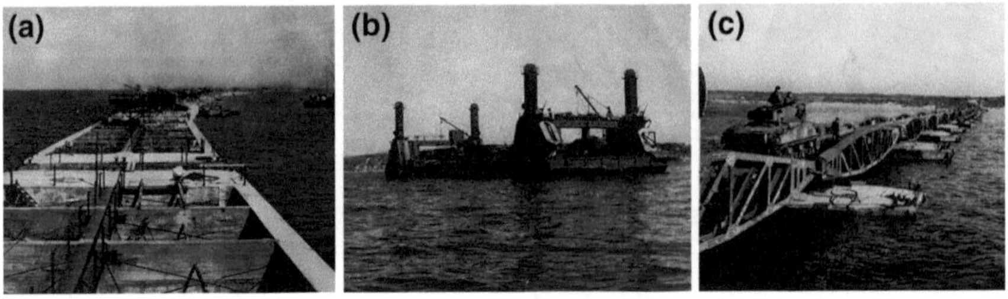

Fig. 17 Mulberry harbour (**a**) breakwater (**b**) pier head (**c**) walkway (*Photo courtesy* The Naval Dept Library, Naval History and Heritage Command, USA, http://www.history.navy.mil/library/online/miracleharbor.htm)

advances in floating structures such as innovations in materials, mooring techniques and measures to protect the structural integrity of these pontoons.

Floating docks were also utilised by the US Bureau of Yards and Docks (1947) at Pearl Harbour. These mobile dry docks had been designed to accommodate submarines, destroyers and other equipment of similar sizes; and were integral to the war efforts following the Kamikaze attacks by Japanese suicide-bombers. While the fleet suffered great damage, the ready availability of the mobile dry docks at nearby advance bases, and the yeoman service rendered by their own crews and the ship repair components at these bases, allowed many of the ships to be saved with minimal time lost to repairs (see Navy Dept online Library, Naval History and Heritage Command, USA).

Modern applications of these concrete docks are seen in Alaska and Japan. The world's first floating prestressed concrete container terminal in Valdez, Alaska[7] is the largest of its kind in the world. It was opened in 1982 and services ships as large as 50,000 tonne-capacity ships on a continual basis. It is a 700 ft long dock, fabricated in two precast water-tight hollow units that were post-tensioned together and anchored on site. As observed in the earlier section on the advantages of these floating structures, the Alaskan facility excels as a high capacity marine terminal given its low maintenance requirements, rapid ability to be deployed, its suitability for use in deep water and poor soil conditions, and most critically, the inherent ability to automatically self-adjust to tidal variations that are critical in a docking scenario.

Japan utilises floating piers in Ujina port at Hiroshima. Built in 1993, these piers allowed the port to extend its waterfront capacity to receive ferry passengers (see Fig. 18 and Chap. 5). Critically, given the Seto Inland Sea has a large tidal change of 4 m, the pontoon berths presented an advantageous ability to match the tides for improved comfort and convenience for small passenger boats. The large scale of these floating piers was adopted to increase stability against rocking by waves, further enhancing the passenger experience.

[7] "Floating Concrete Container Terminal at Valdez, Alaska", *Concrete Construction*, February 1, 1983.

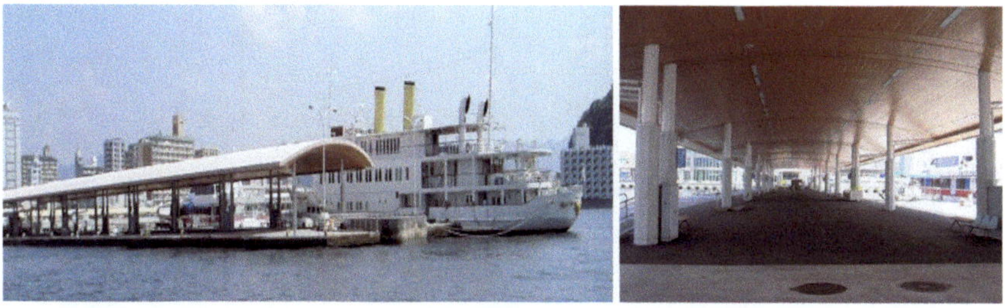

Fig. 18 Ujina floating prestressed concrete pier

4.4 Offshore Floating Structures for Oil Exploration, Extraction and Storage

Another important line of evolution in floating structures is the development of facilities for oil exploration, extraction and storage. Oil has been produced from offshore structures since the late 1940s in facilities that cater to the upstream processes of drilling wells, extraction and processing of the crude oil, as well as temporary storage facilities. There is a variety of oil platforms; ranging from bottom-founded ones to floating platforms to spar type rigs (see Chap. 9).

Soviet construction in the Caspian Sea off the coast of Baku, Azerbaijan, created the first offshore metropolis in the world, Stalin's Atlantis[8] (see Fig. 19). In 1949, Russian engineers discovered a top-quality oil reserve at 1100 m below the seabed while drilling, leading shortly after to the creation of the world's first offshore oil platform, *Neft Dashlari*. A vast web of oil platforms was erected in the middle of the sea to accompany the operations at *Neft Dashlari*, supplying 175 million barrels of crude oil per annum and accounting for 75 % of the country's entire oil production. The development of the area continued with nearly 2000 drilling platforms spreading over a 30 km circle in the sea, joined by a network of bridge viaducts spanning 300 km.

While the initial floating structures associated with the Soviet oil enterprise were fixed platforms and submersible rigs, as oil exploration stretched into deeper environments, the use of semi-submersible floating structures became the only feasible solution. The semi-submersible rig was accidentally invented in 1961 while towing a submersible rig operated on four columns in the Gulf of Mexico.[9] The pontoon used for towing was not buoyant enough to carry the whole structure, which made only half of the structure rise above the water level. The owners of the rig, Blue Water Drilling Company, working for Shell Oil Corporation, noticed that

[8] *"Stalin's Atlantis: This Is What's Left Of The Former Soviet Union's Floating Oil Town", Becket A., November 16, 2012,* www.theblaze.com.

[9] "Offshore Drilling: History and Overview", www.offshoreenergytoday.com, June 25, 2010.

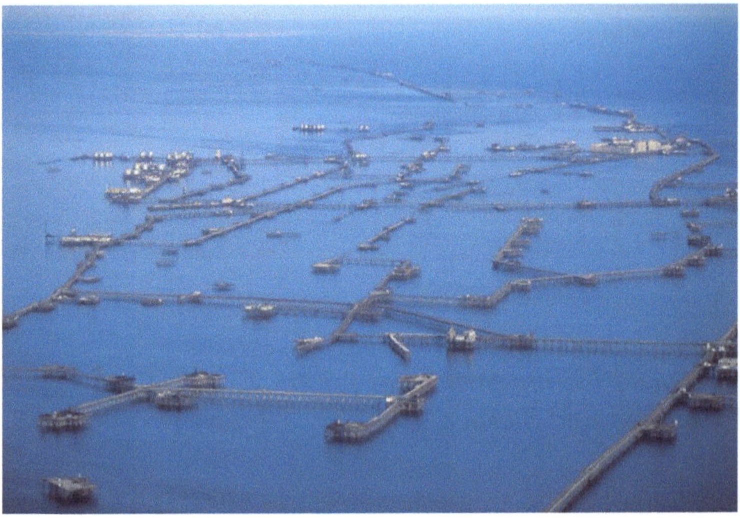

Fig. 19 Stalin's Atlantis (*Photo courtesy* www.spiegel.de)

the motion of the draft at which the structure was towed was small and they jointly decided to keep the rig working in the floating condition.

The first purposely made semi-submersible rig, *Ocean Driller*, was constructed in 1963, which later became the archetypal structure of the oil industry. The concept was quickly adapted and in no less than 10 years, 30 semi-submersible units were built. Today, more than 220 semi- submersible drilling rigs[10] are operating around the world in water depths ranging from 200 to 10,000 feet .

The new generation of semi-submersibles are very large and tall, have a gross weight of approximately 20,000–30,000 tonnes, are outfitted with high pressure pumps, generous high-flow solids controls suits, big bore drill pipe, dual mud systems with upwards of 15,000 bbl pit capacities and automated pipe handling. Most are dynamically positioned, drawing on upwards of 40,000 kW of power plant and boasting a 1,000-metric hook load and either dual activity or significant off line activity. Figure 20 shows Keppel's 6th generation semi-submersible (Merchant 2010).

Another alternative to fixed type and submersible type of offshore rigs was devised before semi-submersible rig was invented and that was the jack-up rig. This is a floating hull with numerous legs which are lowered at the desired location. It can be towed to the location very easily because of its floating platform. Since its advent, over 490 jack-up rigs have been built. As oil production approached greater depths, it gradually became difficult to transfer the products to the shore using pipelines. The oil had to be constantly stored in tanks or transferred. Hence, FPSO (floating production, storage and offloading) units were

[10] Retrieved from: www.shipbuildinghistory.com

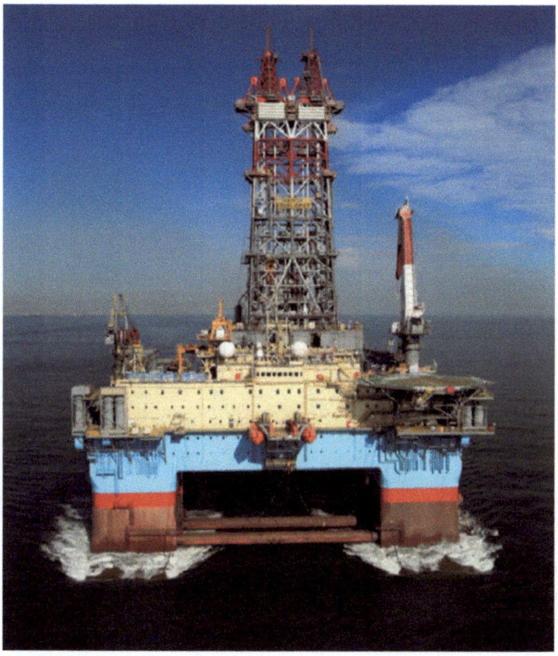

Fig. 20 Keppel-built 6th generation semi-submersible—MAERSK developer (Photo courtesy of Aziz Merchant, KeppelFels)

created. The first oil FPSO was the *Shell Castellon*, built in Spain in 1977. Today, over 200 vessels are deployed worldwide as oil FPSOs. The world's largest FPSO vessel is Kizomba B in Angola near a tension leg platform (see Fig. 21). Another notable large floating structure is the Royal Dutch Shell's Prelude FLNG (floating liquefied natural gas). Measuring 488 m (1,601 ft) long, 74 m (243 ft) wide, made from more than 2,60,000 tonnes of steel and weighing more than 6,00,000 tonnes (five times the weight of the largest aircraft carrier), it is regarded as the largest vessel ever built.

Where the oil industry has begun to embrace large-scale pontoon structures is in the use of floating oil storage facilities. These structures have been built in Japan to stockpile oil against possible oil crisis scenarios. Figure 22a, b show the huge floating oil storage bases in Kamigoto and in Shirashima, respectively (see Chap. 4). In the Shirashima oil storage base, there are 8 large floating steel structures, each measuring 397 × 82 × 25.4 m. One unit holds 7,00,000 m^3 of fuel, equivalent to Japan's oil consumption for a single day. The floating oil facility is equipped to store five different kinds of crude oil as well as refined products, such as gasoline, domestic fuel and jet fuel. The Kamigoto Oil Storage Base is a smaller facility of 5 units of 390 × 97 × 27.6 m (Proceedings of the 16th International Ship and Offshore Structures Congress, Vol. 2).

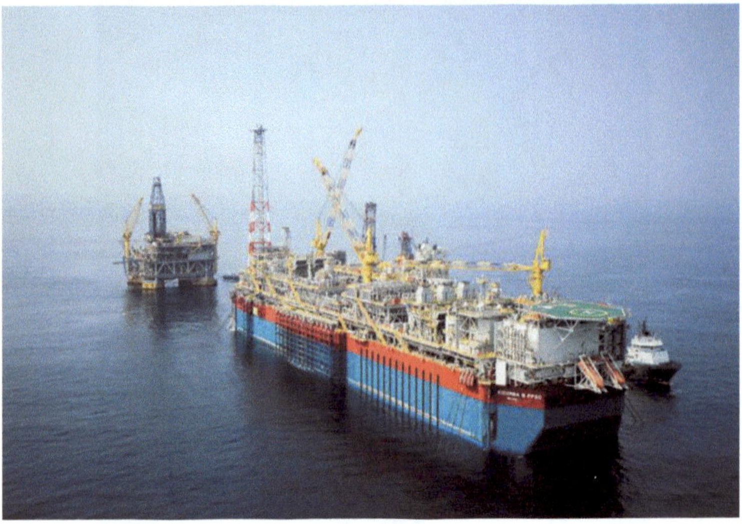

Fig. 21 FPSO vessel, Kizomba B in Angola (*Photo courtesy* energy-pedia news, "Angola: ExxonMobil recognised for industry-leading approach to develop Angola deepwater projects", 2nd May 2011)

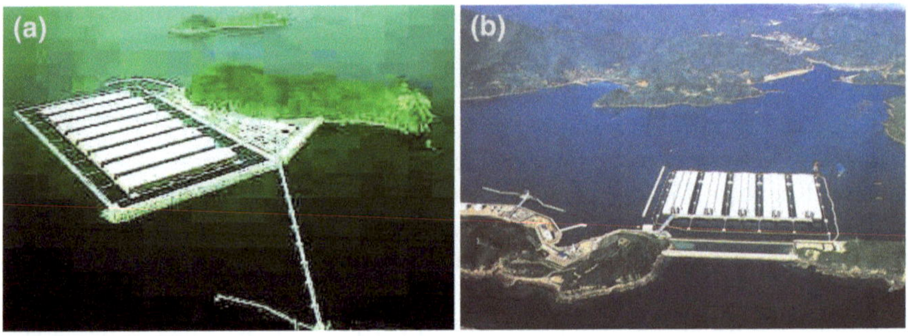

Fig. 22 **a** Shirashima floating oil storage base, Japan (*Photo courtesy* of Shirashima oil storage Co Ltd) and **b** Kamigoto floating oil storage base, Nagasaki Prefecture, Japan

4.5 Floating Entertainment Facilities

As extensions to cities, urban planners have begun to supplement the existing urban environment by programming cultural and recreational functions into floating facilities. These enable a city to expand its ability to cater to the needs of its growing population and enhance coastal living.

Floating structures are popularly employed for use as floating restaurants, giving diners panoramic views of the surrounding sea. An icon of Hong Kong is the famous floating *Jumbo* restaurant, on a 24 × 24 × 3.2 m pontoon (see Fig. 23a).

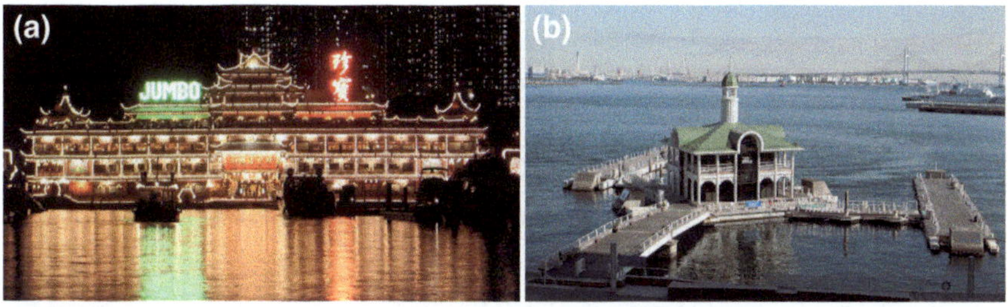

Fig. 23 **a** Jumbo restaurant in Hong Kong and **b** floating restaurant in Yokohoma

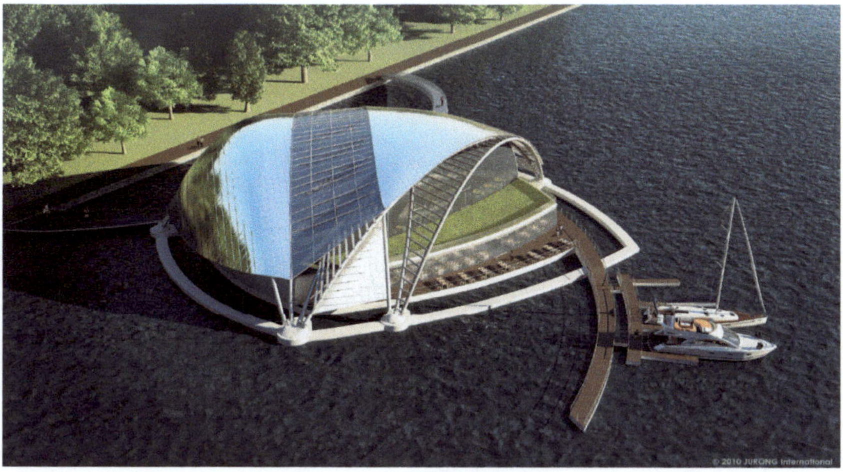

Fig. 24 Prof Wang's proposed floating crab restaurant

Similarly, Japan has a small floating restaurant in Yokohoma that was built in 1992. The pier, next to the restaurant, is also a floating structure (see Fig. 23b). A floating crab restaurant has been proposed by Prof CM Wang of the National University of Singapore (see Fig. 24).

Floating accommodation facilities, such as hotels and dormitories have also been created in recent times. A seven-storey floating hotel, the *Four Seasons,* was built in Singapore and towed to its site at the Great Barrier Reef near Townsville, Australia (see Fig. 25a). Its inherent mobility was used to advantage during cyclone events where it was left anchored at a single point and allowed to move with the wind, reducing resistance and damage. The hotel was later towed to Ho-Chi-Minh, Vietnam and later to North Korea. Canada also has a floating hotel, the *King Pacific Lodge Princess Royal Island,* located in British Columbia (see Fig. 25b).

Japan has an amusement facility that has been created on a large floating island (130 × 40 × 5 m) at Onomichi, Hiroshima. It houses a 3-D theatre, an aquarium and a marina (see Fig. 26).

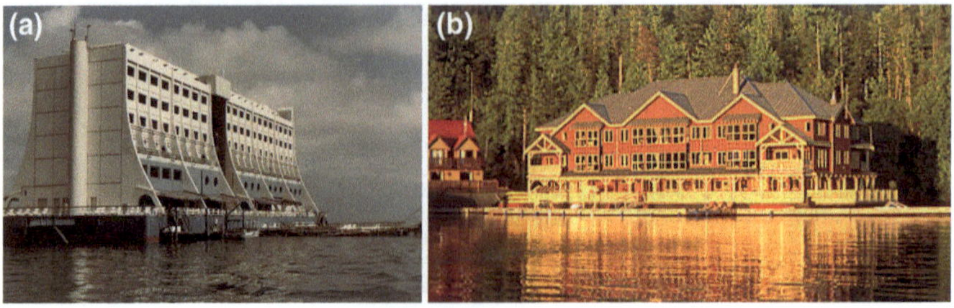

Fig. 25 **a** Four seasons hotel and **b** King Pacific lodge Princess royal Island

Fig. 26 Floating amusement facility at Onomichi, Hiroshima

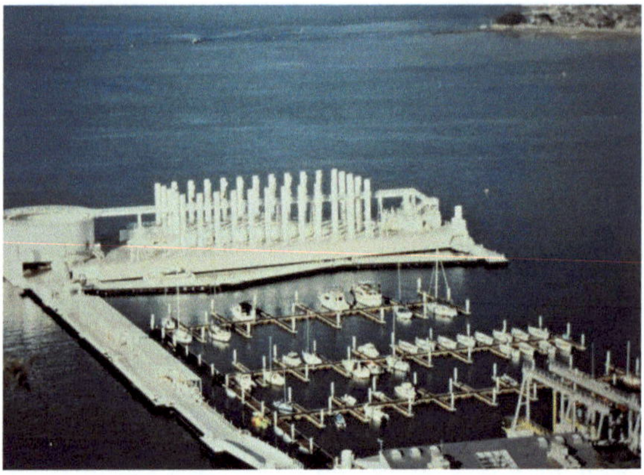

Completed in 2007, the floating platform at Singapore's Marina Bay is the world's largest floating performance stage (see Fig. 27 and Chap. 2). The platform is designed to be a multi-purpose facility on the bay for mass spectator events, sporting activities and cultural performances. The inherent mobility of its component parts has also been incorporated in its design function, ensuring that the structure can be re-configured for use as a mobile dock for water sports and boat shows. Made entirely of steel, the floating platform measures $120 \times 83 \times 1.2$ m, which is 5 % larger than a soccer field. It hosts the nation's National Day Parade, Fireworks Festival and Water Carnival.

The Floating Islands of Han River, Seoul,[11] are part of a man-made archipelago in South Korea comprising three islets, Viva, Vista and Terra (see Fig. 28). The project was developed to enhance the urban relationship of the South Korean capital, Seoul, with the Han River, with the artificial islands housing convention halls, restaurants and other tourist attractions. The design of the three islets exemplifies innovative use of solar-power technology and clever utilisation of the inherent advantage of a floating structure to deal with the problem of varying sea levels.

[11] "Seoul Floating Islands", Haeahn Architecture, H Architecture, *ArchDaily.com*, July 12, 2012.

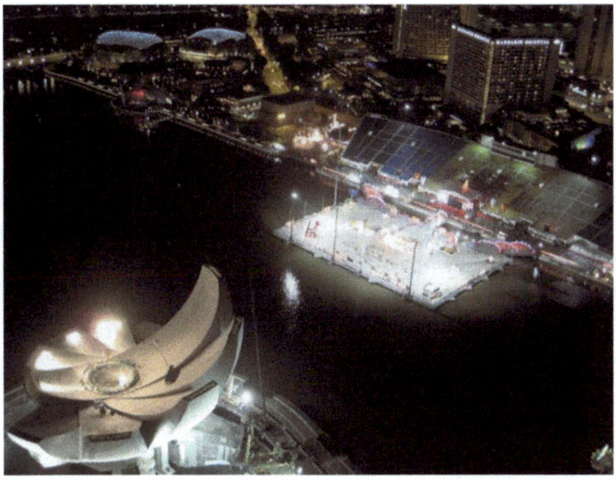

Fig. 27 Performing stage at Marina Bay, Singapore

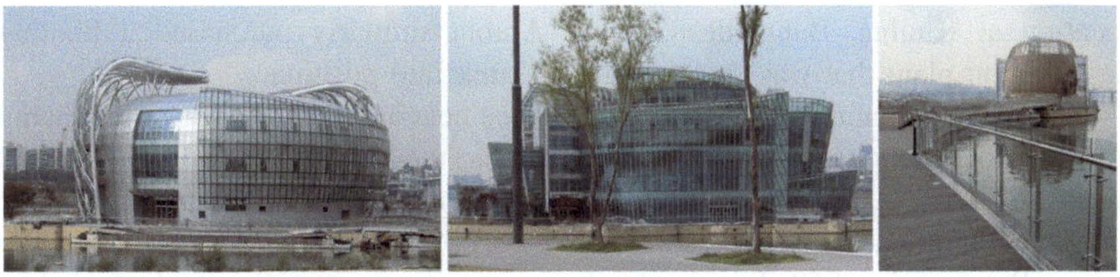

Fig. 28 Floating islands on Han River, Seoul

More recently, Waterstudio.NL designed the *Greenstar* which is a floating hotel for 800 overnight guests and conference centre for 2,000 delegates off the Maldives in the Indian Ocean (see Fig. 29a). Slated to open in 2014, the Greenstar will boost residential capacity for the Maldives. These floating extensions to the city become crucial strategies for island nation-states under threat from rising sea levels. Given the significance of tourism to its economy, the Maldives has also embarked on construction of an ambitious floating golf course, interconnected by underwater tunnels (see Fig. 29b).

4.6 Floating Utility Plants

The mobility of a floating structure has been exploited for industrial purposes since the latter half of the 20th Century. These facilities can be constructed in a particular location, and towed to site and installed as a permanent facility, or moored and towed again to a subsequent site where the need arises. Examples include a floating pulp plant with attached power generator that was built in 1978 in Brazil. It was later towed to its site at Munguba, where it was then converted into a

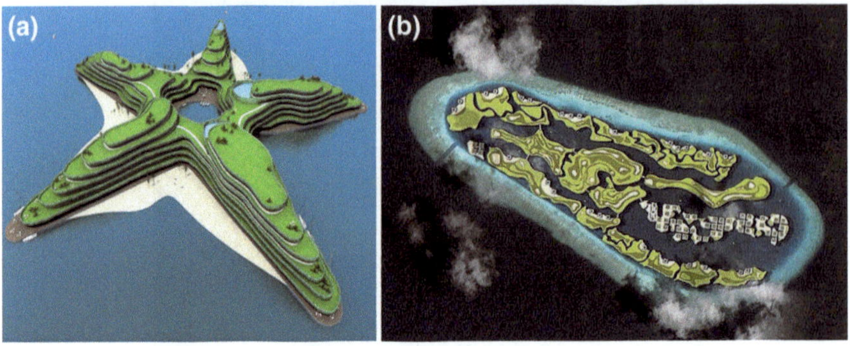

Fig. 29 **a** Greenstar floating hotel and convention centre and **b** 18-hole floating golf course in Maldives

permanent facility and installed on piled foundations. In 1979, Bangladesh purchased a floating power plant that was constructed in Japan, and then towed to location at Khulna, Bangladesh. Saudi Arabia similarly purchased a floating desalination plant that was then towed and sunk into position.

Given that a floating structure can be located near an urban centre without being adjacent to residential and commercial centres, they also provide the additional benefit of being suited to sensitive uses such as nuclear power plants or defence sites. The first floating nuclear power plant was built under the US Army Nuclear Power Program in 1968. Russia also began construction of the floating nuclear power plant, Akademik Lomonosov,[12] in June, 2010 (see Fig. 30). It is anticipated that the power plant will provide up to 70 MW of electricity, enough for a city with a population of 2,00,000 people. The same facility can also be utilised as a desalination plant producing 2,40,000 cubic meters of fresh water per day. The mobile nature of the plant enables the facility to be towed where required in the event of an emergency or natural disaster to provide power.

Japan has built a 70 MW floating solar plant in the Kagoshima Prefecture of Southern Japan (see Fig. 31). Called the Kagoshima Nanatsujima Mega Solar Power Plant, it is the largest solar power plant in Japan and it can generate enough electricity to power approximately 22,000 average households. The plant started operation in November 2013.

Vancouver, Canada has the world's first styrofoam-filled, cellular concrete, floating heliport. Similar floating helipads have been adapted in multiple locations in Japan for use of as emergency rescue bases particularly as these floating structures are immune to seismic forces and, where appropriately designed, can withstand various loads including gravity loads and waves. Figure 32 shows a steel floating emergency rescue base moored at Tokyo Bay.

[12] "Akademik Lomonosov Floating Nuclear Co-generation Plant, Russian Federation", *power-technology.com.*

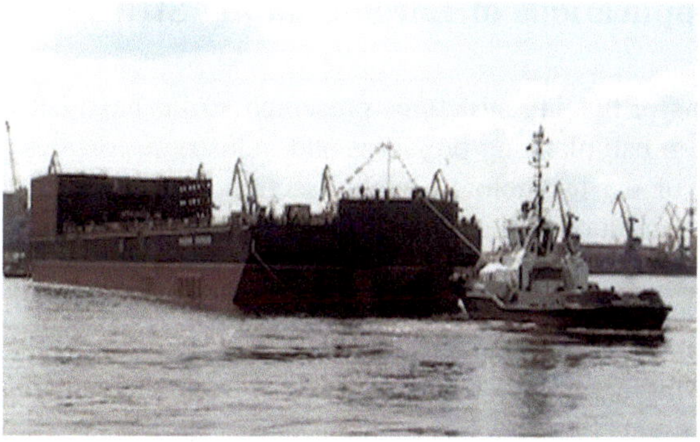

Fig. 30 Floating nuclear power plant, Akademik Lomonosov at its launch in Russia (*Photo courtesy* RIA novosti, Maxim Nichiporenko and Lola Saitmetova)

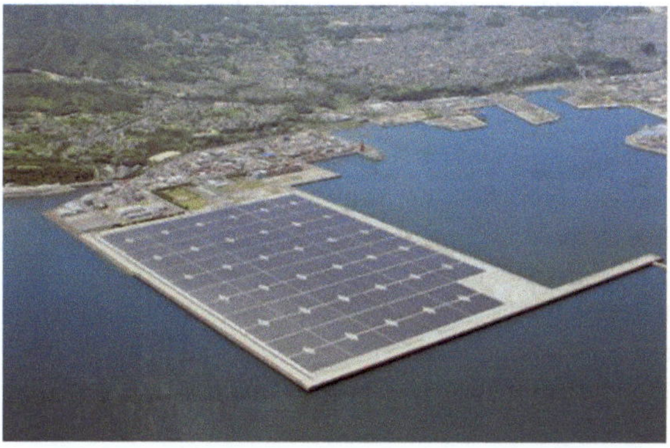

Fig. 31 Floating solar power station in Kagoshima Prefecture http://cleantechnica.com/2013/11/07/kyocera-opens-70-mw-solar-power-station-japan/

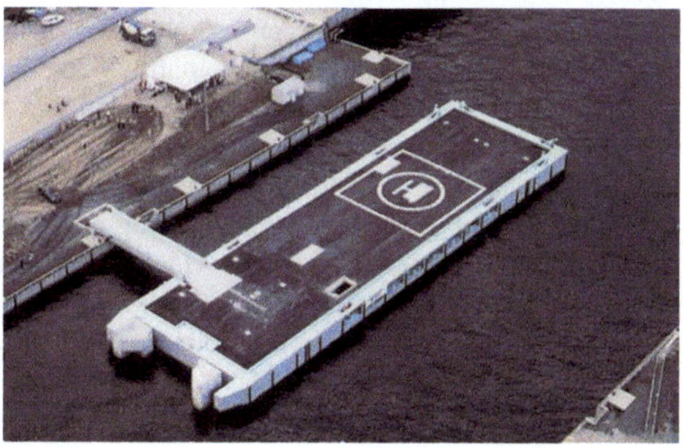

Fig. 32 Floating rescue emergency base at Tokyo Bay (*Photo courtesy* look at the sea. The floating structures association of Japan)

5 Future Applications of Large Floating Structures

The applications of floating structures presented so far have been tied to drivers associated with urban planning pressures and industry necessities. This has led to ubiquitous use of the technology. In this section, we present a snapshot of the speculative and ambitious future applications of large-scale pontoon structures.

One future application of the technology is a revisit of an old idea of a floating airport. This idea may be traced back to Edward Armstrong's 1920 proposal of multiple seadromes (aerodromes located in the sea) positioned as stepping stones for aircrafts flying across the ocean. Owing to technological limits at the time, planes could not travel long distances and needed constant refuelling. In 1943, US Navy Civil Engineers Corps constructed a floating airfield (1810 × 272 ft) consisting of 10,920 pontoons to allow for US Navy airplanes to land in the Pacific Ocean (see Fig. 33). However, the idea lost impetus with the development of non-stop flights (Watanabe et al. 2004).

Test models have been constructed in Japan for a proposed floating airport for *Kansai International Airport*. This project was seminal in triggering research and development of large-scale pontoon structure technology in both academia and its associated industry. Phase 1 of the proposal was a 300 × 60 m semi-submersible type floating structure, proposed in 1974 and Phase 2 was a 1,000 × 60 (to 121) × 3 m pontoon type floating runway, proposed in 1994. Owing to their gigantic size, the term *Megafloat* was coined for them. Technological Research Association of Megafloat (TRAM), established in 1995, pursued the concept of *Megafloat* from 1995–1998 (see Chap. 8). Figure 34 shows Phases 1 and 2 of the project. Although not implemented yet, it is envisaged that a floating civilian airport or a military mobile offshore base (see Fig. 35 and Chap. 11) will be built sometime in the future.

Large-scale floating suburbs and floating island cities are the next evolution of large-scale pontoon technology. These urban forms can either be moored to coastal cities as extensions or free-floating cities in international waters. Architect Vincent Callebaut has designed a model of a floating ecopolis for climate refugees, known as *Lilypad*, as a long term solution to the rising water levels around the world (see Fig. 36 and Chap. 12). The proposed city is envisaged to be self-sufficient, operating fully on renewable energy that will enable it to be independent from the electrical grid. Another example of the proposed floating city is by the Japan Society of Steel Construction that envisaged a floating settlement at the foci of the elliptically shaped Osaka Bay (see Fig. 37). Recently, Waterstudio.NL designed the Citadel (see Fig. 38), a 60-unit floating apartment complex to be built in the Dutch city of Westland, near The Hague. The complex serves as a means of addressing the flood-prone location, a common environmental consideration given the large proportion of land in the Netherlands that lies below sea level. According to architect Koen Olthuistr, the project is scheduled for completion in 2014, and could be Europe's first floating apartment building complex.

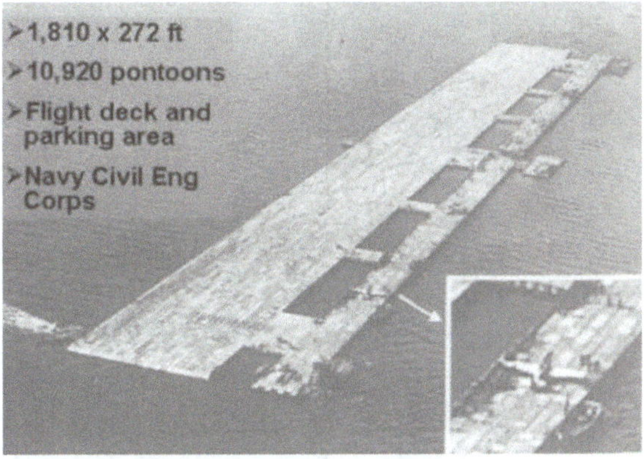

Fig. 33 US floating airfield in the Pacific Ocean during World War II

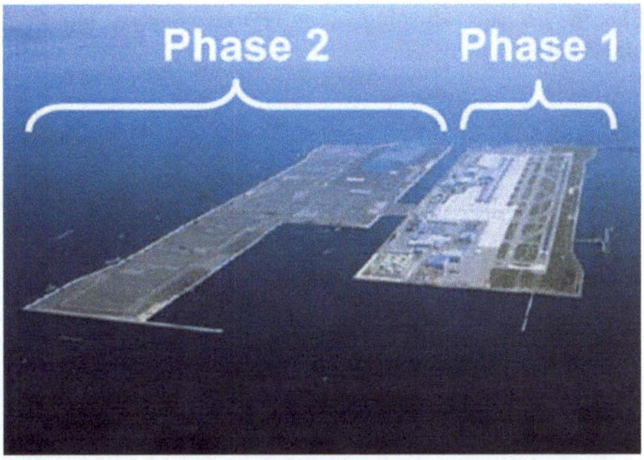

Fig. 34 *Phase* 1 and *Phase* 2 of Megafloat (Suzuki 2005)

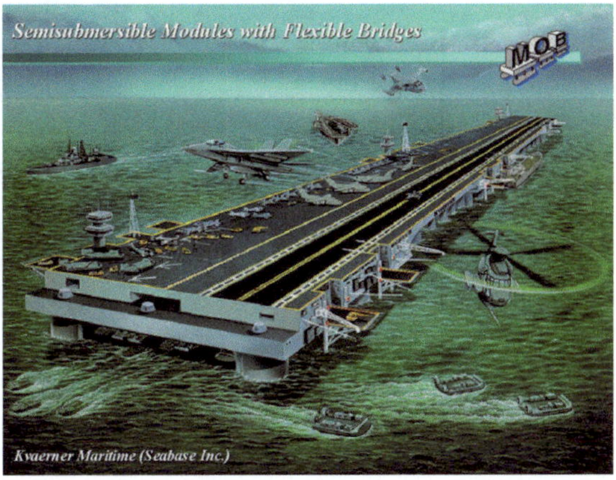

Fig. 35 Conceptual design of a mobile offshore base or a sea base

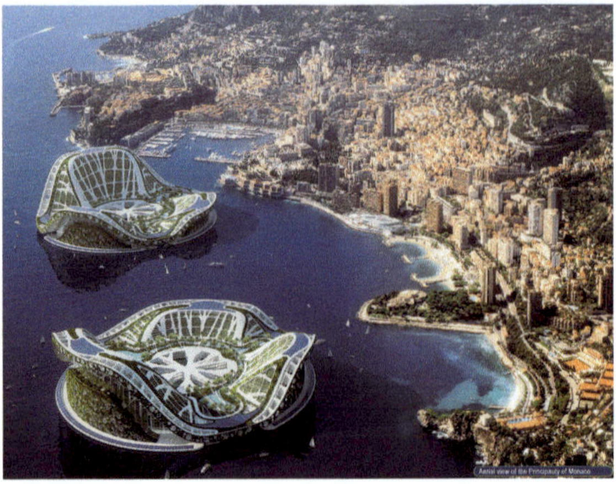

Fig. 36 Lilipads—floating cities (*Photo courtesy* http://vincent.callebaut.org/planche-lilypad_pl31.html)

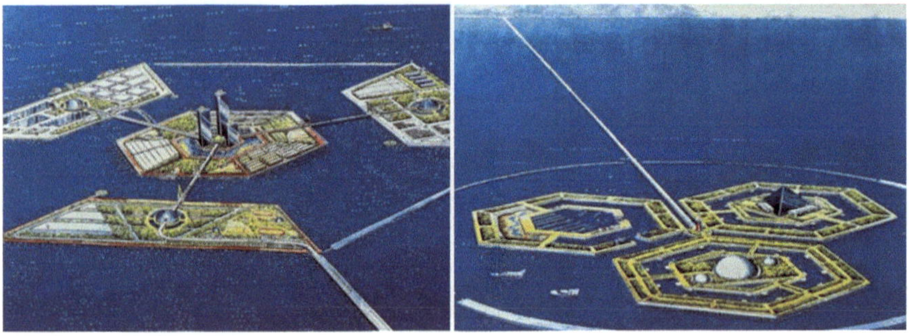

Fig. 37 Proposed floating cities in Osaka Bay, Japan by Japan Society of Steel Construction

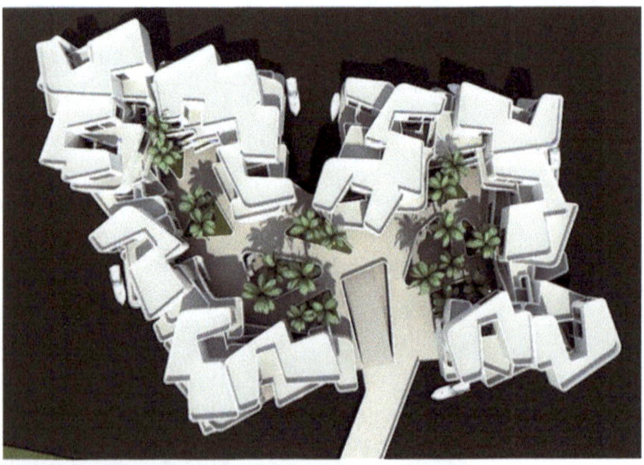

Fig. 38 The Citadel—floating apartment complex designed by Waterstudio.NL (*Photo courtesy* http://www.waterstudio.nl/projects/54)

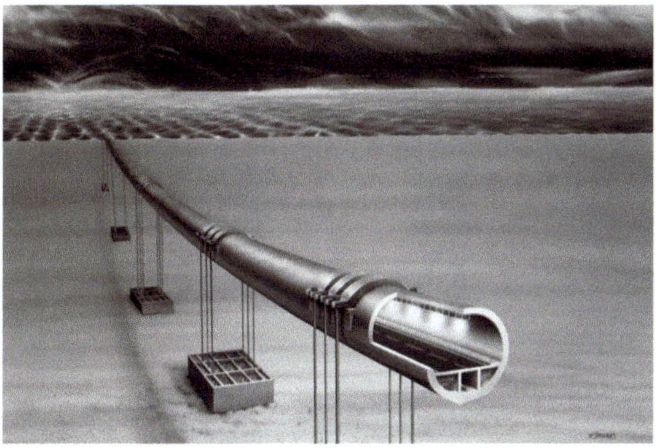

Fig. 39 Proposed submerged floating tunnel across Sognefjord in Norway (*Photo courtesy* Norwegian public roads administration)

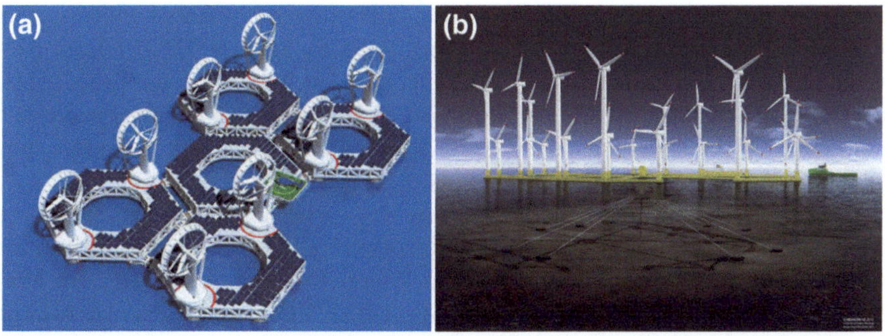

Fig. 40 **a** Proposed 'wind lens' project by Kyushu University, Japan utilizing both wind and solar power (*Photo courtesy* Main(e) international consulting LLC) **b** Hexicon, a Swedish design for wind energy farm (*Photo courtesy* Main(e) international consulting LLC)

Another future application is a planned submerged floating tunnel (SFT), or *Archimedes Bridge*. This application of a floating structure is perhaps the beginning of a new line of evolution for the technology in taking an innovative approach to waterway crossings. The design calls for a tubular structure, fully immersed and stabilized by appropriate mooring which exploits buoyancy force to carry traffic loads. The first SFT model is to be constructed in Qiandao Lake (China), though it was first proposed in Italy in 2007 (Mazzolani et al. 2008) (see Fig. 39). Other countries like Norway, Italy, Indonesia and Japan are engaged in further research and development of this technology.

The growing concern over climate change has brought the development of renewable energy technology to the forefront of innovation (Fraenkel 2000) and this impact has also been observed in the area of floating structures. Floating wind turbines have gained recognition worldwide after its successful installation in various European sites (Main(e) International Consulting LLC 2013; Wang et al. 2010).

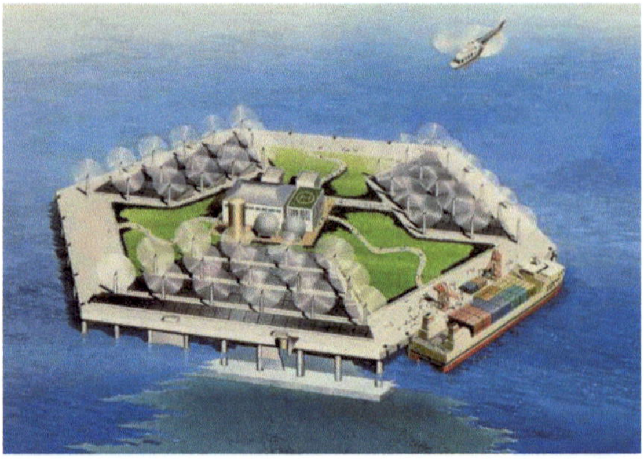

Fig. 41 Concept design of a clean energy plant by floating structure association of Japan

Following the events of the nuclear accidents at Fukushima, Japan is considering plans to set up a pilot semi-submersible type floating wind farm off the Fukushima coast (see Fig. 40). The Fukushima FORWARD project of offshore wind farming will have a potential of 1.2 TW compared to the capacity of 0.2 TW from conventional sources.[13] Future floating clean energy plants, such as the one proposed by the *Floating Structure Association of Japan* (see Fig. 41), offer potential to create a new industry for governments to secure their energy needs and provide a new exportable good.

In the near future, the authors anticipate that large floating structures will be deployed as Ocean Thermal Energy Conversion (OTEC) platforms to harvest the solar energy trapped in the water surface (Wang et al. 2011). By using the honeycomb concrete framing system, some of these OTEC platforms can be as large as 356 m long × 70.7 m wide × 25 m deep, and are designed to withstand severe hurricane, wind and wave conditions (see Fig. 42 and Chap. 10). This permits application of these structures in both coastal and deep ocean sites. A number of such projects have been proposed to various local governmental agencies in several jurisdictions to highlight the feasibility of building and deploying OTEC platforms of various sizes to produce base load electrical power, fresh water, and liquefied hydrogen. The technology for these OTEC systems is based on integrating commercially established components including concrete platforms developed by Dr. Alfred Yee of Precast Design Consultants and EHS, the Kalina Cycle by Recurrent Engineering and EHS, the cold water pipe by EHS and Intecsea, and the integrated engineering design by EHS and Worley Parsons.

The proposals have produced a myriad of design solutions, such as: (a) a 10 MW OTEC platform project to supply power and fresh water for Diego Garcia currently under consideration with participation by the U.S. Navy and the United Kingdom; that is likely to be constructed in Singapore; (b) a 125 MW OTEC

[13] "Fukushima Floating Offshore Wind Farm Demonstration Project (Fukushima FORWARD)",Prof *Takeshi Ishihara*, University of Tokyo, Retrieved from: www.japan.ahk.de.

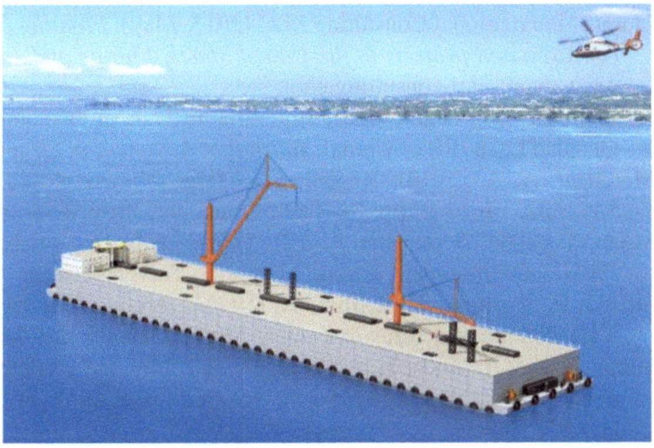

Fig. 42 125 MWE OTEC Plant designed by Dr Alfred Yee

platform in the Republic of the Marshall Islands (RMI) to supply power and water to the local population and to the U.S. military facility on Kwajalein, with accompanying facilities to supply liquid hydrogen for the world market (c) a 60 MW size OTEC platform for Guam to supply much of the power requirements for the U.S. military; and (d) a series of large-scale open ocean OTEC platforms to produce freshwater and liquid hydrogen for Singapore.

References

Bureau of Yards and Docks (1947). The continental bases, building the Navy's bases in world war II. *Chapter 9—floating Drydocks, Vol. I, part II, history of the bureau of yards and docks and the civil engineer corps* 1940–1946. Retrieved from www.history.navy.mil.

Clauss, G., Lehmann, E., & Ostergaard, C. (1992). Conceptual design and hydromechanics. *Offshore structures* (Vol. 1). Berlin: Springer-Verlag. (Translated by M.J. Shields).

Fraenkel, P. (2000). Renewable option 5: Marine power—harvesting the power of the sea. *PeopleandPlanet.net.* Retrieved June 2004, from (http://www.peopleandplanet.net/doc.php?id=452).

Hara, S., Yamakawa, K., Hoshino, K., & Yukawa, K. (2004). At-sea towing of a mega-float unit. *Journal of Marine Science and Technology, 8*(3), 138–146.

Hosam El-Din, H. S., Alaa, M. R, Basil, A. S. (2010). Durability and strength evaluation of high-performance concrete in marine structures. *Journal of Construction and Building Materials, 24*(6), 878–884.

Koekoek, M. (2010). *Connecting modular floating structures—general survey and structural design of a modular floating pavilion.* Master thesis. The Netherlands: T.U. Delft and Engineering Company Public Works, Rotterdam.

Koh, H. S., & Lim, Y. B. (2008). The mega floating platform at Marina Bay. *Proceedings of the Conference on Iconic Structures in Singapore and Asia* (pp. 145–160). Singapore: IStructE Singapore Regional Group.

Maeda, H., Rheem, C. K., Washio, Y., Osawa, H., Nagata, Y., Ikoma, T., & et al. (2001). Reduction effects of hydroelastic responses on a very large floating structure with wave energy absorption devices using OWC system. *Proceedings of 20th Offshore Mechanics and Arctic Engineering Conference.* Rio de Janeiro, Brazil.

Main(e) International Consulting LLC (2013, May). *Floating offshore wind foundations: Industry consortia and projects in the United States, Europe and Japan: An overview*. Retrieved from www.maine-intl-consulting.com.

Mazzolani, F. M., et al. (2008). Structural analysis of the submerged floating tunnel prototype in Qiandao Lake (PR of China). *Journal of Advances in Structural Engineering, 11*(4), 439–454.

Merchant, A. A. (2010). Keppel mega semi-submersibles. *Proceedings of the Conference on Structural Marvels* (pp. 81–110). Singapore: IStructE Singapore Regional Group.

Moan, T. (2004, September 12–17). Safety of floating offshore structures. *Proceedings of 9th PRADS Conference*. Germany: Luebeck-Travemuende.

Suzuki, H., Yoshida, K., & Iijima K. (1996a). A consideration of the structural design of a large-scale floating structure. *Journal of Marine Science and Technology, 1*(5), 255–267.

Suzuki, H., & Yoshida, K. (1996b). Design flow and strategy for safety of very large floating structures. *Proceedings of International Workshop on Very Large Floating Structures*. Hayama, Japan.

Suzuki, H. (2005). Overview of Megafloat: Concept, design criteria, analysis and design. *Marine Structures, 18*, 111–132.

Takahashi, S. (1996). Design of vertical breakwaters. *Reference document No. 34*. Japan: Port and Harbour Research Institute, Ministry of Transport.

The Hetch Co. (1945, May). *Miracle harbor—we'll take our harbors with us*. British information services. Retrieved from www.history.navy.mil.

Wang, C. M., Utsunomiya, T., Wee, S. C., & Choo, Y. S. (2010). Research on floating wind turbines: a literature survey. *IES Journal Part A: Civil and Structural Engineering, 3*(4), 267–277.

Wang, C. M., Yee, A. A., Krock, H., & Tay, Z. Y. (2011). Research and developments on ocean thermal energy conversion. *IES Journal Part A: Civil and Structural Engineering, 4*(1), 41–52.

Watanabe, E., Wang, C. M., Utsunomiya, T., & Moan, T. (2004). Very large floating structures: Applications, analysis and design. *CORE Report No. 2004–02*. Singapore: National University of Singapore.

Wong, L. H., Tan, H. S., Wang, C. L., Lim, H., Ho, H. C., Wang, C. M., et al. (2013). Floating wetlands at Punggol. *The IES Journal Part A: Civil and Structural Engineering, 6*(4), 249–257.

Floating Performance Stage at the Marina Bay, Singapore

H.S. Koh and Y.B. Lim

Abstract The Marina Bay was identified as the venue for the National Day Parade for 2007 and subsequent years. The concept was to create a large usable space on water for mass spectator events and to design it as a multi-purpose facility to be used for sporting activities, mega festivities, cultural performances and water activities. The pontoon-type floating platform technology was adopted to create the large-scale floating performance stage. The Marina Bay floating platform measured 120 m long, 83 m wide and had a depth of 1.2 m and was built by assembling fifteen modular steel pontoons. This floating stage, completed in April 2007, is believed to be the world's largest floating performance stage.

1 Introduction

The Marina Bay was identified as the venue for the National Day Parade (NDP) for 2007 and the subsequent years. The concept was to create a large usable space on water for mass spectator events and to design it as a multi-purpose facility to be used as an alternate site for sporting activities while the new Singapore Sports Hub[1] was being developed as well as for mega festivities, cultural performances and water activities.

[1] The old National Stadium at Kallang was demolished in 2006.

H.S. Koh (✉)
Naval Systems Programme Centre, Defence Science and Technology Agency,
1 Depot Road, Singapore 109679, Singapore
e-mail: khocksen@dsta.gov.sg

Y.B. Lim
Land Systems Programme Centre, Defence Science and Technology Agency,
1 Depot Road, Singapore 109679, Singapore
e-mail: lyokeben@dsta.gov.sg

C.M. Wang and B.T. Wang (eds.), *Large Floating Structures*,
Ocean Engineering & Oceanography 3,
DOI 10.1007/978-981-287-137-4_2

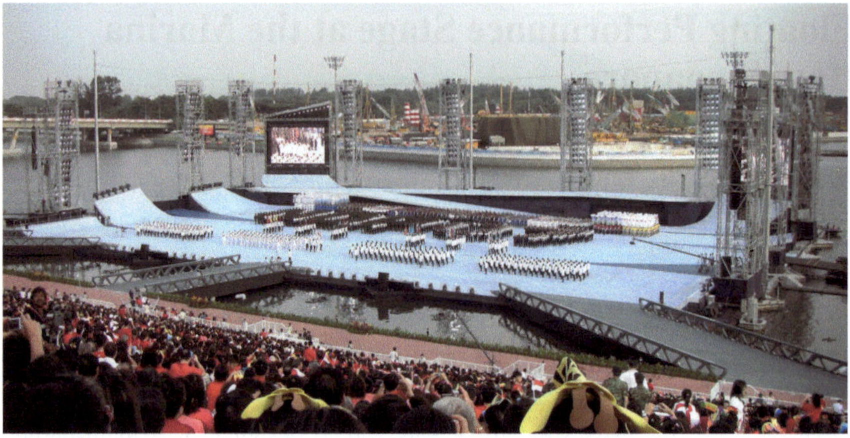

Fig. 1 National day parade held on the floating performance stage on 9 August 2007

The pontoon-type floating platform technology was adopted to create the large-scale floating performance stage. The Marina Bay floating platform measured 120 m long, 83 m wide and 1.2 m depth and was built by assembling modular pontoons made entirely of steel. This floating stage, completed in April 2007, is believed to be the world's largest floating performance stage. It hosted the National Day Parade (NDP) in August 2007 (see Fig. 1). A 27,000 seating capacity gallery was built facing the floating platform and functions as an integrated facility. It allowed the spectators on shore to view the various events on the stage as well as on the water against the backdrop of the Singapore city skyline (see Fig. 2). The integrated floating performance stage is now a landmark in the Singapore cityscape. Besides hosting the annual NDP, the floating platform also provides a wonderful opportunity for all kinds of activities such as concerts, sports events and activities on and around the water. An artificial turf was laid on its large surface and turned the floating platform into a temporary sport field.

This chapter discusses the pontoon-type floating structure technology and presents the engineering challenges and considerations involved in building the floating platform, design concepts, technical analysis as well as the construction, assembly and verification testing.

2 Engineering Challenges and Considerations

The Marina Bay floating platform, slightly larger than the size of a football field, is one of the most technically challenging floating platforms of her size, in view of the many unique considerations, constraints and innovative designs. The main challenge was to design and create a large usable space on water to carry a high load comprising 9,000 people, 200 tons of stage props and three 30-ton vehicles.

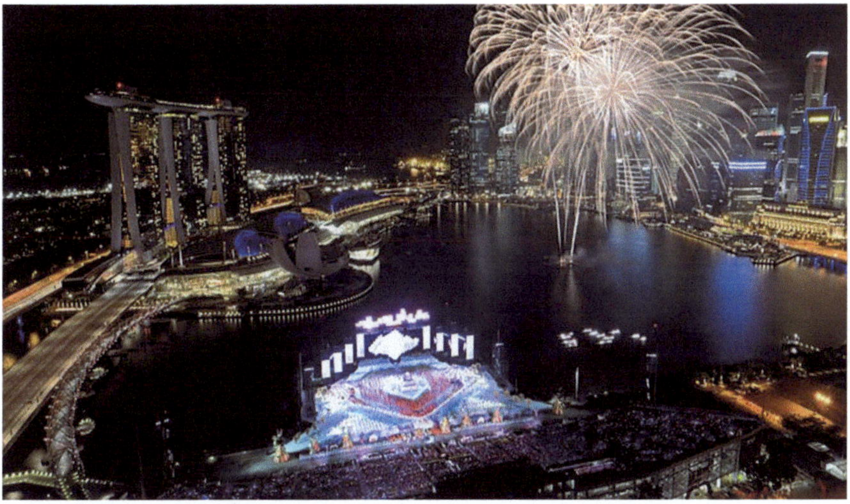

Fig. 2 Performing stage at Marina Bay, with backdrop of Singapore city skyline on 9 August 2013 (*Photo courtesy* of The Straits Times)

The way in addressing this challenge was to design a mega floating structure using pontoon-type floating technology. Pontoon-type, very large floating structures are rare in the world (Suzuki 2005).

Singapore's first application of the mega floating structure technology is the construction of the floating performance stage at the Marina Bay (Koh et al. 2007, 2008, 2009). The floating installations offered significant advantages, in particular the reduced development and construction period, ease of assembly as well as concurrent construction and installation. It only took 13 months to deliver the floating stage at the Marina Bay, meeting all the requirements.

2.1 Versatility and Mobility in Deployment

Most large floating structure applications reported by Watanabe et al. (2004) have been designed as a "single piece". The design of the floating platform at Marina Bay has to address the key requirements of versatility and mobility in deployment. To achieve this, the floating platform is uniquely designed to be modular and is constructed from assembling fifteen steel pontoons with a one-of-its-kind connecting system to allow for dismantling and re-assembling of the floating platform. Given the modular nature of the design, the floating pontoons can be easily relocated within the bay area or re-configured into various shapes and sizes to meet different event requirements.

2.2 Rigidity

With a modular design, one of the greatest challenges is connecting all the multiple floating modules into one strong single integral platform to create the large working surface in a manner that keeps the relative motions of the connected pontoons within the allowable limits for the intended purposes. These pontoon-type floating modules take advantage of the inherent free buoyancy to achieve floatation to take the gravity loads. The connecting system is the key component for the functionality and robustness of the integral platform and is specially designed to inter-lock the pontoons to ensure rigid connection and to withstand heavy loads.

2.3 Stability

One of the key design criteria is the stability requirement of the floating structure. The platform (and her pontoons) has to be stable in water and unaffected by waves, winds and tidal currents and at the same time safe for holding mass performances on it.

2.4 Relationship to Water Surface

The freeboard must not allow sea waves to get onto the surface and yet the depth cannot be increased, which is desirable to prevent tilt effect. The floating structure needs to be moored to permit vertical movement of the structure to follow the changing sea level. For given payloads, the surface of the floating structure thus remains a constant distance from the water surface despite the large tidal variations.

2.5 Consideration of Water Environment

The other key challenge faced in the design included having to contend with the water environment. The shallow water at the site limits the platform draft, and consequently the depth while the changing tides put constraints on both the positioning of the platform with respect to the shore and the gradient of the access bridges that link the platform structure to the land. There is a need to ensure that the presence of the floating structure would not create disruption to the water flow from the Singapore River and affect the marine ecology in the bay.

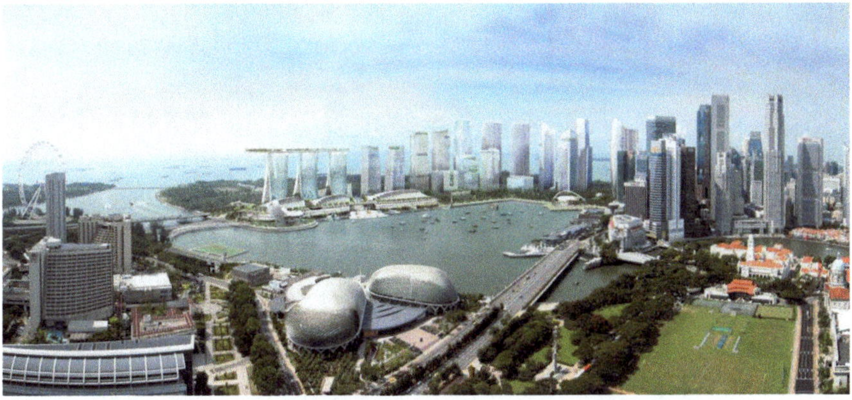

Fig. 3 Location of floating platform at Marina Bay (*courtesy* of URA)

2.6 Statutory Requirements

The design had to address the non-existence of established design rules and standards for connecting multiple floating pontoons rigidly as an integral platform. The design has to be compatible with the urban master-planning for the bay area (see Fig. 3), which was designated the new downtown and to be implemented within the environment and regulatory constraints imposed as the Marina Bay was designated as a fresh water reservoir. This requires the design to architecturally blend with the surrounding infra-structure developments and to address the environmental specifications for a reservoir.

2.7 Rapid Installation and Ease of Assembly

To achieve the desired (tight) construction deadlines, the modular pontoons are constructed at shore-based shipyards in multiple locations with concurrent on-site installation of supporting structures. When completed, each pontoon module is capable of floating so that they can be floated or towed to the bay site and assembled on water to the final built-form with minimal on-site work. On the other hand, filling the seabed and waiting for it to settle often requires time, anywhere around 2–5 years, before construction may occur.

3 Design Concepts

The design concepts are developed based on considerations in response to these challenges, as well as low acquisition and maintenance cost.

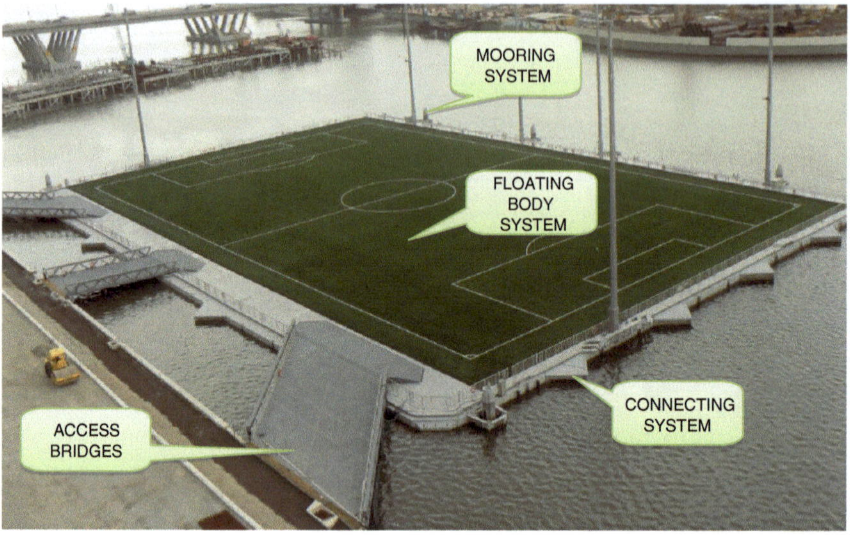

Fig. 4 System description of floating performance stage at Marina Bay

3.1 System Description

The floating platform at the Marina Bay is characterised by four main components,[2] as shown in Fig. 4, that need to be well integrated to achieve the desired outcome:

- Floating body
- Connecting system
- Mooring system
- Access bridges or linkways

3.2 Floating Body

The floating body or platform comprises the fifteen identical pontoons of dimensions 40 m × 16.6 m × 1.2 m each (see Fig. 5), which works on Archimedes' principle of buoyancy. The pontoon is of welded mild steel construction owing to the material's high tensile strength and elasticity to the heavy loads. They float on water, and as such, easier to move around.

Each pontoon is a box-type structure based on two longitudinal and two transverse watertight bulkheads to subdivide it into nine watertight compartments

[2] The breakwater was eventually provided by the Marina Barrage erected to create a reservoir in the Marina Bay.

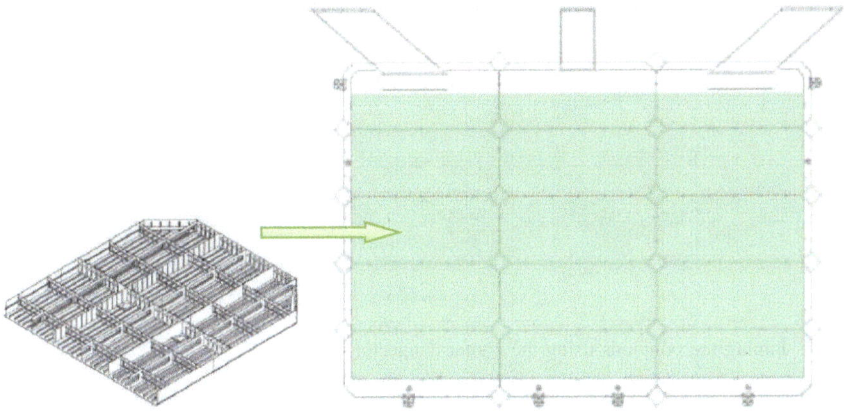

Fig. 5 Plan view of floating platform consisting of 15 pontoons (each measuring 40 m × 16.6 m × 1.2 m) connected together and three access bridges

to keep it stable in the event of flooding. Three non-watertight longitudinal full-depth web frames and several transverse frames were installed. The pontoons are characterised by their low depth-to-length ratios. The self-weight of the floating platform is about 2,800 ton.

The floating structure is subjected to heave (up/down) movement and rotational motions (about the horizontal axes) when load is applied. The buoyancy force acting from below the structure stabilises heaving movements and oscillations caused by gravity and dynamic loads. However, eccentric loads and/or moments can cause tilt movements or deflections in the floating structure, with one of its sides rotating deeper into the water.

3.3 Connecting System

The principal considerations for the connecting system include the need to satisfy the structural requirements that address the operating conditions, structural strength, serviceability, durability and safety standards.

The connecting system ensures adequate strength and rigid connection to the floating platform when the pontoons are inter-locked, and yet shall be easily dismantled and re-configured. These pontoons are joined at the corners using a floating corner connector and along the mating edges with side connectors (five along the longer edge and two along the shorter edge). These are made of high tensile steel and designed to withstand the heavy static loads and dynamic forces resulting from personnel movement and wave motion.

The corner connecting system utilizes the full depth of the platform and consists of a male member at the corner of each pontoon and a female member of a 4 m by 4 m diamond-shaped floating connector unit (see Fig. 6). The hollow edges of the

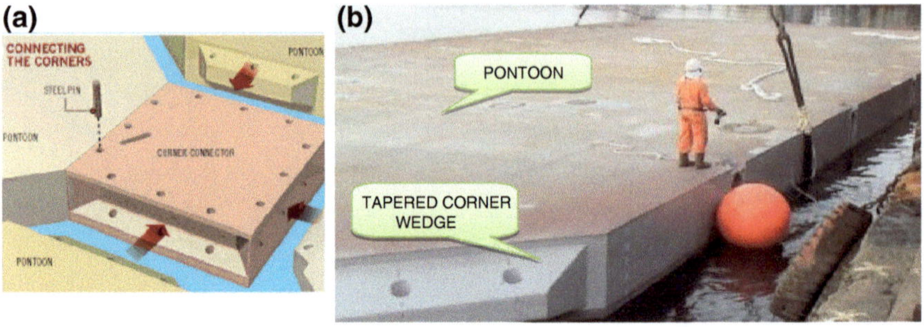

Fig. 6 Joining the pontoons using the corner connector

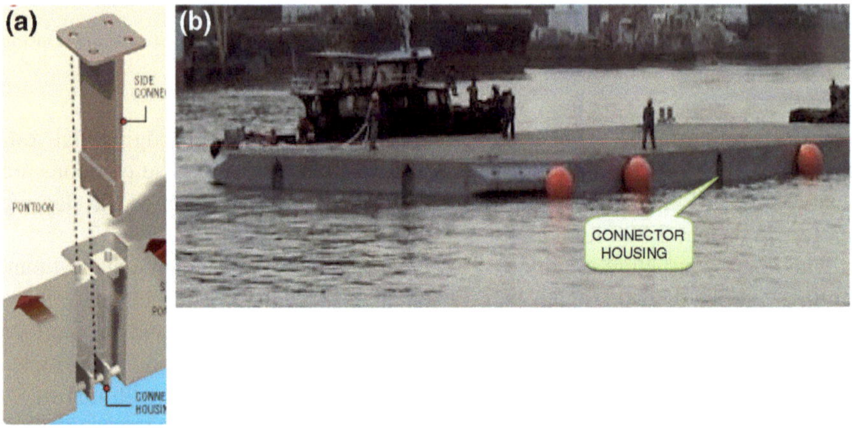

Fig. 7 Joining the pontoons using the side connector

corner connectors slip into the tapered wedges of the pontoons. The coupling members are kept in place by detachable steel locking pins. Twenty corner connectors are used to assemble the stage.

The pontoons are locked adjacently using blade-type engagement to bear the tensions and bending moment between adjacent pontoons (see Fig. 7). These side connectors lock the top involving steel stubs and bottom of the adjacent pontoons together to form a rigid connection so as to ensure that there is no differential displacement which may render the platform non-operational. Eighty side connectors are used to assemble the platform.

3.4 Mooring System

A mooring system is used to hold the floating platform in position considering key factors of the current and waves, and environmental constraints of shallow water

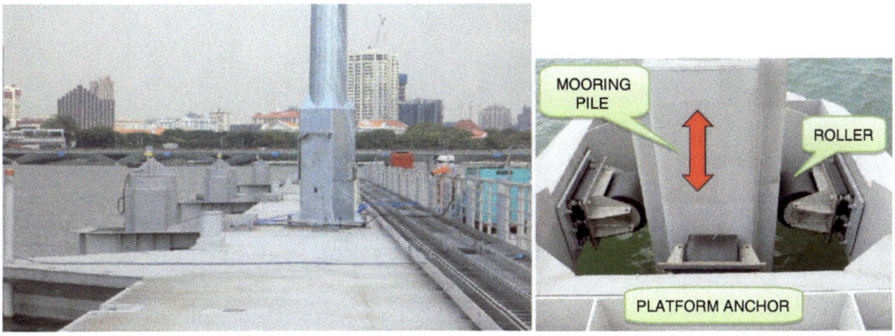

Fig. 8 Detachable mooring system

and tides as well as wind. Conventional mooring systems involving cables and chains were found not suitable for the floating platform because the tidal range was relatively large (of up to 4 m) when compared to the shallow depth of the platform. The water depth in the middle of the Bay varies between one metre and seven metres while the water channel between the floating platform and the opposite embankment is about 100 m. The heavy frame-guide dolphin mooring method and caisson-type mooring system were also unsuitable because the floating platform has to be moved to another site from time to time. Therefore, a dolphin mooring system of six piles (4 on the long side and 1 each on the short side) with the aim of restraining the lateral and yaw movements, and easing the process of installation was devised. But these dolphins are designed to be detachable just above the seabed to allow a clear water surface for sea-sporting activities in the bay.

Each detachable mooring system comprises four components, namely the mooring pile, the platform anchor, the casing pile and the fender rollers (see Fig. 8). The casing pile is rammed 16–20 m into the sea bed and protrudes about 50 cm above the seabed to provide a footing where the upper removable mooring pile is secured by bolting at mating flanges. The top of the pile is linked to the sides of the floating platform via fender rollers using the platform anchor attached. The loads and moments exerted by the floating platform are transferred to each pile via the rubber fender rollers. The rollers guide the stage up and down as the current and tide change.

3.5 Access Bridges/Linkways

Three wide access bridges connect the floating platform to the land and seating gallery for the access by vehicles and mass movement of people onto the platform. Evacuation and rescue are also important design considerations.

The centre bridge is 8 m wide and 23 m long and can carry 30-ton vehicle. The two side bridges are 10 m wide and 40 m long for people and light vehicle. These bridges have to be designed in two segments. The first segment is a fixed bridge

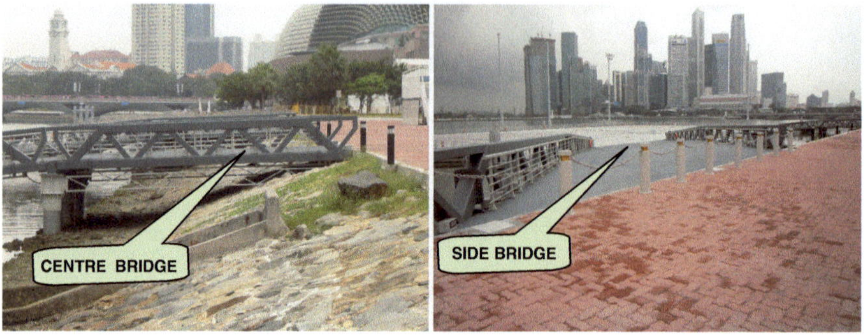

Fig. 9 Access bridges

supported by foundation piles. The second segment is a gangway, articulated at one end with hinged connections that help it move with the tide so that the change in the gradient of the gangway allows unaffected access at all times despite the tidal variations in the bay of 4 m (see Fig. 9). The other end of the gangway moves horizontally on the deck resulting from the tidal change. The access bridges have the ability to match the tides for serviceability and comfort.

4 Design Analysis

The analysis and design of floating structures need to have special considerations when compared to land-based structures.

4.1 Structural Analysis

The floating platform is of welded mild steel construction and has been designed according to the American Bureau of Shipping (ABS) Rules for Building and Classing Steel Vessels for Service on Rivers and Intracoastal Waterways, as well as applicable industry standards and codes. This was complemented with building construction rules for the mooring system design and access bridges.

The primary hull structure of the pontoon has been designed with a uniform deck loading and 30-ton wheeled vehicle loads, which resulted in the top plate thickness of 12 mm. The bottom and side shells are of 8 mm thick plates. For the global response, as there were no established rules for the design of thin flexible plated structures and for connecting multiple pontoons as an integral platform, a first-principle approach was adopted for the structural design. This requires extensive finite element analysis to predict the static bending structural responses of the floating platform (including the connectors).

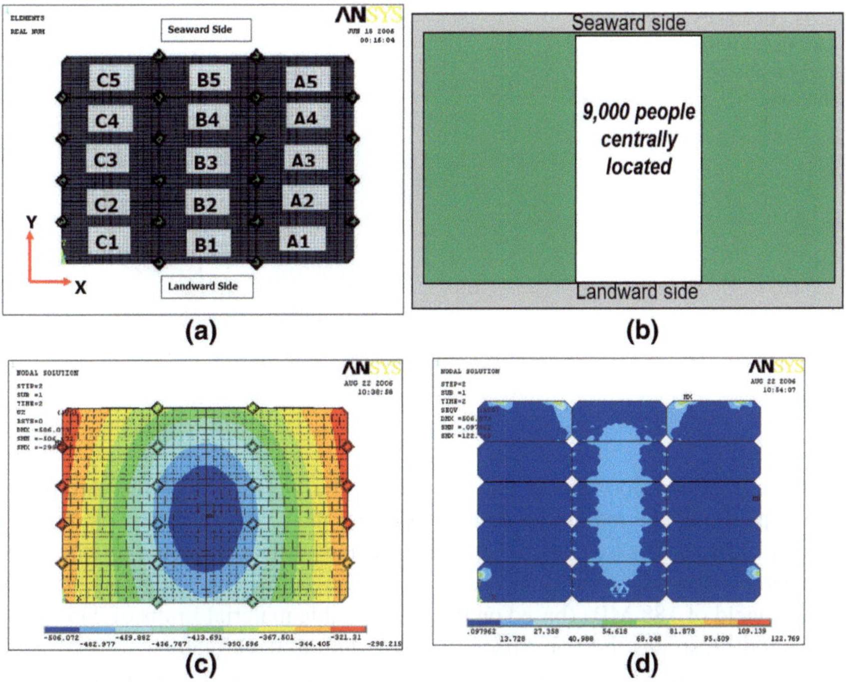

Fig. 10 **a** Finite element model of floating platform. **b** Load case: live load centrally located. **c** Plots of vertical deflections. **d** Plots of von-Mises stress on top deck

The response of the global platform to various loading scenarios was analysed using a three dimensional (3D) finite-element model of the complete platform, consisting of the fifteen pontoons, the connectors and the three access bridges. For the finite-element model of the pontoon, plate and beam element elements were used. Supporting the platform model are linear springs that simulate the buoyancy force. In-plane restraints were applied at the six mooring piles locations. Figure 10a shows the computer model used for the engineering analysis.

For the static analysis, the floating platform under static dead weight (including its own weight, stage props or towers) and different live load scenarios including eccentric patterns (due to personnel movement) (for example, Fig. 10b) were studied to determine the stress distributions and deflections (tilt movements) of the platform as well as the maximum loads at the connecting system. For structures (such as multi-media screens) that are subjected to wind loading, the loads transmitted to the top deck of the platform have to be considered.

The stresses are checked against material strength requirements while the vertical deflections were checked against serviceability requirements. Figure 10c shows the vertical deflection contours in the plan view of the floating platform under static dead weight and the live load scenario depicted in Fig. 10b. The platform is not flat and is in sagging condition. In this example, the platform displaces a maximum of about 500 mm towards middle of the platform to a

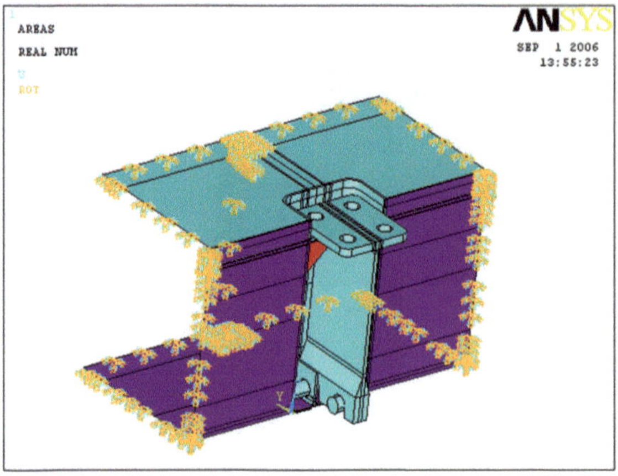

Fig. 11 Boundary conditions for localized finite element model of side connector and its housing

minimum of about 300 mm at the sea side of the platform. For the thin-plated floating platform, the maximum deflection is less than *L*/400, where *L* is the span of the platform. Stresses in all steel plates and stiffeners owing to the most unfavourable loading conditions, have to remain below the allowable factored material yield of 188 MPa (based on 0.8 × allowable material limits of 235 MPa). Figure 10d illustrates the von-Mises stress on the top deck plate.

From the results of the global response analysis, the local stress response at both the corner and side connectors under combined loading was evaluated using detailed structural modelling and static analysis. Figure 11 shows the local structural model for the analysis of the stress response of the side connector with its housing. Figure 12 shows the results of the finite element analysis of the corner and side connectors in terms of von-Mises stresses.

In areas of high stress concentration, such as pipe sleeve or connector pin, high tensile steel (355 MPa) is used for the corner connector. The side connector is constructed of the high tensile steel material to keep stresses within their allowable tensile stress limits.

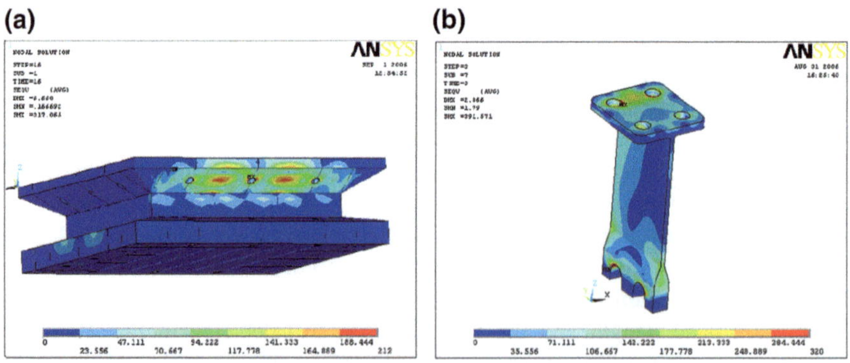

Fig. 12 Plots of von-Mises stress. **a** Corner connector and **b** side connector

For the mooring system, the design of the piles was performed considering the operational and survival environmental conditions: wind speed (20–30 knots), water current (2 knots) and significant wave height (0.5–1 m). For the sizing analysis, the floating platform was considered as a rigid body since it has large flexural rigidity in the horizontal direction. The extreme scenario where the forces are assumed to be co-linear and act simultaneously in the same direction both perpendicularly to the long side and parallel to the long side of the floating structure was simulated. These forces are calculated using the force pressure balance to derive the lateral loads, which are then distributed among the piles. The wave force (using Morison's formula) and current acting directly on each pile are added to its load carrying capacity. The horizontal displacement of the pile is adequately controlled. The upper and lower column of the piles has an outer diameter of 914 mm and is bolted together by 20 bolts, each of allowable tensile strength of 450 MPa.

For the access bridges, the two outer bridges were designed for performers or personnel and use by pedestrians. The centre bridge is assessed by both vehicle (one 30-ton vehicle) and people loading (400 kg/m^2). Based on the loading assumptions and the structural arrangement, the required minimum section modulus for the bridge structural members (truss girders, longitudinals and transverses) and the required minimum deck plate thickness have been assessed in accordance with the Canadian Institute of Steel Construction (CISC) beam formula and ABS "Rules for Building and Classing Steel Vessels for Services on Rivers and Intracoastal Waterways" (1997) respectively. The Lloyd's Register's "Rules and Regulations for the Classification of Linkspans" (1998) was used to define the load cases and to establish the maximum allowable stress and deflection criteria, which were used in defining and evaluating Finite Element (FE) model. The FE analysis was conducted to assess the distribution of stresses for the bridge truss girders in response to the maximum loading conditions as defined by the Lloyd's Register "Linkspans" rules. The Lloyd's Register rules define load conditions for vehicle loading, and for people loading.

4.2 Buoyancy and Stability Analysis

When the floating platform is fully loaded it sinks to a draft of 0.5–0.6 m, allowing adequate freeboard to prevent any water from getting onboard.

The robustness of the floating platform and each single pontoon is checked for its stability. The design of the pontoons and the floating platform meets the International Maritime Organisation's (IMO) guidelines for intact and damaged stability, with each pontoon having at least two-compartment subdivision status. The intact and damage stability computations of the single pontoon and platform are based on static and calm water conditions. The assessment of the stability was carried out in accordance with IMO's guidelines for stability criteria, Resolution A.469(XII) "Guidelines for the Design and Construction of Offshore Supply Vessels". Different loading conditions were considered and the "worst" deck loading generally corresponds with the maximum deck loads, resulting in the least freeboard.

For the intact stability, the metacentric heights (GM) for the different loading cases are way above the stated criteria of 0.15 m. Comparing with the computed limiting VCG curve, the stability of the pontoon and platform for all loading conditions are well within acceptable limits.

The floating platform is assessed according to "two-compartment" damaged stability. For each floating platform configuration, there are more than 10,000 unique two-compartment flooding combinations to be analysed. Given that the IMO damage criteria as applicable to the floating platform is defined in terms of submergence of the margin line (i.e. deck edge) and positive stability, several factors were considered to reduce the number of computational cases to a manageable number, for example:

- Symmetry in configuration and compartment sizes;
- Similarity in the righting arm curves for light and loaded conditions;
- Maximum draft condition is always the limiting load condition;
- Maximum reduction in freeboard occurs when the compartments that are damaged are those which create the largest heel and trim.

The critical damaged condition occurs when damaging the largest adjoining compartments whose joint centroid of buoyancy is furthest from that of the intact floating platform. The damaged stability was thus restricted to the largest draft condition in each configuration and the damaged compartments that gave the worst angles of heel or trim by virtue of the distance from the centre of buoyancy. For all cases, the damaged stability for the floating platform and pontoons meets the IMO criteria by wide margins. Owing to the large horizontal dimension of the floating platform, it remains stable, even in the event of flooding of a few compartments.

4.3 Frequency Analysis

An important parameter for hydrodynamic analysis of floating structure is the natural frequency of the structure. To avoid resonance and any response enhancement, the natural frequency of the floating platform should be significantly less than the frequency of exciting forces. The anticipated forces include effects of the marching contingency or performers during the NDP, jogging or running. Figure 13 shows the first two vibration modes of the floating platform and the natural frequencies are 0.99 and 1.02 Hz respectively. These were low frequency motion and less than the typical exciting frequencies of 2–3 Hz.

Besides ensuring that the access bridges were designed with adequate strength, deflection and vibration (resulting from human movement) analyses were performed to determine the required structural stiffness. The natural frequency of each of the two segments were computed and measured. These were found to be above 4 Hz.

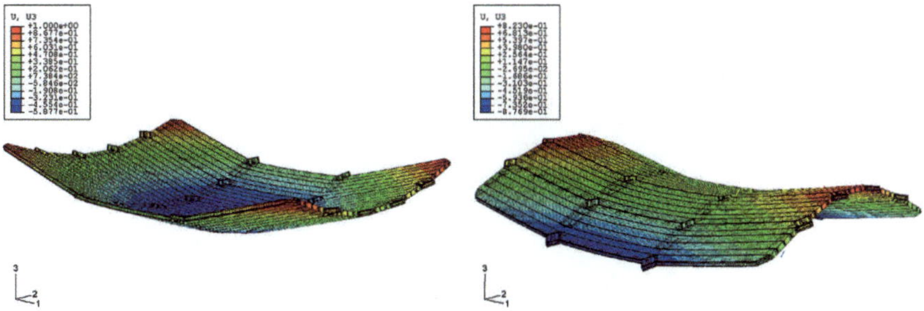

Fig. 13 Vibration frequencies and mode shapes. **a** First mode: 0.99 Hz and **b** second mode: 1.02 Hz

4.4 Hydroelastic Analysis

As the floating performance stage is relatively flexible due to its low profile in relation to the length dimensions, it exhibits elastic behavior similar to a thin plate when subject to wave actions (Fujikubo 2005). Waves caused heave oscillations, roll and pitch and variances in buoyancy along the length and width. Hence there was a need to perform hydroelastic analyses to estimate the structure and connector loads when the floating platform was exposed to the action of waves found in the Marina Bay (Wang et al. 2008a).

The challenge posed in the hydroelastic analysis was the modelling of an equivalent plate that captures the dynamic flexural behaviour of the floating stage. The floating platform was modelled as a thin plate and numerical technique was used to solve the fluid-structure interaction to determine the vibration frequencies and model stresses of the floating structure. This is done by determining an equivalent solid plate model (having the same length and overall depth dimensions as the actual structure), but its Young's modulus and Poisson's ratio were adjusted to furnish the same flexural stiffness and deflection behaviour as the floating platform. To do this, a free vibration of the floating platform modelled by the detailed finite-element model was performed. By adjusting the Young's modulus and Poisson's ratio so as to match the free vibration frequencies and mode shapes of the equivalent plate model with the results of the detailed finite-element model of the floating stage, the Young's modulus and Poisson's ration were determined for the hydroelastic analysis. To account for the flexibility at the connecting lines of the pontoons, plate strips (1.2 m in depth) along the boundaries were adopted to simulate the connecting system (see Fig. 14).

The analysis shows that the hydroelastic response of the floating structure is relatively mild due to the calm waters in the Marina Bay when compared to the static bending response from imposed loads (Fig. 15).

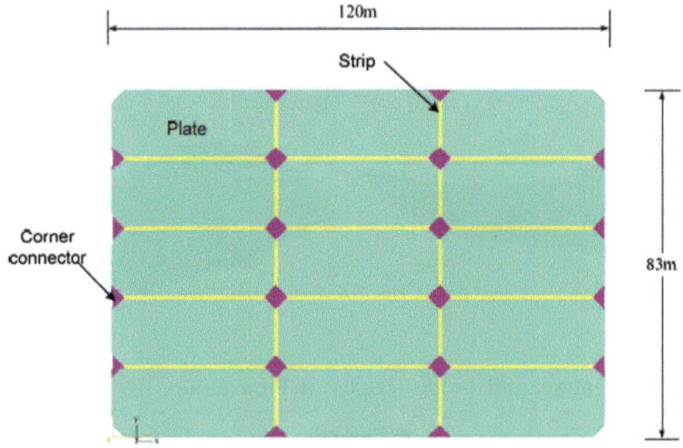

Fig. 14 Proposed equivalent plate model

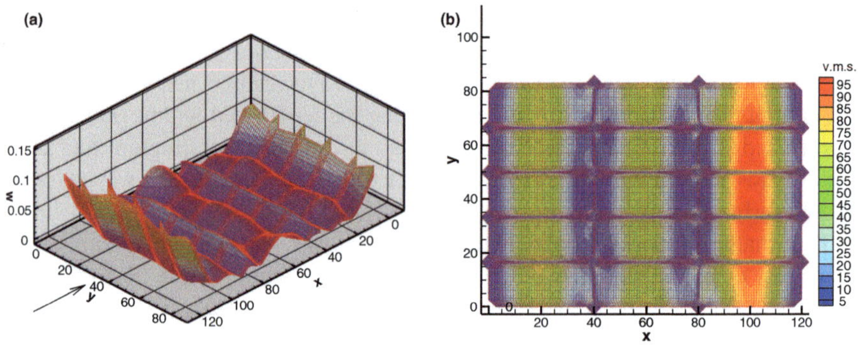

Fig. 15 Deflection and corresponding von-Mises stress in the bottom plate of floating platform under head sea of wave period = 2.4 s and water depth = 0.8 m

4.5 Heave Analysis

Another important area to study is the dynamic behaviour of the large floating platform in a heaving harmonic motion when some 9,000 performers or people simultaneously jumping in a mass event such as the National Day Parade. The objective is to estimate the heaving response of the floating structure so as to establish whether the participants will be comfortably at ease during their performances (Wang et al. 2008b).

By assuming the platform as a rigid body, the added mass and damping coefficient are first estimated. This requires the evaluation of a three-dimensional velocity potential governed by Laplace's equation. The Green's function method was used to solve the boundary value problem for frequencies ranging from 0.01 to 0.50 Hz to obtain the added mass and radiation damping coefficient. The added mass decreases with increasing frequency values, with rapid decrement at the lower frequency values. For high frequency values, the added mass becomes relatively small. The radiation damping coefficient first increases with respect to increasing frequency values and reaches a peak value at about 0.05 Hz, and then it decreases finally to a zero value for a very high frequency value.

Based on the estimated added mass and radiation damping, the finite element software NASTRAN was employed to solve the transient dynamic problem for the displacement and acceleration responses. The impulse force used for the transient problem is based on experimentally obtained ground reaction force (due to a human jumping) as measured by using a force plate.

Figures 16a, b show the heaving displacement with responses (positive sign indicates downward direction) for 2 and 50 s durations, and Fig. 16c, d present the acceleration response (positive sign indicates downward direction) for 2 and 50 s for the full-size platform when the performers jumped up and down once. These results are the same for more than one jump. As anticipated, owing to the relatively high natural period for heaving and critical damping factor, the transient displacements (maximum value of 0.30 mm) and accelerations (maximum of 0.0105 m/s^2) are relatively small. These transient responses will not be perceived by the performers. The large size floating platform provides a much greater inertia (due to greater added mass and radiation damping coefficient) in resisting motion. This advantage cannot be realized when the platform is supported at its edge and suspended in air. This human response analysis provided the assurance that the floating stage is suitable for mass performance and sporting activities.

4.6 Hydrodynamic Model Tank Tests

Model tank tests were performed to study the behaviour of the floating platform in water. Several combinations of parameters were investigated, including loading conditions, tidal variations, wave heights and water speeds. Measurement data concerning the motions and mooring loads in both operational and extreme environmental conditions were collected for engineering analysis. In particular, the experimental results obtained allowed the calibration of the wave loading applied to the global finite element analysis. Figure 17 shows the hydrodynamic model test to examine the performance of the floating platform.

5 Construction, Assembly and Verification Testing

To minimise on-site welding, the pontoons, connectors and access bridges were pre-fabricated off-site and transported to the Marina Bay.

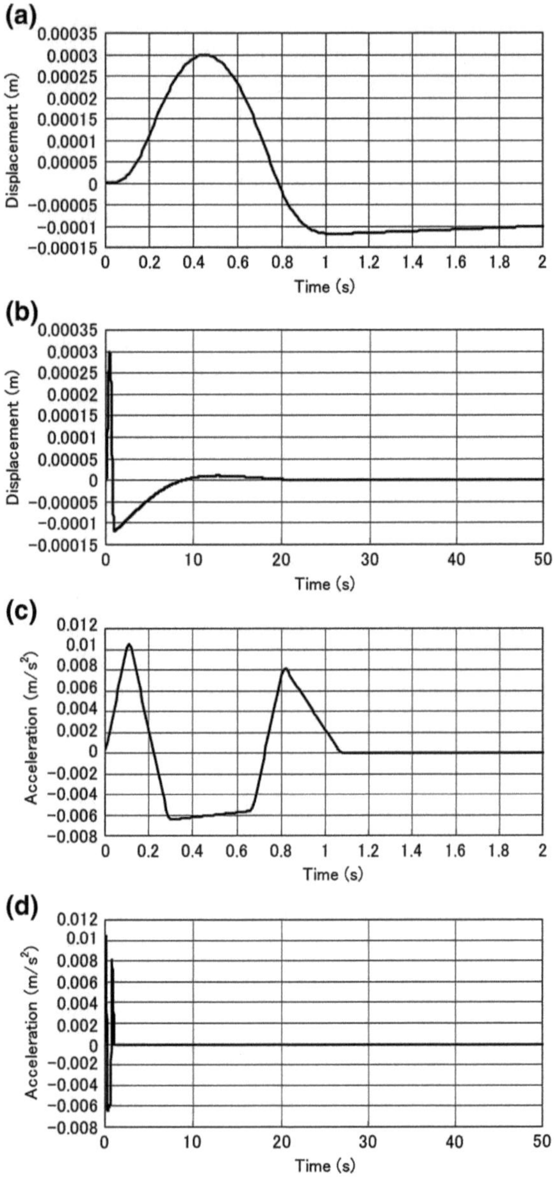

Fig. 16 Heaving displacement response for full-size platform **a** until 2 s and **b** until 50 s. Heaving acceleration response for full-size platform **c** until 2 s and **d** until 50 s

5.1 Installation of Fixtures at Site

Concurrent with the pre-fabrication off-site, the supporting piles for the three access bridges and the mooring piles were installed at the Marina Bay site (Fig. 18).

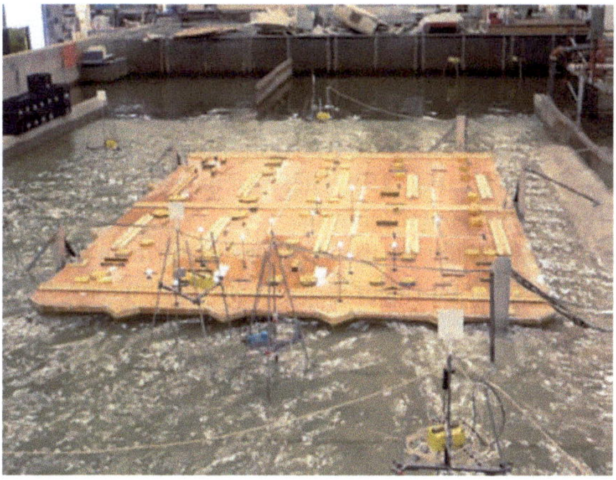

Fig. 17 Hydrodynamic model test

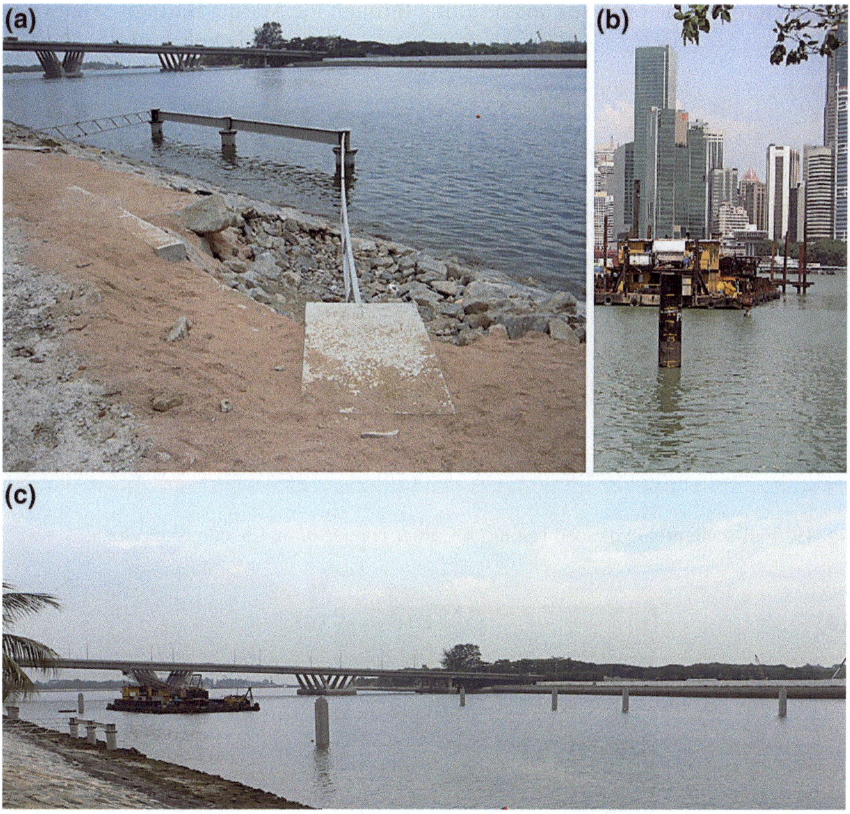

Fig. 18 **a** Installing supporting piles for access bridge. **b** Installing mooring piles. **c** Completed fixtures in Marina Bay

5.2 Component and Full-Load Testing

In addition to numerical computations, prototype model of the corner and side connectors were constructed and load-tested in the workshop to verify the stresses and elastic deformation (see Fig. 19). These were measured and the results were compared with the finite element analysis.

Before the final installation of the floating platform in the bay, the mooring piles were also subjected to pull load test to verify its holding strength and elastic deformations (see Fig. 20).

5.3 Transportation to Site and on-Site Assembly

The pontoons were transported or towed individually by tugboats from the shipyard where they were fabricated, to the Marina Bay. The journey included a passage through the open sea and a sailing through the Marina Barrage, which was

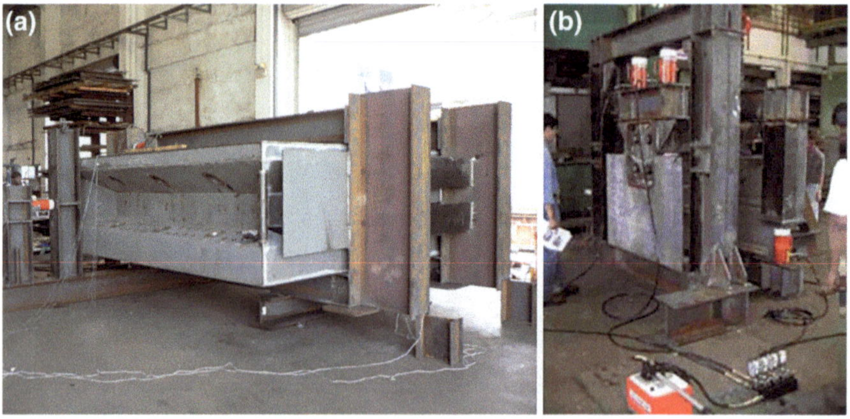

Fig. 19 Full-scale prototype load testing. **a** Corner connector and **b** side connector

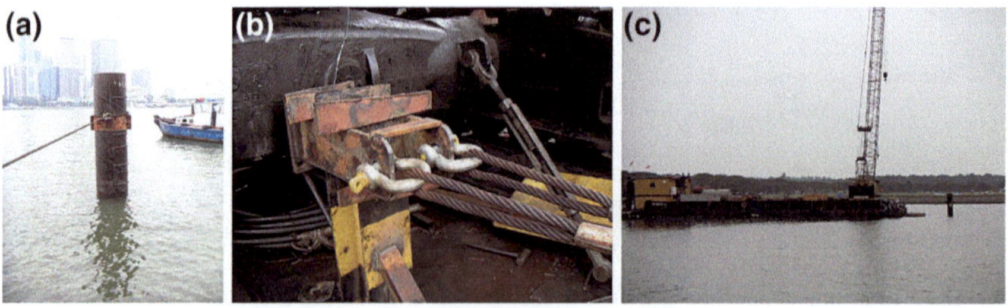

Fig. 20 Load testing of mooring piles. **a** Pull cable attached to pile. **b** Pull cable attached to winch onboard barge and **c** pull load test set-up with barge

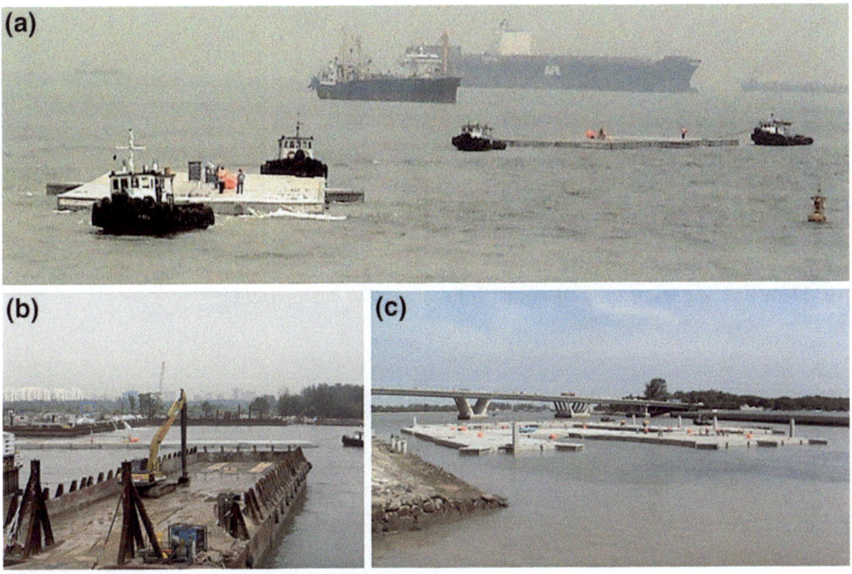

Fig. 21 **a** Pontoons being towed by tug boats. **b** Passing through the Marina Barrage. **c** Arriving and assembly of floating pontoons on-site

not fully closed, at the mouth of the Singapore River (Figs. 21a, b). This is possible owing to the modular design. Once at site, the pontoons were assembled using the connecting system to form the floating platform, which was then secured to the mooring piles (Fig. 21c).

5.4 Erecting the Access Bridges

The access bridges were then erected and assembled. Because the components were to be transported by truck, the access bridges components were limited to transportable size of approximately 2.5 m by 10 m. At site, the primary structure was bolted together by way of boltable flange connections and installed in place. Once the primary structure was in place, the deck panels were installed (Fig. 22).

5.5 On-Site Testing

Extensive full-scale load tests were conducted on the platform at the site to validate the floating platform design and to ensure that the platform could withstand the large loads of performers, vehicles and stage props (Fig. 23).

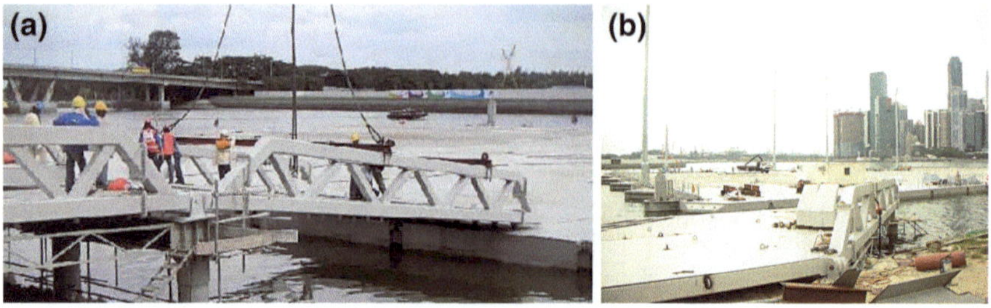

Fig. 22 **a** Erecting the truss girders and **b** installing the deck panels

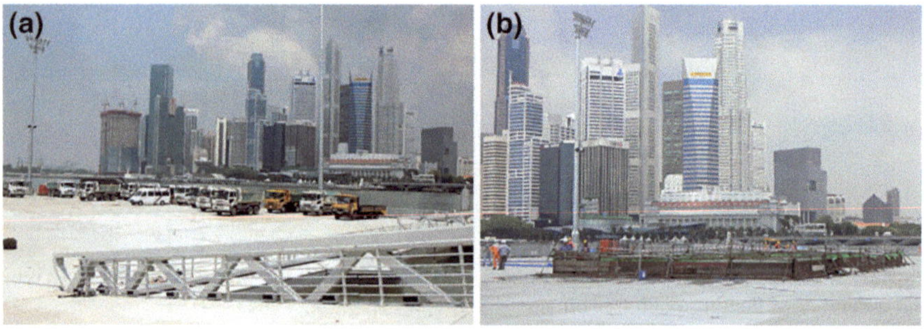

Fig. 23 Load testing of the floating platform **a** extreme vehicle load test **b** water containment load test

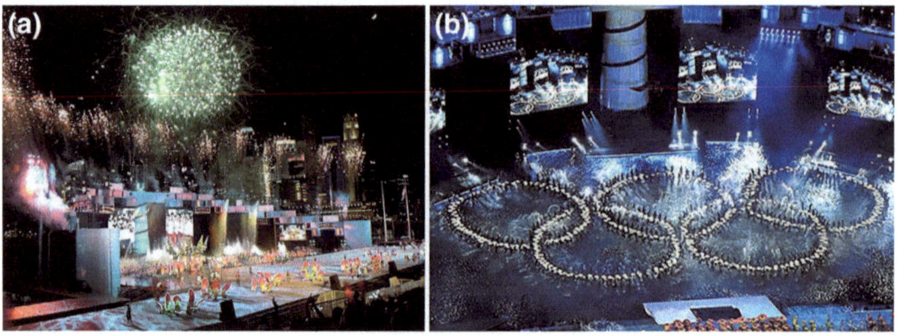

Fig. 24 Staging the opening and closing ceremonies of the inaugural Youth Olympic Games 2010 on the Marina Bay floating platform (*Photo courtesy* en.Wikipedia.org and flickr.com)

6 Conclusion

The Marina Bay floating platform is a large-scale pontoon structure and is unique and iconic from the structural engineering perspective. It has been staging the annual NDP celebrations and hosted the Opening and Closing ceremonies of the inaugural Youth Olympic Games in 2010 on water (see Fig. 24), and is now a landmark on the Singapore water. It serves to open up new frontier for space creation on water in land-scare countries and complement land reclamation, construction of high-rise buildings or development of underground caverns. It will certainly inspire many applications of the floating structure technology in floating storage facilities, mobile offshore platform, floating islands or cities and floating airports or industrial facilities in the future.

References

Fujikubo, M. (2005). Structural analysis for the design of VLFS. *Marine Structures, 18*, 201–226.

Koh, H. S., & Lim, Y. B. (2007). The floating platform at the Marina Bay: The challenges. In *Proceedings of IStructE Asia-Pacific Forum on Structural Engineering: Innovations in Structural Engineering,* Singapore.

Koh, H. S., & Lim, Y. B. (2009). The floating platform at the Marina Bay, Singapore, *Structural Engineering International: Journal of the International Association for Bridge and Structural Engineering (IABSE), 19*(1).

Koh, H. S., Lim Y. B., Seow T. H., & Thapar, A. (2008). The Floating performance stage @ Marina Bay Singapore–New possibilities for space creation. In *Proceedings of the International Conference on Offshore Mechanics and Arctic Engineering*, Portugal.

Suzuki, H. (2005). Overview of megafloat: Concept, design criteria, analysis and design. *Marine Structures, 18*, 111–132.

Wang, C. M., Song J. H., Utsunomiya, T., Koh, H. S., & Lim Y. B. (2008a). Hydroelastic analysis of floating performance stage @ Marina Bay Singapore. In *Proceedings of the International Conference on Offshore Mechanics and Arctic Engineering*, Portugal.

Wang, C. M., Utsunomiya, T., & Koh, H. S. (2008b). Heaving response of a large floating platform. *IES Journal Part A: Civil and Structural Engineering, 1*(2), 97–105.

Watanabe, E., Wang, C. M., Utsunomiya, T., & Moan, T. (2004). Very large floating structures: applications, analysis and design, *CORE Report No. 2002–2004*, Singapore: National University of Singapore.

Yumemai Floating Swing Arch Bridge of Osaka, Japan

E. Watanabe, T. Maruyama, S. Ueda and H. Tanaka

Abstract The Yumemai Bridge is a floating swing arch bridge constructed across a water channel and supported by two hollow steel pontoons in the Port of Osaka. It connects two reclaimed islands to the regional road network and can swing around a pivot axis near one end of the girder with the assistance of a tug boat. The design of the bridge specifically addressed the action of waves, wind and earthquakes, the required swinging mechanism and the issue of durability. The pontoons are made corrosion-free with their side lined with titanium plates and with the other wet surface by cathodic protection. The mooring system consists of dolphins with movable reaction walls and rubber fenders. The reaction walls are located on the fixed dolphins with piles driven in the seabed. The structure is designed to be strong and stable enough to withstand typhoon-level winds and waves.

1 Introduction

The City of Osaka had undertaken the "Techno port Osaka" project to develop in its waterfront area a new metropolitan center with advanced features for the twenty first century. This project covers the construction of three reclaimed islands, Maishima, Yumeshima and Sakishima, in the waterfront area where "shima"

E. Watanabe (✉)
Osaka Regional Planning Institute and Kyoto University, Toyonaka, Osaka and Kyoto, Japan
e-mail: watanabe@rpi.or.jp

T. Maruyama
Komaihaltec Inc., Osaka, Japan

S. Ueda
IDEA Consultants Inc. and Tottori University, Tokyo and Tottori, Japan

H. Tanaka
Yoshida-Gumi, Osaka, Japan

C.M. Wang and B.T. Wang (eds.), *Large Floating Structures*, Ocean Engineering & Oceanography 3,
DOI 10.1007/978-981-287-137-4_3

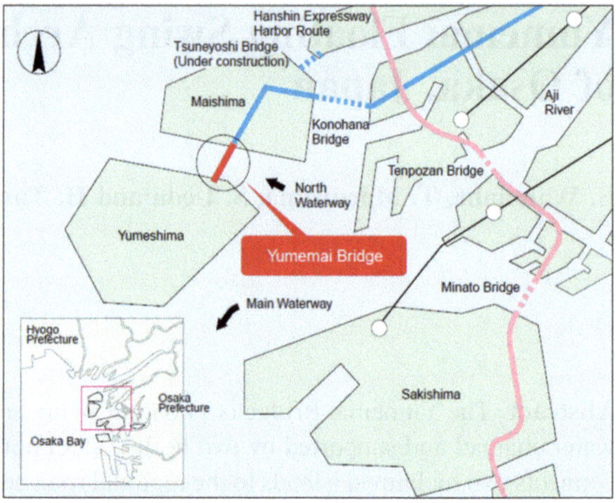

Fig. 1 Construction site of Yumemai Bridge

implies island; whereas "Mai", "Yume" and "Saki" implies dance, dream and bloom, respectively. The location of these islands is shown in Fig. 1. The artificial islands of Maishima, Yumeshima and Sakishima were planned to be reclaimed either for residential, commercial and various amenity facilities. At the time of construction stage, Yumeshima was the last island to be reclaimed and in fact, it was not fully reclaimed at the time of completion of the Yumemai Bridge. Furthermore, Sakishima has been completed much earlier than the other two islands.

Under these circumstances the City of Osaka planned the construction of the Yumemai Bridge, with a view not only to contribute to accelerated development and improvement of these reclaimed islands, but also to play an important role in transportation access to the waterfront area.

The channel between Yumeshima and Maishima, called "North Waterway," as a subsidiary to the main waterway located between Yumeshima and Sakishima had been providing a passage mainly for small crafts. These two waterways are the only routes via which ships and boats can access to major facilities in the Port of Osaka. If the main waterway becomes unusable due to an accident or for other unforeseeable reason, the North Waterway needs capacity as an international passage through which even large vessels can navigate.

To ensure marine passage in emergency, it was decided to construct a movable bridge over this North Waterway. Compared with tunnels and ordinary fixed bridges with large vertical navigation clearance, a movable bridge is considered to be much advantageous in terms of costs, construction time, and land use.

The Yumemai Bridge was constructed as a floating swing arch bridge, the world's first type of movable bridges (Fig. 2). It comprises a floating bridge over the waterway, transitional girder bridges at both ends of the floating bridge, and approach bridges on both grounds of Yumeshima and Maishima. This floating bridge, a large arch bridge structure floating on two steel pontoons (58 m × 58 m × 8 m), is horizontally supported by two mooring dolphins with rubber fenders. When positioned in normal service, the floating bridge

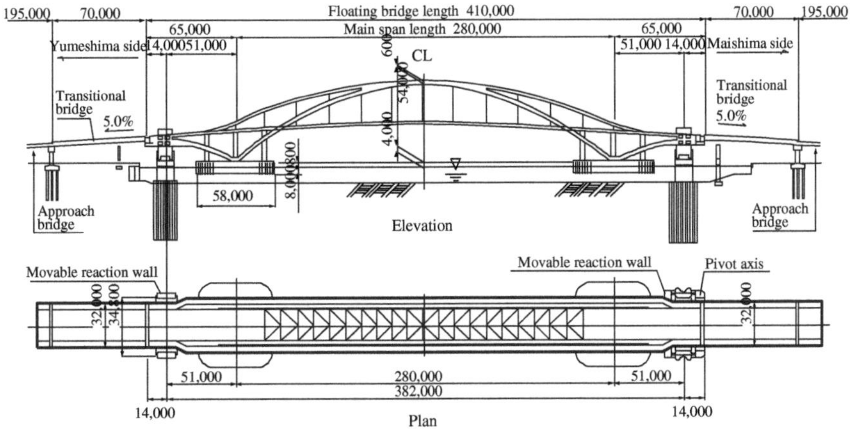

Fig. 2 Plan and elevation of Yumemai Bridge, a floating swing arch bridge

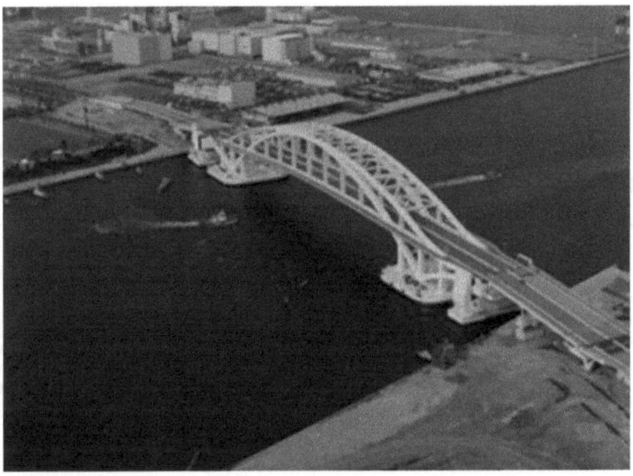

Fig. 3 Panoramic view of Yumemai Bridge after the on-site installation, courtesy of Osaka construction industry times

accommodates a navigation passage width of 135 m (see Fig. 3). In emergency, when the main waterway is out of service, the entire floating bridge is swung by tugboats to widen the passage width (200 m or more) enabling the passage for large vessels (Maruyama et al. 1998a–d; Watanabe et al. 1998).

2 Selection of Bridge Type

Normally, lift bridges, swing bridges, retractable bridges, and bascule bridges are typical types of movable bridges. These movable bridges, excluding the bascule bridge, were subjected to a comparative study for suitability to the Yumemai Bridge under a condition of crossing a relatively wide (200 m) navigation passage in emergency. During the preliminary investigation stage, a floating swing type

Table 1 Comparison of various movable bridges

	Lift bridge (Curved chord truss through bridge)	Floating bridge (Pool-type upper side supported bridge)	Floating bridge (Swing type bridge)	Swing bridge (Cable-stayed bridge)	Retractable bridge (Abacus bridge with lifts)
Schematic (Unit: m)	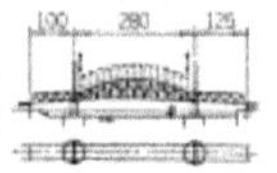	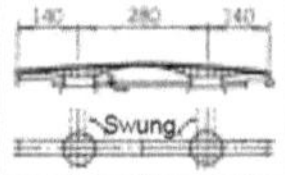	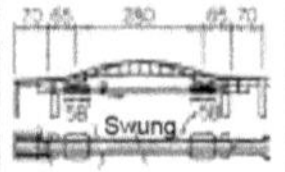	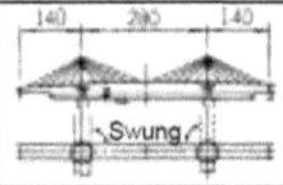	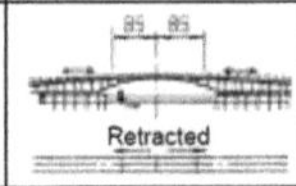
Economy	×	×	○	○	×
Structural rigidity	○	△	△	○	×
Construction ease	△	△	○	△	△
Movable performance	○	△	△	△	△
Ground displacement	×	○	○	△	△
Serviceability	○	△	△	○	×
Maintenance ease	△	△	○	△	△
Overall rating	△	×	◎	○	×

Note ◎: excellent; ○: feasible; ×: infeasible; △: fair

was studied in particular detail from various aspects, including mooring method. Table 1 summarizes the result of the comparison. A floating swing bridge and a cable-stayed swing bridge were found to have economic advantage over the other types. Closer investigation of these two candidates led to choose the floating bridge of swing type, for the following reasons:

(1) The bridge over the North Waterway will not have to be opened so frequently unless something unexpected happens. When it becomes necessary to be opened, the floating bridge can be swung by tugboats, requiring very little power and minimum driving equipment. In addition, opening by towing is accurate.
(2) Yumeshima was still in the process of reclamation at the time the construction started, and ground displacement and subsidence by consolidation are inevitable. With a floating bridge, the influence of ground displacement and subsidence on the bridge and bridge driving system can be minimized.
(3) The bridge is erected and assembled at a large dockyard and towed to the installation site. Since the superstructure and substructure can be constructed simultaneously at different locations, a floating bridge can substantially save the construction time.

For the superstructure of the floating bridge, single-rib arch, double-rib arch, and truss designs were proposed. As the result of comparison, the double-rib arch design was selected because of its superior overall rigidity and the potential for minimizing wave influence and local distortion by uniformly distributed flexural and torsional rigidity along the bridge axis. The aesthetic effect as a symbol of Osaka City was also taken into account (Maruyama et al. 1998a–d).

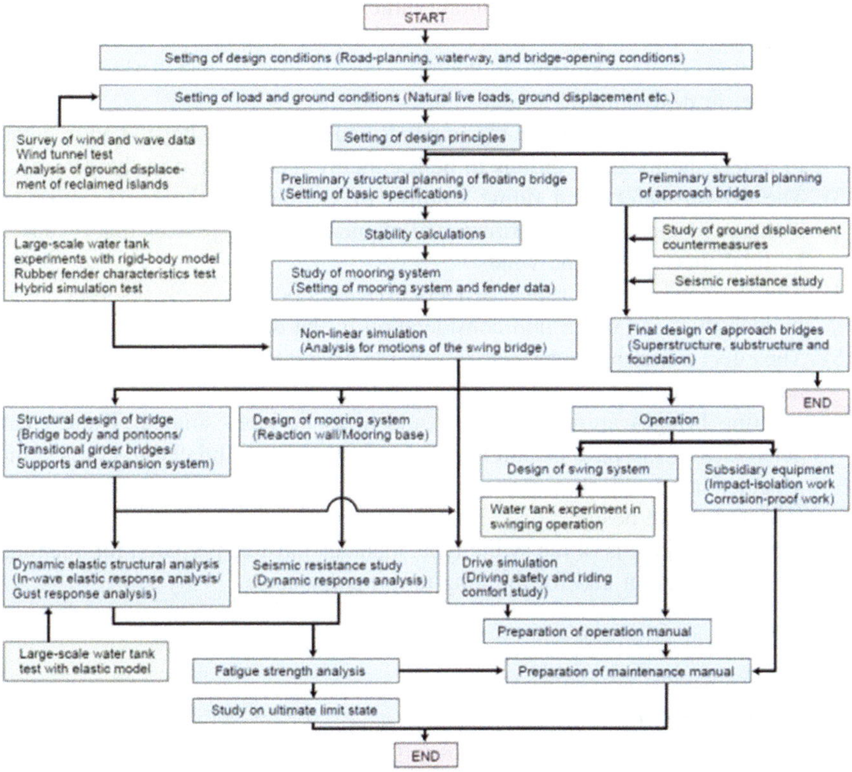

Fig. 4 Flow chart of Yumemai Bridge

3 Technical Challenges

"The Japanese Specifications for Highway Bridges (JRA 2012)" must be used for bridge design in Japan at the present time but the design was made at that time referring to the older version (JRA 1996). In designing this floating bridge, however, these standards alone were not sufficient. It was necessary to establish new design techniques and a new concept of design safety factors. Floating structures have been one of the main focuses of investigation in various occasions, e.g., at the time of planning the Kansai International Airport, and in the "Mega-Float Project" led by the Ministry of Transport. Among large floating structures now in service in Japan are oil storage terminals at Kami-goto (Nagasaki Pref.) and Shirashima (Kita-kyushu City). With reference to the data and experience thus accumulated, various technological challenges had to be made in the process of designing the Yumemai Bridge. Figure 4 shows the flow chart showing this process. The major technical challenges are listed below (Maruyama et al. 1998a–d; Watanabe et al. 1998, 1999, 2001a, b, 2002, 2003a, b; Watanabe and Utsunomiya 2001, 2003; Watanabe 2003, 2004):

(1) Since a floating bridge is far more vulnerable to meteorological and oceanographic conditions than a conventional fixed bridge, proper environmental conditions that are suitable to the characteristics of the installation site must be met in designing the bridge.
(2) The motion of a floating bridge in winds and waves must be studied in detail to reflect the local conditions to be accurately incorporated in the design.
(3) The driving safety and riding comfort of vehicles on the bridge must be maintained against the deformation of alignment of bridge deck that is caused by tidal flux, waves and winds.
(4) The characteristics of rubber fenders used as mooring shock absorbers must be identified and taken into consideration in the design.
(5) The effect of ground displacements on the bridge structure must be evaluated and taken into consideration in the design.
(6) Floating bridges are generally considered to be base-isolated. These characteristics of the base-isolation must be proved and validated to confirm the seismic safety so that the seismic performance must also be desirably reflected in the design.
(7) Since this bridge is of the swing type, manuals must be prepared for opening/closing operation and for the long-term maintenance.

4 Design Conditions

4.1 Meteorological and Oceanographic Conditions

The basic design wind velocity (V_{10}: wind velocity at height of 10 m) was set to be 42 m/s with a 100-year return period, based on wind velocity data obtained near the installation site and the observation records in the past from 1931 to 1995 supplied by the Osaka District Meteorological Observatory. For bridges near the installation site, the regulation prescribes the traffic safety wind velocity limit at $V_{10} = 20$ m/s. This value was adopted as the design condition. The wind velocity limit for safe bridge opening/closing was set to be $V_{10} = 15$ m/s, on the basis of the marine operation standard for the Port of Osaka. As for design tidal level, a tidal fluctuation between DL + 4.8 m (design high tide) and DL − 0.52 (ultra-low tide) was assumed, the design datum level (construction datum level) being CDL + 0 m. The design wave was set to be the significant wave height of $H_{1/3} = 1.4$ m, based on typhoon and gale data for the past 40 years from 1956 to 1995. Moreover, the wave diffraction calculations for the waterway and the large-scale water tank experiments were conducted. Based on the wave spectrum observation at the wave observation tower in Osaka Bay, the simulation of bridge drift in winds and waves was conducted by employing the Bretschneider-Mitsuyasu type wave spectrum.

The design tidal current velocity under ordinary conditions was set to be 0.2 m/s on the basis of existing data; that under storm condition, in spite of the lack of data, was set to be 0.5 m/s by estimation taking into account topography of the bridge site.

4.2 Earthquake

"Expected earthquake," taking into account the influences of the active faults, topography, geology and ground condition of the bridging site, was used to determine the bridge's seismic requirements. Specifically, the design considered two types of expected earthquake waves: one based on the Tohnankai (East South Sea)-Nankai (South Sea) Seismic Fault Line model, which corresponds to a level II type I earthquake (earthquake of Plate boundary type) as provided in the Specifications for Roadway Bridges and Commentary (JRA 1996) and the other based on the Uemachi Active Fault Line model, which corresponds to a type II earthquake (near-field earthquake).

4.3 Tsunami

With respect to tsunami, the height (the maximum change of water level) of tsunami and its maximum flow velocity of 2.4 m/sec were considered respectively based on the result of a numerical simulation on the Osaka Bay. Furthermore, after the Tohoku district great earthquake which occurred on March 11, 2011, the predicted value of the tsunami height due to Tohnankai-Nankai earthquake was calculated and the tsunami height in this area in Osaka is predicted to be 3–4 m. Nevertheless, the more effort must be made in near future to obtain more reliable result at the site.

4.4 Combined Loads, Allowable Stress Increment Factors, and Safety Factors

The design of the Yumemai Bridge is essentially based on existing design standards based on the allowable stress design. However, for combined loads, which are not covered by any of existing design standards, allowable stress increment factors were set using the safety evaluation techniques proposed by the Japan Society of Civil Engineers. Table 2 lists the allowable stress increment factor for each load combination on the floating structure and the mooring dolphins. For each

Table 2 Combined loads and allowable stress increment factors

(a) Floating structure (superstructure, pontoon and pivot pin)

Loading status	Combination of loads	Allowable stress increment factor
Standard	D + U + L + 1	1.00
Temperature	D + U + L + 1 + T	1.15
Storm	D + U + W + WP	1.20
Earthquake	D + U + EQ	1.50
Swinging	D + U + W + WP + DR	1.25
Construction and towing	D + U + W + WP + ER	1.25

(b) Mooring dolphin (including reaction wall, reaction wall supporting beam, reaction wall anchor frame, RC dolphin, steel pipe sheet pile well foundation, and floating structure fender installing section)

Loading status	Combination of loads	Allowable stress increment factor
Standard	D + U + GD	1.00
Temperature	D + U + T + GD	1.15
Storm	D + U + W + WP + GD	1.50
Earthquake	D + U + EQ + α* GD (α coefficient)	1.50
Swinging	D + U + W + WP + DR + GD	1.25

D Dead load
L Live load
I Impact
U Uplift of buoyancy
E Earth pressure
W Wind load
T Effect of temperature fluctuation
EQ Effect of earthquake
WP Wave pressure
PD Tidal force
GD Effect of ground displacement
SD Effect of supporting point displacement
DR Bridge driving toad
ER Load during erection
CO Ship collision load
TU Effect of tsunami

major component, the margin to ultimate collapse was determined and incorporated in the safety factor.

5 Study on Stability

For the Yumemai Bridge, a long-span floating bridge supported on two pontoons, floating stability is a very important concern. Since this bridge must secure a large vertical navigation clearance of 26 m under the bridge girders, the bridge's center

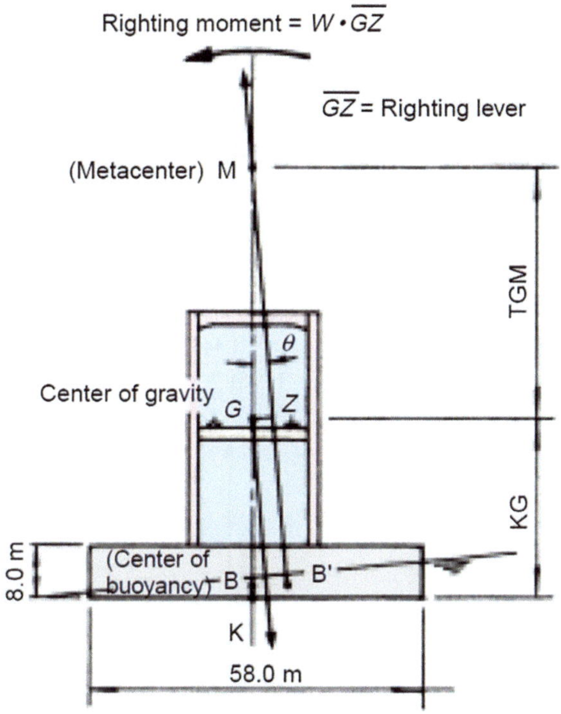

Fig. 5 Centers of buoyancy, gravity and metacenter

of gravity and the wind loading point are necessarily positioned high. Therefore, to ensure its stability careful attention is required. For the initial righting moment to secure a satisfactory static stability of the floating structure, it is essential that the vertical distance between the center of gravity and the transverse metacenter (i.e., TGM), shown in Fig. 5, are positive. The greater the TGM value, the greater the stability of the floating structure.

Using the basic design dimensions of pontoons, static stability calculations were carried out for three cases: without live load (S1), with biased live load (S2), and with full live load (S3). Table 3 shows the results. For all three cases, the calculated values are larger than those for conventional marine structures and ships, verifying that this floating bridge is extremely stable.

Dynamic stability was evaluated using the following formula: Area $(A + B) \geq 1.4 \times$ Area $(B + C)$ in which areas A, B and C are as schematically shown in Fig. 6. To satisfy this formula, the righting moment must be at least 1.4 times the inclining moment. In this manner, the ratio of the required righting moment to the inclining moment was determined to secure the satisfactory dynamic stability. Table 3 also shows the ratio for a 58m × 58m × 8 m pontoon, for each of the above-mentioned three cases.

Table 3 Results of static and dynamic stability calculations

	Case S1 and D1	Case S2 and D2	Case S3 and D3	Condition
Mean draft d (m)	4.80	5.08	5.30	
Displacement Δ(t)	31.445	32.287	34.700	
Height of center of gravity KG (m)	26.38	26.58	26.72	
Transverse metacenter TGM (m)	30.09	27.03	24.92	TGM > 0
Lateral inclination θ (deg)	0.00	1.14	0.25	
DSR = $(A + B)/(B + C)$	1.44	4.97	4.50	DSR ≥ 1.4

(Specific gravity of seawater: 1.025)

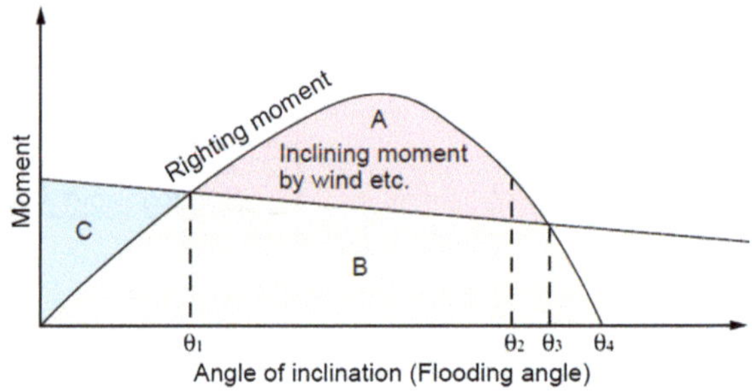

Fig. 6 Dynamic stability study

6 Mooring Method

The floating bridge is supported vertically by the buoyancy of seawater. It must also be supported horizontally to resist such lateral forces as wind, wave, earthquake and tsunami. Horizontal support is achieved by mooring. The following three different mooring methods were compared and assessed for applicability to this floating bridge:

- Anchor chain mooring
- Submersible mooring
- Rubber fender mooring

Comparison revealed that the rubber fender mooring not only the most effectively restricts the bridge motion but also it can lead to the best choice economically. Therefore, the focus of our attention was on the two rubber fender mooring methods: reaction wall and link damper. Figure 7 schematically shows the two mooring methods, and Table 4 compares their characteristics. The reaction wall method has been adopted, due to its superior bridge motion prevention characteristic, and convenience in bridge swinging operation. The rubber fenders used for mooring this bridge have nonlinear reaction characteristics as shown in Fig. 8.

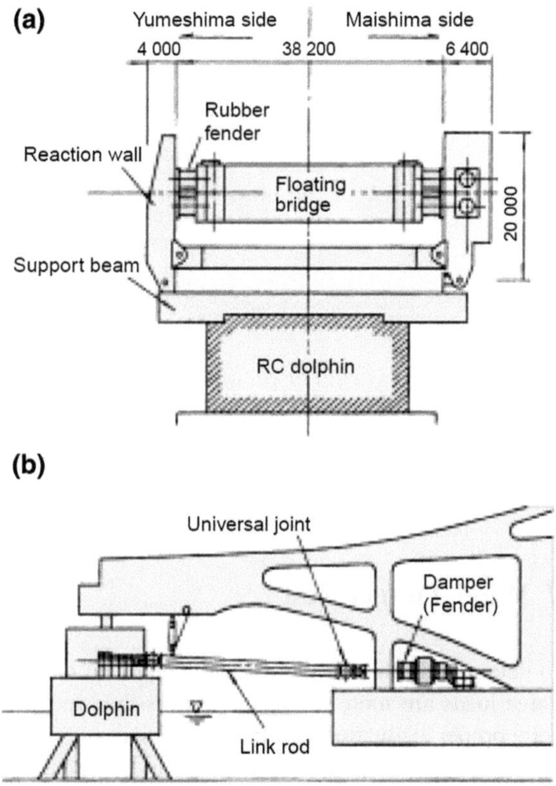

Fig. 7 Schematic of mooring systems. **a** Reaction wall method. **b** Link damper method

Table 4 Comparison of mooring methods

Type / Item	Reaction wall method	Link damper method
Motion of bridge body	Motion in wind and waves is relatively small, since mooring point is at the same level as the center of gravity	Motion in wind and waves is relatively large, since mooring point is at the same level as the pontoon
Opening/ closing operation	Release from the mooring system and positioning of the bridge are relatively easy, since they involve only moving the reaction walls	Rod connection/disconnection requires labor and involves operation of bridge position retaining mechanism
Technical problem	Steel 20 m high movable reaction wall	Link mechanism rod connection/ disconnection mechanism
	Reaction wall operating mechanism, and fixing pin insertion/removal mechanism	
Dolphin (foundation)	Load-acting point is high, resulting in a large moment	Load-acting point is low, resulting in a small moment

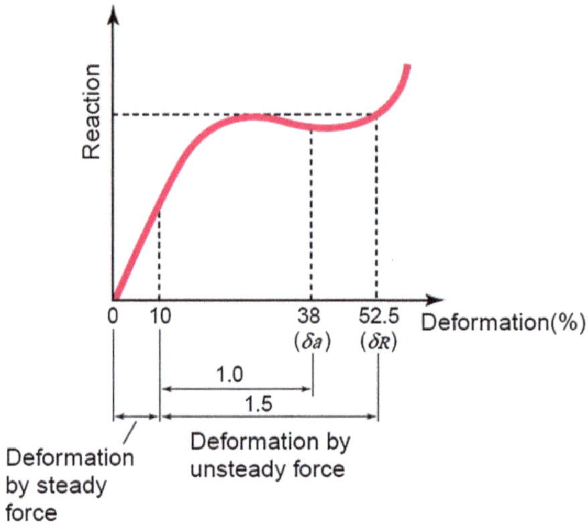

Fig. 8 Relationship between reaction force and deformation of a rubber fender

7 Wind Tunnel Test

In determining the design of each component of a floating bridge, storm wind and wave loads are more influential factors than in the case of a fixed bridge. Therefore, proper evaluation of wind load is necessary. If wind load can be reduced by a relatively simple measure, the cost can be saved. In view of this, the static wind load characteristics (mainly drag coefficient) of this floating bridge was investigated by conducting a wind tunnel test (Maruyama et al. 1995) using a rigid 3D model, and effective wind load reduction measures were sought (see Fig. 9).

The test revealed that the following measures are effective in reducing wind load:

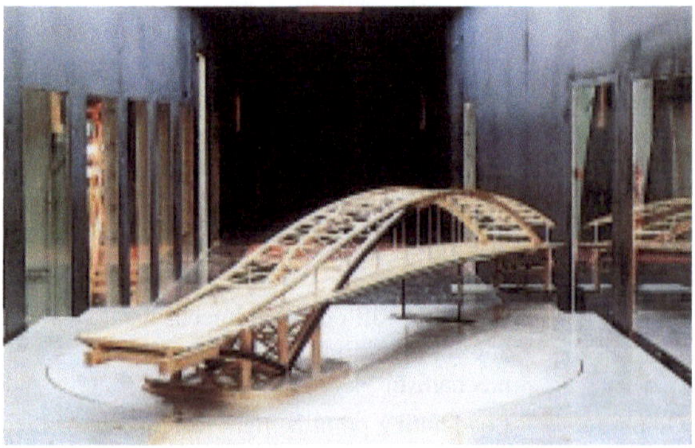

Fig. 9 A wind tunnel test of Yumemai Bridge

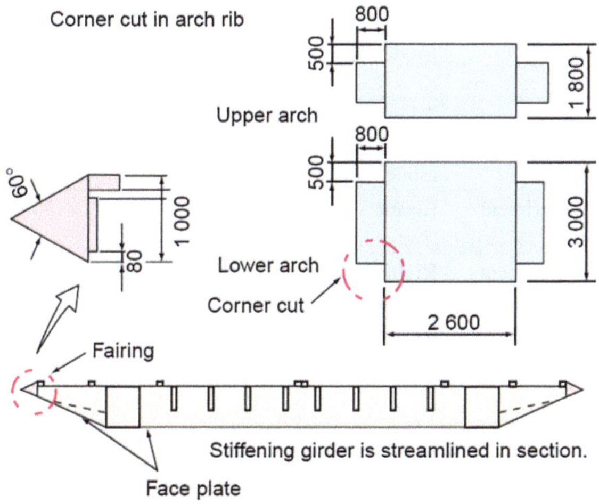

Fig. 10 Measures for reduction of wind load

(1) A corner cut is formed in each side face of the upper and lower arch ribs, so that the aspect angle becomes approximately 30°.
(2) A fairing is provided on both ends of the stiffening girder, and the bottom face of the stiffening girder is closed to form streamlines around the profile of the girder.

These measures reduce the drag force by about 20 %. Consequently, the measures shown in Fig. 10 were applied to the basic box cross section of the stiffening girder of this bridge.

8 Large-Scale Water Tank Experiments and Non-linear Computer Simulations

In designing a floating bridge, it is essential to clarify the bridge's drift characteristics in wind and waves, and to obtain accurate drift characteristic values. Since the Yumemai Bridge would be moored by rubber fenders, a new analytical technique had to be developed that takes into account the non-linear characteristic of the rubber fenders. It was also necessary to determine the effect of the relatively flexible bridge structure's elasticity and to investigate opening/closing safety of the mechanical system.

For designing the floating bridge, a structural analysis program was developed, and large-scale water tank experiments were conducted to verify the appropriateness of simulation-based calculations. Hybrid simulation testing was also carried out to clarify the actual behavior of constant reaction rubber fenders with high nonlinearity. Three different large-scale water tank experiments were conducted, and drift simulation results were validated using three different programs which contain the same basic formulas but were improved for consistency among respective experiments. Table 5 outlines the large-scale water tank experiments and the hybrid simulation test.

Table 5 Large-scale water tank experiments and a hybrid test

	Model	Scale	Water tank size (length × width × depth)	Site of experiment
Experiment 1	Topographic model, rigid model	1/80	50 m × 40 m × *	Tsuchiura, Ibaraki pref
Experiment II	Elastic model	1/40	190 m × 30 m × *	Nagasaki, Nagasaki pref
Experiment III	Rigid model, bearing model	1/80	100 m × 5 m × *	Akishima, Tokyo
Experiment IV	Rubber fender model	1/ 12.5	–	Totsuka, Yokohama, Kanagawa pref

*Depth conforms to water depth at site
For experiment II, height was varied

8.1 Experiment I

Rigid model experiment: a topographic model of 5.125 m length S: 1/80 as shown in Fig. 11

The objectives of the experiment were (Nagata et al. 1999):

(1) To clarify the oceanographic conditions at the bridging site, taking into account the influence of wave diffraction and interference, by accurately reproducing the waterway between Yumeshima, Maishima and the seawall structures
(2) To obtain data regarding the motion of the entire floating bridge, deformation of rubber fenders, and to develop simulation techniques and data that could represent such motion and deformation.

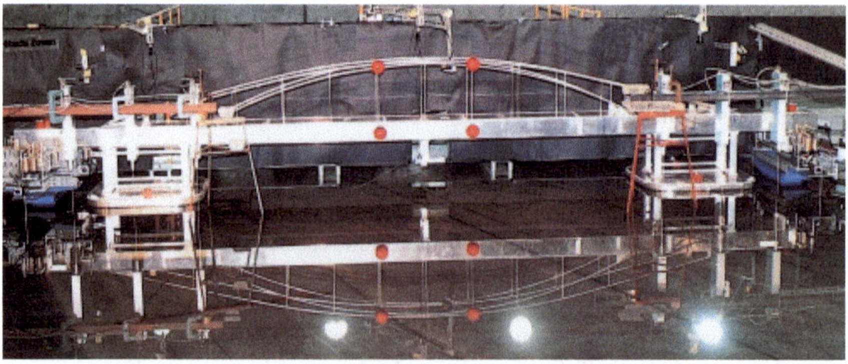

Fig. 11 Rigid model experiment on the in-plane motion of a moored floating structure

8.2 Experiment II

In-wave elastic model experiment: a model of 10.25 m length S: 1/40 as shown in Fig. 12

The objectives of the experiment were (Ueda et al. 1999):

(1) To experimentally investigate the elastic response of the floating bridge in waves
(2) To experimentally verify in-wave elastic response simulation's applicability to the structural design

8.3 Experiment III

Swinging operation experiment with a model of 5.125 m length S: 1/80 as shown in Fig. 13

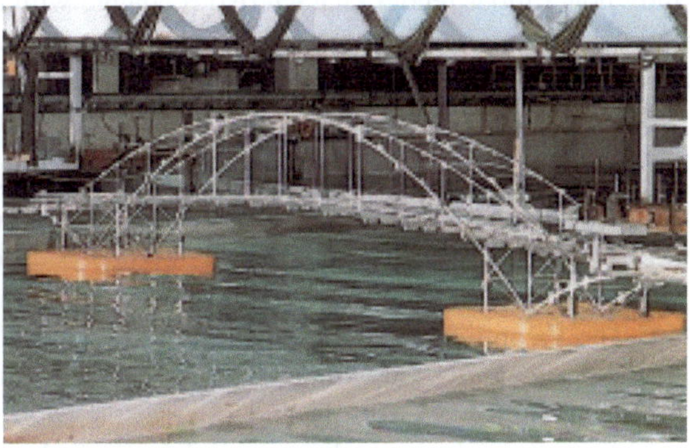

Fig. 12 Elastic model experiment on the in-plane elastic response

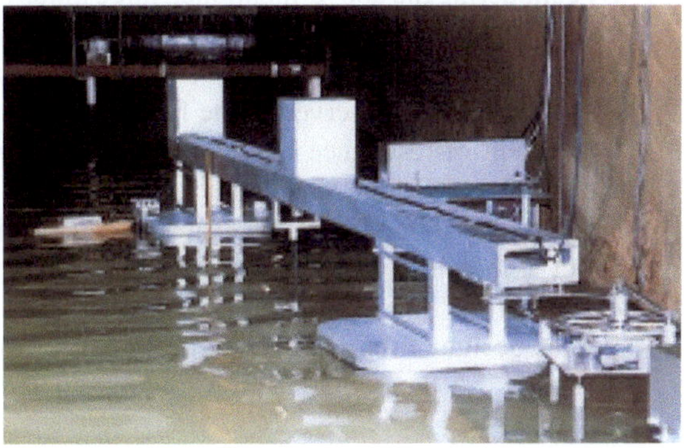

Fig. 13 Experiment for swinging and temporary mooring operations

The objectives of the experiment were (Kobayashi et al. 1999):

(1) To confirm bridge swinging performance, and check loads imposed on the swing mechanism during swinging operation
(2) To obtain data on swinging operation, such as the tug thrust of a tugboat
(3) To confirm the temporary mooring force
(4) To validate the analytical program by comparing the analytical results with analytical results

8.4 Experiment IV (Hybrid experiment)

This floating bridge is horizontally supported by mooring with rubber fenders (Oda et al. 1998). Various experiments and simulation analyses were carried out to clarify the motion of the floating bridge in wind and waves. It is known that the reaction characteristic of rubber changes when the rubber is subjected simultaneously to different deformations other than compression. The hysteresis of deformation also changes with loading patterns. Therefore, it was necessary to study how rubber fender characteristics change with deformation and to verify the applicability of the formula for simulation calculations. To these ends, hybrid experimentation was carried out using a scale model of the rubber fender.

The following items were proved by these experiments and hybrid simulations.

(1) The diffraction experiment by a local geographical feature model proved that change of wave height depends significantly on the conditions of a construction point of site. This change has been experimentally grasped with sufficient accuracy by experiment.
(2) The theoretical predictions through the simulation were compared with the experimental verifications regarding the dynamic motions of the floating bridge and fender displacement under the actions of wind and the wave. The discrepancy between the proposed theoretical prediction and the experimental values was found to be within the error of 20 %.
(3) By conducting the above-mentioned three kinds of dynamic experiments on the floating bridge, the response characteristics by the wind and wave of a floating bridge were clarified. Moreover, it must be added that any other peculiar response was not observed in the field condition of the Yumemai Bridge.
(4) Coefficient-of-resistance CD = 1.58 of the sea water at the time of towing of the Yumemai Bridge was conjectured.
(5) The viscous damping of the water due to rolling motion was found to be about one half of the damping due to waves caused by radiation and diffraction.
(6) For the design of the roll bearing and the pivot, their safety may be secured by assuming 1.5 times the reaction force predicted by the simulation.

(7) Since the design of a mooring mechanism has a large nonlinearity, such as mechanical characteristic changes of fender material and an expansion gap, its design may be rather difficult.

However, by the hybrid simulation, the action and reaction force of the mooring part became clear, and the validity of the design was proved.

9 Driving Safety and Riding Comfort Simulation

The vertical load of this floating bridge is supported by the buoyancy of seawater, and the horizontal load by the mooring system. Therefore, in addition to deflection, which is a general problem with ordinary fixed bridges, each of the following changes had to be studied from the viewpoint of vehicle driving safety and serviceability:

(1) Change in longitudinal gradient of the transitional girder bridge decks due to tidal change
(2) Change in longitudinal and transverse gradients of the floating bridge deck due to wind and waves
(3) Change in draft of each pontoon due to live loading

It was necessary to confirm that these changes would cause no problem in regard to driving safety and riding comfort. At present, there is no regulation or standard specifying the requirements regarding the riding comfort of vehicles on bridges. Therefore, driving on this floating bridge was simulated, and a questionnaire survey was conducted as to the vibration feeling and riding comfort on existing bridges in Osaka. The relations between the simulation and the questionnaire survey results were used as data for relative evaluation of floating bridge riding comfort (Kawatani et al. 1997).

To evaluate driving safety, vehicular lateral and vertical accelerations were calculated by simulation. The result showed that these accelerations would cause no problem in driving safety, considering the long oscillation period of the bridge. A large bus carrying 36 passengers was run at a speed of 30–60 km/h on existing long-span bridges, viaducts in the urban area and ordinary roads in Osaka. Vibration acceleration was measured in the bus, and the 36 passengers were asked to fill in a questionnaire on riding comfort, to obtain the correlation between vibration acceleration and riding comfort. Riding comfort was rated in five grades as shown in Table 6.

As an example evaluation result, Fig. 14 shows the correlation between riding comfort index and mean or maximum vertical vibration acceleration as measured on the bus floor over the front wheels. Here a_1 and c_1 designate the calculated accelerations at M.S.L. (Mean Sea Level) under good surface condition whereas a_2 and c_2 designate the ones at L.W.L. (Low Water Level) under good surface condition. Similarly b_1 and d_1 designate the calculated accelerations at M.S.L. under bad

Table 6 Rating of riding comfort on the floating bridge

1: No peculiar vibration is felt
2: Some vibration is felt which causes no problem
3: Obviously peculiar vibration is felt
4: Vibration is considerably large and uncomfortable
5: Vibration is extremely large, uncomfortable and uneasy

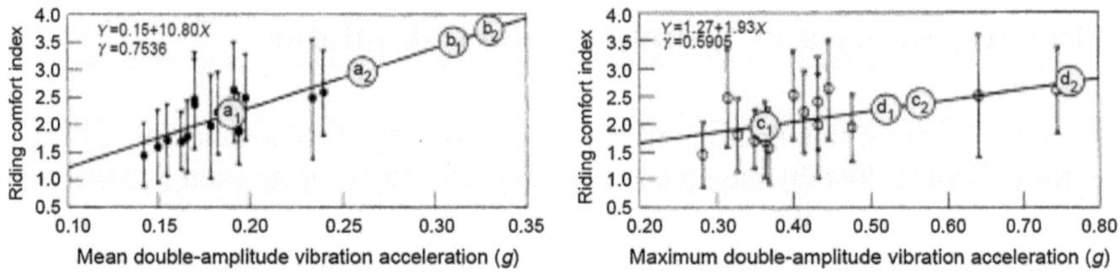

Fig. 14 Correlation between dynamic acceleration and riding comfort index

surface condition whereas b_2 and d_2 show the ones at L.W.L. under bad surface condition. The correlation shows that although passengers may feel somehow peculiar in the worst case but may not be annoyed by the vibration. Consequently, it may be reasonable to conclude that passengers on the vehicles can ride comfortably on the floating bridge.

10 Design of Superstructure

In designing the superstructure of this bridge, the sectional force was calculated based on the results of static and dynamic analyses. Figure 15 shows the models used for static analysis. The buoyancy working on each pontoon was evaluated at the vertical spring set at each node of the pontoon. A floating bridge is subjected

Fig. 15 Models for the static analysis. **a** Vertical load model. **b** Horizontal load model

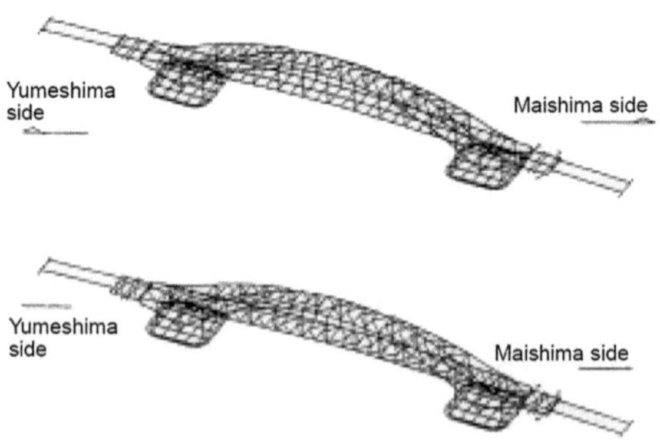

simultaneously to wind and wave loads. To study the influence by the elastic response of the floating structure in wind and waves, dynamic gust response in wind and in-wave elastic response analyses were carried out to obtain the sectional force (Utsunomiya et al. 2003).

The sectional forces due to dead load, drift force of waves, tidal force, and lateral inclination caused by winds and waves, thus obtained based on the static and dynamic analyses were evaluated superposed over each other, to design the superstructure. The intersection between each support and inside arch forms a corner and generates complex stress. The design section was therefore studied by 3D FEM analysis, to confirm structural safety. Each arch rib has been designed as a beam-pillar member that receives axial force and in-plane and out-of-plane bending moments. The ultimate load resistance of the design arch rib was estimated in order to confirm that the arch rib collapse load is sufficiently high compared with the load working on the rib, leaving a sufficient safety margin.

11 Design of Pontoons

Considering the shallow waterway at the bridging site and the long span of the floating bridge, pontoons of PC structure would become huge in size, making it difficult for the bridge to allow safe passage of boats. In addition, large thin-wall PC structures are difficult to be constructed. In view of these aspects, steel pontoons were adopted for this floating bridge. Figure 16 shows the pontoon internal structure. The outermost frame of the pontoon is of double-hull structure comprising outer wall and water-tight inner wall, as a failsafe measure against possible water leakage in the event of damage to the outer wall. The water-tight inner wall is installed 3 m inside the outer wall. For safety in case of ship collision, the

Fig. 16 Structural details of a pontoon

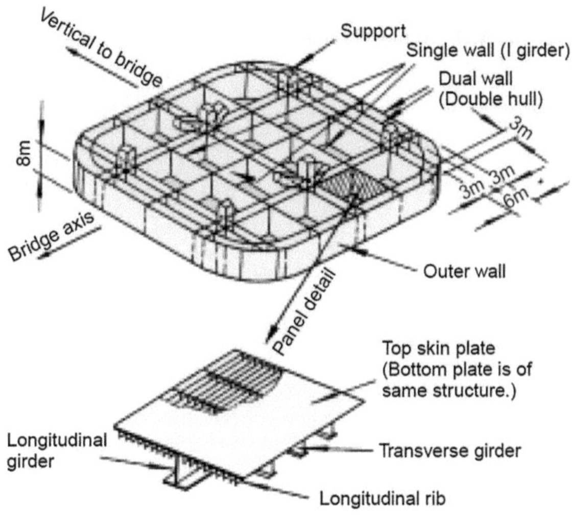

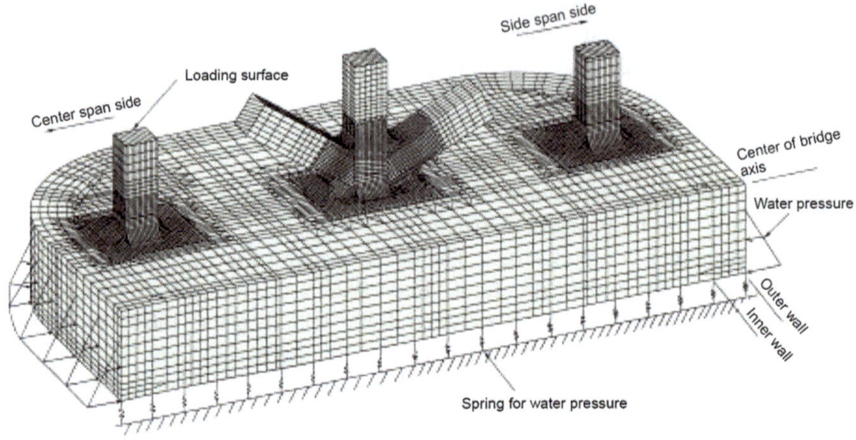

Fig. 17 FEM model of a pontoon

outermost construction limit was set at 6 m inside from the outer wall. Superstructure supports are positioned within this limit. The stress generated at the base of each pontoon support is too complex to be determined by skeleton analysis only. Stress flow was therefore clarified by analyzing the FEM model of the entire pontoon shown in Fig. 17.

A Ship collision with a floating pontoon was simulated to confirm safety. Dynamic 3D FEM model analysis (LSDYNA3D) was used. Figure 18 shows the FEM model used for this analysis (Maruyama et al. 2000). Figure 19 shows the analytical results. The maximum outer wall deformation was approximately 1.7 m, and the deformed outer wall did not reach the water-tight inner wall that is located 3 m inward. This proves that even if the outer wall is partly damaged, water will not enter the inner wall, so traffic on the bridge will not be affected.

Fig. 18 An FEM model for a ship collision

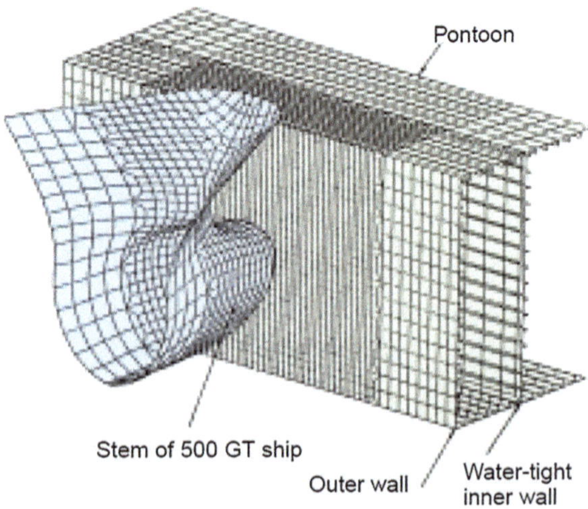

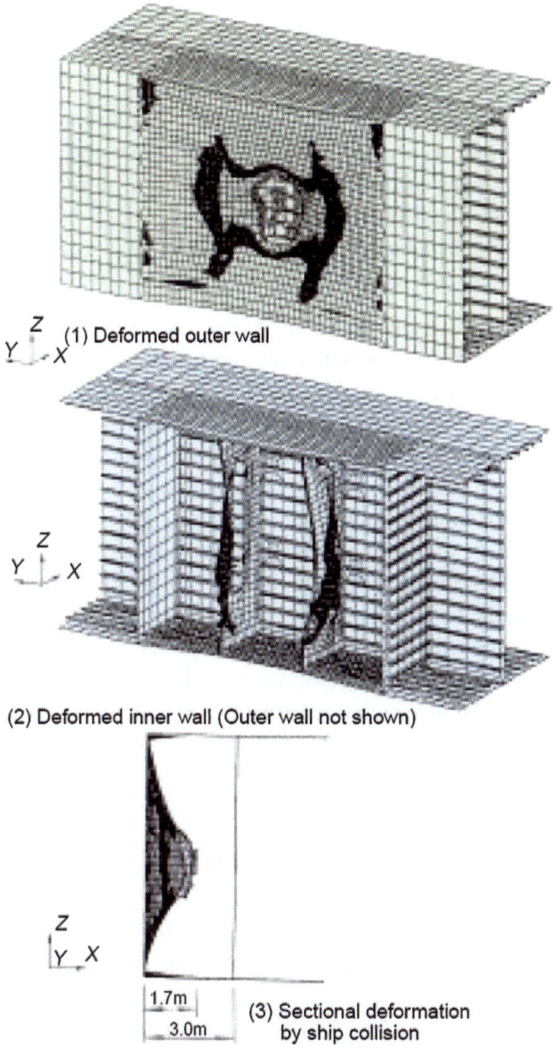

Fig. 19 A result of an FEM analysis of a ship collision

12 Corrosion-Proof Measure

For a floating bridge, the sufficient reservation for the water tightness is one of the most important issues. When the pontoons are made of steel, the specifications must carefully specify the corrosion-proof measure. For the superstructure, the structural elements classified as 'inside' are painted with a modified epoxy resin paints while the external surface is painted with fluoride resin paint. About the component peculiar to a floating bridge, such as a pontoon, the following paint was used in the case of the operation of the Kansai International Airport connecting bridge, or Tokyo Aqua Line and the conventional marine structures.

Pontoon and reaction wall components

(1) Inside of the sea: Inorganic zinc rich paint + Tar epoxy resin paint + electric anticorrosion
(2) Splash and ebb-and-flow zone: Lining method by titanium clad steel
(3) Side wall upper part: Urethane elastomer paint + fluoride resin paint
(4) Upper surface: Epoxy resin paint + fluoride resin paint
(5) Inside: Tar epoxy resin paint
(6) Rubber fender attachment side of a reaction force wall: Steel lining by stainless steel clad steel

Corrosive environment is the severest and also maintenance management is difficult for splash and an ebb-and-flow belt. Consequently, the metal lining by the titanium clad steel was proposed to be adopted here in view of its desirable advantage of being free from rust and strong shock resistance.

Since titanium clad steel is very expensive, it was decided to limit its application to vertical range of splash zone of 2.4 m, namely, from 1 m beneath the sea surface to the significant wave height position of 1.4 m above the water surface.

While the internal surface of the reaction wall may touch the floating bridge due to its rigid body motion of the bridge, the prevention from the corrosion was decided by applying the lining by stainless-clad steel. The design life of the aluminum anode in the cathodic protection system was designed to be 30 years and the anodes are going to be replaced depending on the necessity. The required current of electric anticorrosion was calculated and the 48 aluminum alloy anodes were attached for each of the two pontoons.

13 Construction Procedures

The superstructure of the floating bridge with two pontoons was constructed at a dockyard about 10 km away from the bridge installation site. Construction began in March 1998 and was completed in July of 2000. The dock, which measures 62 m wide by 408.3 m long by 12 m deep, could precisely accommodate the two pontoons, with the superstructure girder end protruding by about 5 m outside the dock. Each superstructure block (average weight: 60 t, maximum weight: approx. 110 t) was mounted by the temporary support method, using two 120 t suspension jib cranes installed on both sides of the dock. Figure 20 shows the construction procedure (see Figs. 21, 22, 23, 24).

The temporary bents used in constructing the superstructure can be grouped into three types: the center- and side span stiffening girder supporting bents, which are set on the foundation installed at the dock bottom; center arch supporting bents, which are set on the main stiffening girder structure; and bents for supporting arch members and stiffening girder blocks on the pontoon. The tallest bent used was 36 m tall. The total weight of all temporary bents used was about 4,500 t.

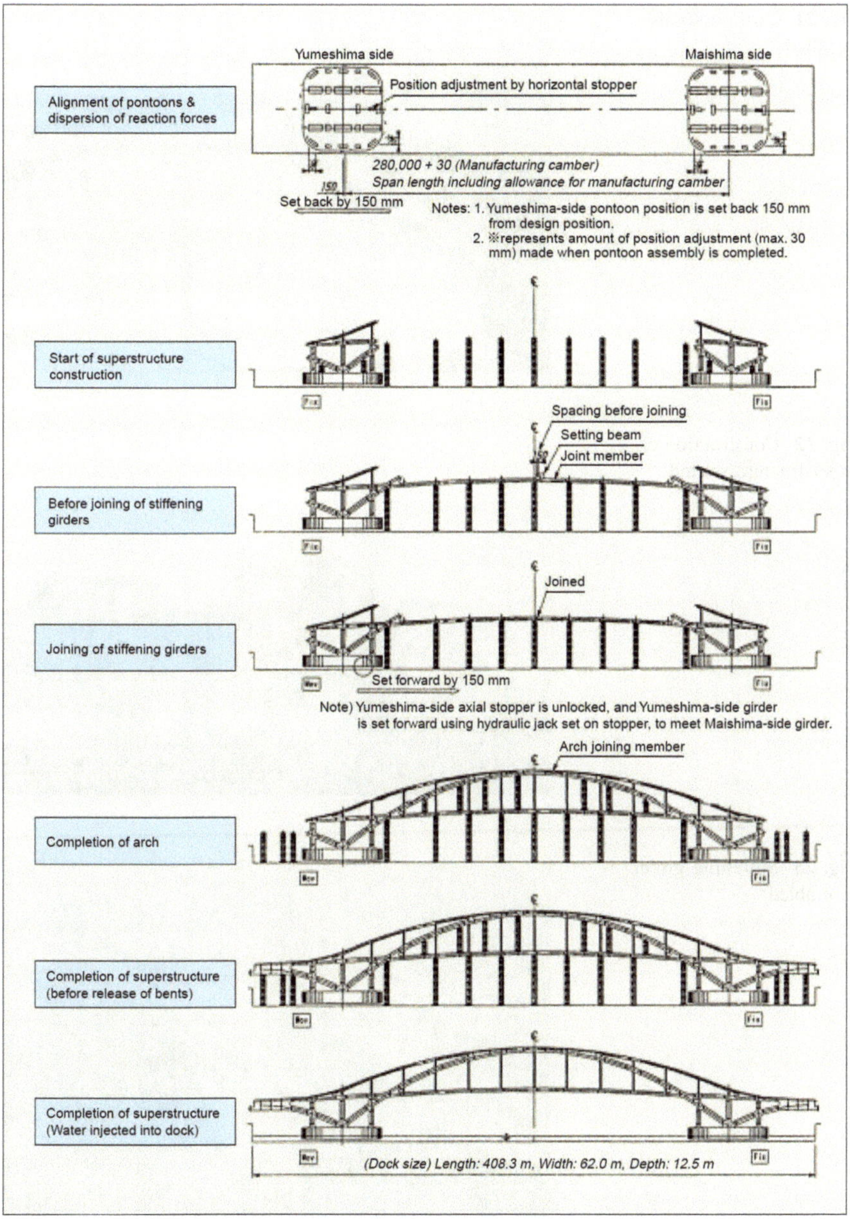

Fig. 20 Construction sequence of pontoons and superstructure at dockyard

While center-span stiffening girders and structural members were constructed over each pontoon, the pontoons were exposed to sunlight and could warp excessively due to the temperature differential between the upper and lower sides. Before construction, therefore, dimensional measurements were taken during day

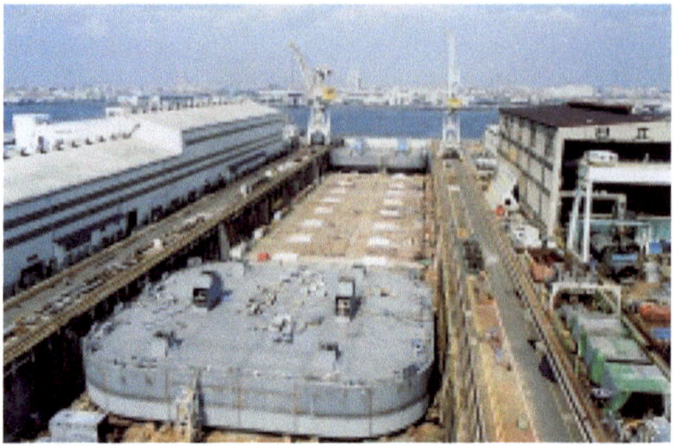

Fig. 21 Two pontoons installed

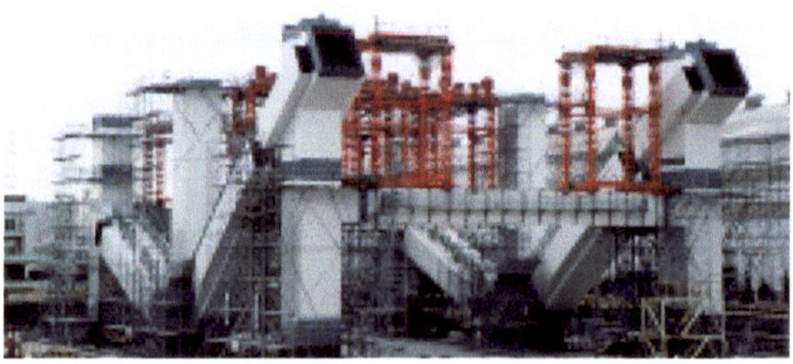

Fig. 22 Construction of superstructure started

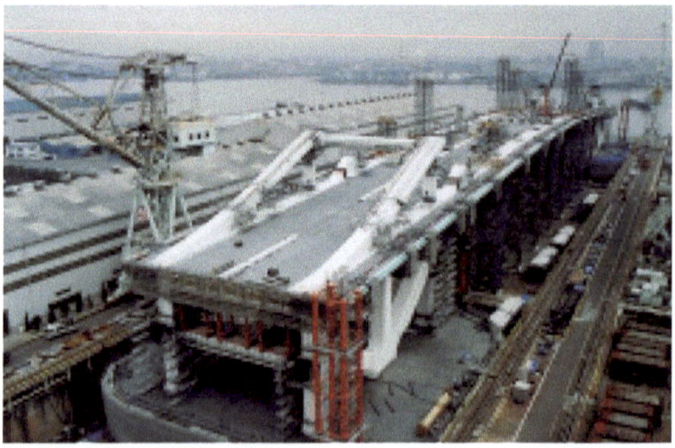

Fig. 23 Stiffening girder assembled

and night, and adjustments were determined, to secure construction accuracy. Upon installation at the site, the center-span stiffening girders between the two pontoons are also affected by temperature fluctuation. Therefore, the two pontoons were set about 150 mm further apart from each other than the design distance, and center-span stiffening girder blocks were arranged from each end toward the

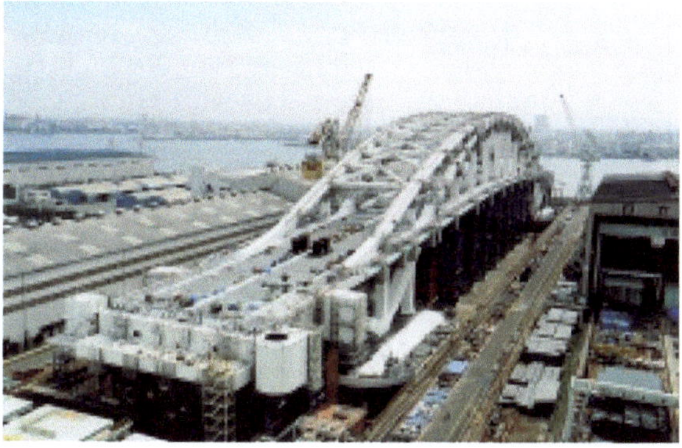

Fig. 24 Full view of completed superstructure

center. After the final block was installed, the Yumeshima-side pontoon was set forward by about 200 mm (=150 + 50 mm for contraction during winter) to join the stiffening girders.

14 Pulling Out of Dock, Towing and Installation

Upon completion, the floating bridge was pulled out of the dock and towed by tugboats to the site on July 9, 2000, where it was successfully installed at the installation site on the same day. Figure 25 outlines these operations (see Figs. 26, 27 and 28).

14.1 Pulling Out of the Dock

The floating bridge has two pontoons spaced 280 m apart. After either of the pontoons came out of the dock, the most demanding task was to control the positions of the two pontoons. The position of the trailing pontoon was controlled by operating the dockyard winches and carriages connected to the winch on the pontoon. The position of the leading pontoon was controlled by operating tugboats on both sides of the floating bridge.

14.2 Towing

The floating bridge, thus pulled out of the dock, was towed at a speed of about 3 knots by a formation of eight 3,600 HP tugboats, over the 9 miles to the installation site. The towing took about 3.5 h.

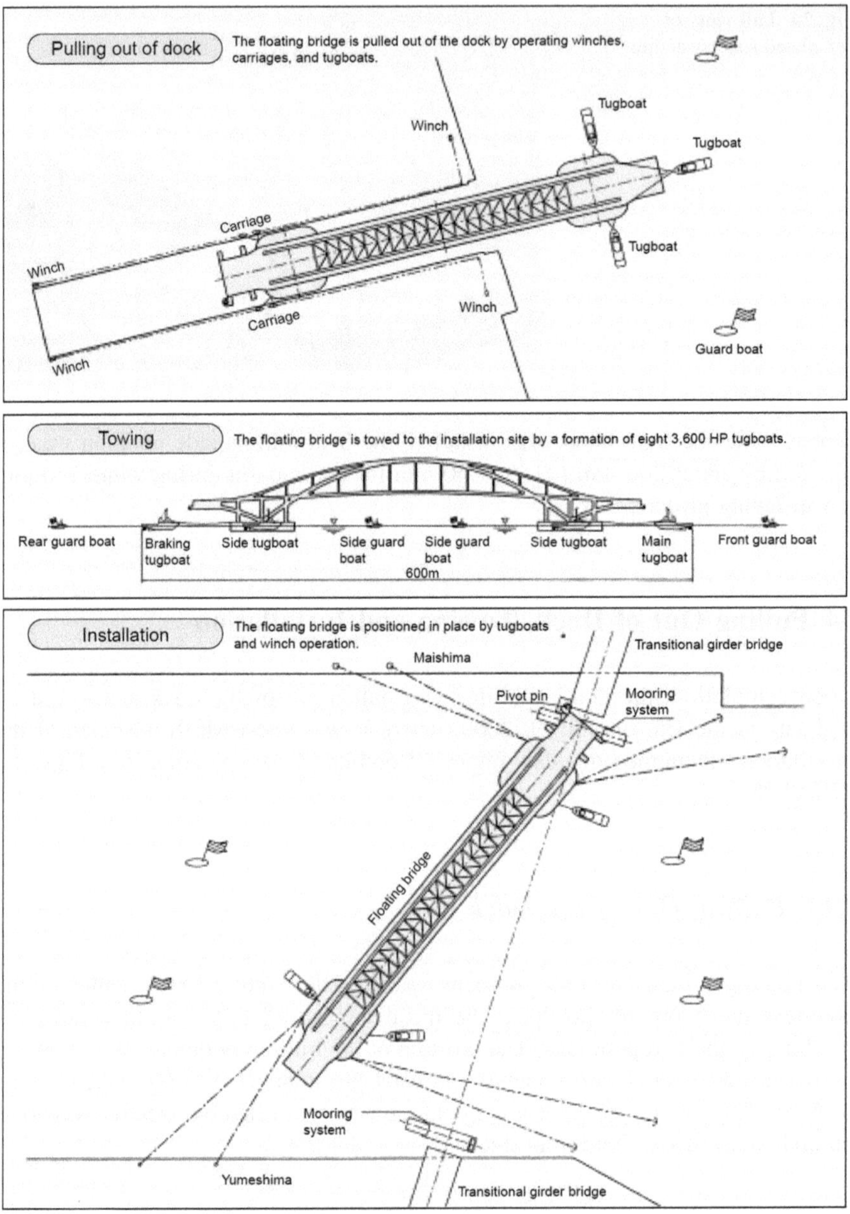

Fig. 25 Pulling out of dock, towing and installation at the destination

Fig. 26 Floating bridge being pulled out of dock

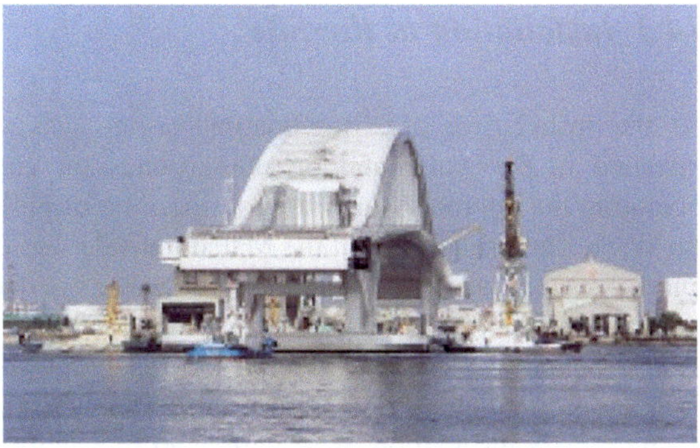

Fig. 27 Floating bridge being towed

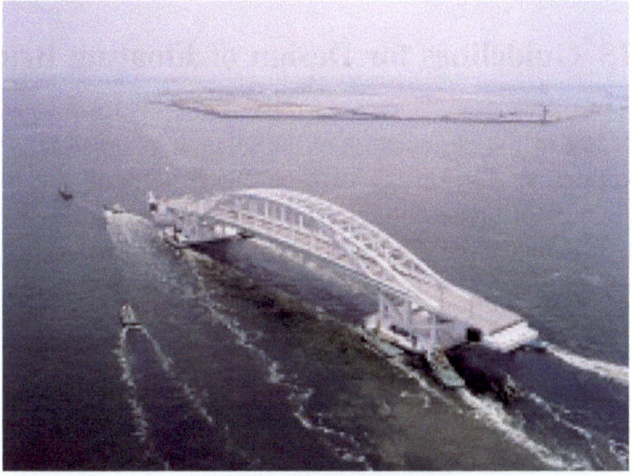

Fig. 28 Floating bridge being installed at the site

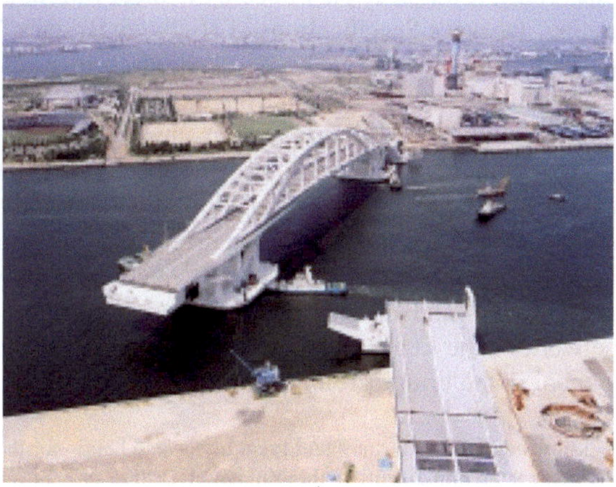

14.3 Installation at the Site

At the installation site, the floating bridge was wired to anchors previously installed in the water and on the grounds of Yumeshima and Maishima, by operating the pontoon winch used at the time of pulling the floating bridge out of the dock. The bridge end on the Maishima side was drawn to the mooring system, and a pivot pin was inserted in the same way as in opening/closing the bridge. The other end of the bridge was then rotated by tugboats and connected to the mooring system on the Yumeshima side. Finally, reaction walls were raised to complete the installation. The entire operation, from pulling the floating bridge out of the dock to the on-site installation, was completed in 1 day.

15 Guidelines for Design of Floating Bridges

Many of the engineers had the chance to become members of a subcommittee of Guidelines for Design of Floating Bridges in the Steel Structures Committee of JSCE. Fortunate enough, the guidelines for design of floating bridges were published under the auspices of JSCE in the format of performance-based Design in March 2006 (JSCE 2006).

16 Conclusions

The Yumemai Bridge is the world first floating swing arch bridge. For the design of this bridge, it was necessary not only to meet various existing design standards, such as the "Japanese Specifications of Highway Bridges," but also to solve many technical problems that had not been solved. The bridge has successfully been installed on the site, thanks to cooperation from the academic sector, and from various industrial fields, including the shipbuilding, machinery and electric industries.

The Yumemai Bridge was opened to traffic except for the side walk part in the autumn of 2000. After completion, container quays etc. were improved by reclaimed land and traffic has been drastically increasing. According to the statistics in 2012, the traffic was about 12,000 vehicles a day with the majority being large-sized trailers and the bridge has been playing the important role of the traffic network of the Osaka port area. Moreover, the white and beautiful bridge is one symbol in the bridge group of the Osaka harbor.

Acknowledgments Writers feel great honor and happy for their having been able to take part in the project of implementing the Yumemai Bridge project and fully appreciate the efforts and enthusiasm of all of the people who participated in the project of Yumemai Bridge, particularly the people from the Ministry of Land, Infrastructure, Transport and Tourism, MLIT of the Government of Japan, Municipal Government of Osaka City, bridge and ship engineers of bridge and shipyard companies.

References

JRA Specifications for Highway Bridges and Commentary. (1996). Part I Common design principles, Part II Steel bridges and Part V Seismic design. Maruzen (in Japanese).

JRA Specifications for Highway Bridges and Commentary. (2012). Part I Common design principles, Part II Steel bridges and Part V Seismic design.Maruzen (in Japanese).

JSCE. (2006). Guidelines for design of floating bridges. In E. Watanabe (Ed.), *Steel structures series 13, floating bridges subcommittee, committee of steel structures*. Tokyo: Japan Society of Civil Engineers, Maruzen.

Kawatani, I., Maruyama, T., Kawamura, Y., & Kishimoto, T. (1997). Vibration serviceability of floating bridge. In *Proceedings of Structural Engineering J.S.C.E.* (Vol. 43A, pp. 757–764) (in Japanese).

Kobayashi, M. et al. (1999). Model experiments on swinging of a movable floating bridge in waves. In *Proceedings of VLFS'99* (pp. 546–554). Hawaii, USA.

Maruyama, T., Ogawa, K., & Shimodoi, H. (1995). Wind-tunnel test for the Maishima–Yumeshima bridge. In *Proceedings of Wind Engineering of J.S.C.E.* (pp. 161–162).

Maruyama, T., Watanabe, E., & Tanaka, H. (1998a). Floating swing bridge with a 280 m span, Osaka. *Structural Engineering International*, *8*, 174–175.

Maruyama, T. et al. (1998b). Planned design of Yumeshima-Maishima bridge (1)—movable floating bridge. *Bridge and Foundation*, *32*(2), 15-24 (in Japanese).

Maruyama, T. et al. (1998c). Plan and design of Yumeshima-Maishima bridge (2)—movable floating bridge. *Bridge and Foundation*, *32*(3), 15-24 (in Japanese).

Maruyama, T., Watanabe, E., Utsunomiya, T., & Tanaka, H. (1998d). A new movable arch bridge in Osaka Harbor. In *Proceedings of the Sixth East Asia–Pacific Conference on Structural Engineering and Construction*, EASEC6 (pp. 429–434). Taipei, Taiwan.

Maruyama, T. et al. (2000). A study on the strength of floating bridge pontoon in small ship collision. In *Proceedings of Structural Engineering J.S.C.E.* (Vol. 46A) (in Japanese).

Nagata, S. et al. (1999). Motions of a movable floating bridge in waves. In *Proceedings of VLFS'99* (pp. 358–366). Hawaii, USA.

Oda, K., Maruyama, T., Tanaka, H., Nagata, S., & Yamase, S. (1998). Hybrid simulation for a new movable floating bridge, In *Proceedings of the Sixth East Asia-Pacific Conference on Structural Engineering and Construction*, EASEC6 (pp. 435–440). Taipei, Taiwan.

Ueda, S. et al. (1999). Experimental study on the elastic response of a movable floating bridge in waves. In *Proceedings of VLFS'99* (pp. 766–775). Hawaii, USA.

Utsunomiya, T., Watanabe, E., Murakoshi, J., Fumoto, K., & Tanaka, H. (2003). Development of dynamic response analysis program for floating bridges subjected to wind and wave loadings. In *Proceedings of International Symposium on Ocean Space Utilization Technology, Ministry of Land, Infrastructure and Transport, Ship and Ocean Foundation. National Maritime Research Institute* (pp. 417–424).

Watanabe, E., Maruyama, T., Kawamura, Y., & Tanaka, H. (1998). A new movable floating bridge project in Osaka city. In *Proceedings of the IABSE Symposium Long-Span and High-Rise Structures,Kobe* (pp. 155–160).

Watanabe, E., Maruyama, T., & Tanaka, H. (1999). Design and construction of a floating swing bridge in Osaka. In *Proceedings of the Third International Workshop on Very Large Floating Structures* (pp. 888–897). Honolulu, Hawaii, USA.

Watanabe, E., Maruyama, T., Kawamura, Y., & Tanaka, H. (2001a). Why is a floating swing arch bridge built in the Port of Osaka? In *Proceedings of New York City Bridge Conference, Journal of Bridge Engineering, ASCE.*

Watanabe, E. et al. (2001b). An Osaka floating swing arch bridge towed to the site from dockyard. In *Proceedings of the Fourth Symposium on Strait Crossings 2001, Bergen* (pp. 293–299).

Watanabe, E., & Utsunomiya, T. (2001). Wave response of a floating bridge with separate cylindrical pontoons. *In Proceedings of the Fourth Symposium on Strait Crossings 2001, Bergen* (pp. 301–308).

Watanabe, E., Utsunomiya, T., Okafuji, T., Murakoshi, J., & Fumoto, K. (2002). Development of wave response simulation program of floating bridges and a benchmark test. In *Proceedings of Second International Symposium on Steel Structures, ISSS'02, Seoul* (pp. 67–78).

Watanabe, E. (2003). Floating bridges: past and present. *Structural Engineering International,2*, 128–132.

Watanabe, E., & Utsunomiya, T. (2003). Analysis and design of floating bridges. *Progress in Structural Engineering and Materials, Wiley Interscience,5*(3), 127–144.

Watanabe, E., Ueda, S., Maruyama, T., & Takeda, S. (2003a). Konstruktion der Yumemai-Brücke schwimmende Bogenbrücke in Osaka. *Stahlbau,5*, 323–330.

Watanabe, E., Utsunomiya, T., Okafuji, T., Murakoshi, J., & Fumoto, K. (2003b). Development of response simulation program of a floating bridge subjected to irregular waves. *Journal of Structural Engineering, JSCE,49A*, 661–668.

Watanabe, E. (2004). Floating bridges: Past and present (translated version of Structural International, SEI2/2003, by IABSE), Zurich, Switzerland. Russian Journal: Bridge Construction Abroad, MOCKBA 2004 (pp. 9–13) (in Russian).

Floating Oil Storage Base

S. Ueda

Abstract A floating oil storage system has been adopted at two bases, Kamigoto and Shirashima, Japan. Each floating oil storage base holds in reserve approximately 5 million kilolitres of crude oil in steel-made storage reservoirs measuring about 400 m in length, 100 m in width, and 30 m in depth. In this chapter, an outline of those bases, design concept and conditions, mooring system in special fender characteristics, wind and wave conditions and the motions of floating structures, the numerical simulation and factors considered are described. The design standard for oil storage base as established by the Ministry of Transportation is briefly introduced as well as protective facilities and ancillary facilities.

1 Introduction

Japan depends on imports for nearly all of its energy and therefore has established petroleum stockpiling to maintain economic stability in the event of a supply disruption. The Japanese government started petroleum stockpiling in 1978. As of 31 March 2012, stockpiles of crude oil and oil products stored in public and private facilities totaled 185 days of domestic consumption, of which 55 % is stockpiled in public facilities.

Japan's national petroleum stockpiling bases employ four types of reservoir systems according to the geographic conditions at the site: above-ground reservoirs, in-ground reservoirs, floating oil reservoirs, and underground rock cavern reservoirs. The floating reservoir system is unique, comprising large steel floating oil reservoirs moored to concrete caisson type dolphins through the use of high performance large rubber fenders. The floating reservoirs type was adopted for both the Kamigoto Oil Storage Base and the Shirashima Oil Storage Base.

S. Ueda (✉)
Institute of Environment Informatics, Yokohama, Japan
e-mail: ued21010@ideacon.co.jp

C.M. Wang and B.T. Wang (eds.), *Large Floating Structures*,
Ocean Engineering & Oceanography 3,
DOI 10.1007/978-981-287-137-4_4

Table 1 Properties of floating oil storage bases

		Kamigoto base	Shirashima base
Total oil storage capacity		4,400,000 kl	5,600,000 kl
Number of oil reservoirs		5	8
Maximum capacity of oil reservoir		880,000 kl	700,000 kl
Size of oil reservoir	Length	390 m	397 m
	Breadth	97 m	82 m
	Depth	27.6 m	25.4 m
Draft	Full laden	24.5 m	22.7 m
	Half	14.6 m	13.4 m
	Ballasted	4.75 m	4.1 m

Table 1 shows the properties of floating oil storage bases. The bases for oil stockpiling are designed for a long working life with design loads of natural and environmental conditions adopted for a one-hundred-year return period.

2 Floating Oil Storage Base and Design Standard

2.1 General Geographic Conditions

The Kamigoto Oil Storage Base is located at the Kamigoto islands, about 60 km north-west of Nagasaki City, Nagasaki prefecture. The site is surrounded by several islands such as Nakatorishima Island, Shuge Island, Orishima Island and Kashiwajima Island. This forms a sheltered basin that is further protected through two main breakwaters. The first has been constructed in response to west-ward waves between Orishima Island and Kashiwajima Island, and the second, at the northern side of the basin. The main facilities of Kamigoto Oil Storage Base include five huge steel floating oil reservoirs and an offshore terminal for loading and unloading of crude oil, and oil fences to prevent oil diffusion in the basin. Each reservoir measures 390 m in length, 97 m in width and 27.6 m in depth and the maximum crude oil reservation capacity is 880,000 kL. The total oil storage capacity is 4,400,000 kL. Figure 1 shows an aerial view of the Kamigoto Floating Oil Storage Base.

The Shirashima Oil Storage Base is located at the Shirashima Islands, about 10 km north-west of Kitakyushu City. The basin is located on the eastern coast of Ojima Island, and the basin is protected from waves by breakwaters on the remaining three sides surrounding the basin. The main facilities of Shirashima Oil Storage Base include eight huge steel floating oil reservoirs and an offshore terminal for loading and unloading of crude oil. The facility is ringed with oil fences to prevent oil diffusion into the basin. Each reservoir measures 397 m in length, 82 m in width and 25.4 m in depth and the maximum crude oil storage capacity is

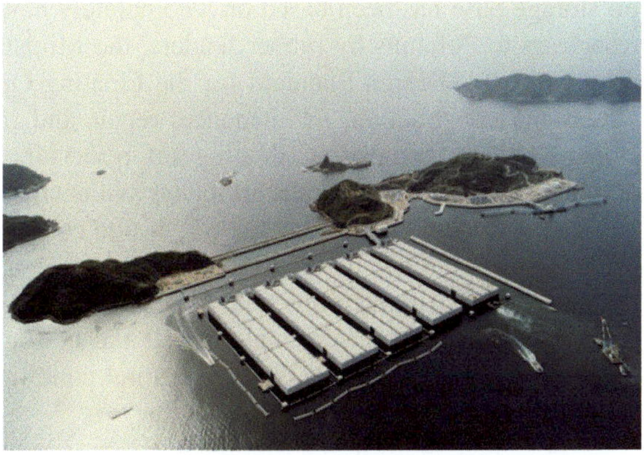

Fig. 1 Aerial view of the Kamigoto Floating Oil Storage Base

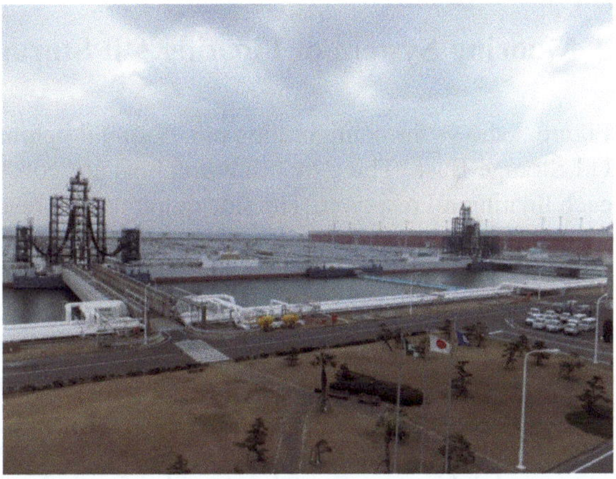

Fig. 2 View of the Shirashima Floating Oil Storage Base

700,000 kL. The total oil reservation capacity is 5,600,000 kL. Figure 2 shows a view of the Shirashima Floating Oil Storage Base.

As this floating reservoir type oil stockpiling requires a robust mooring system, rubber fenders have been employed in conjunction with mooring dolphins. A more detailed description on the characteristics of these rubber fenders will be provided further in this chapter along with the procedure recommended to estimate the motions of the floating oil reservoirs and the deformation and reaction forces of the rubber fenders.

2.2 Design Standard for Floating Oil Storage Base

The large floating oil reservoirs described above must continue to be safely moored to dolphins in the extreme wind and wave conditions at the site and remain safe and serviceable during earthquakes. Given that the construction of a floating oil storage base has been based on a completely new technology to moor huge oil reservoirs to dolphins by rubber fenders, the Ministry of Transport, Japan had to develop a new Design Standard for the Floating Oil Storage Base. The standard covers the construction, maintenance, repair and improvement of a floating oil storage facility comprising floating oil reservoirs, working dolphins, mooring dolphins, deepwater terminals, oil fences and ancillary facilities (The Japan Port & Harbour Association 1999). A sp ecial design code was also developed for floating oil reservoirs that are moored to dolphins by the use of high performance large rubber fenders. Although this kind of rubber fender is commonly used at mooring dolphins of deep water terminals, it was the first attempt to use the material for permanent mooring even in extreme natural conditions. Several characteristics of rubber fenders shall be examined herein

3 Mooring System of Floating Oil Storage Base

Figure 3 shows the configuration of mooring dolphins and fenders of the Kamigoto Oil Storage Base. Three mooring dolphins are constructed at the side and end of each floating oil reservoir. The side dolphins will restrain lateral motions by rubber fenders and the end dolphin will restrain both lateral motions and longitudinal motions. The longitudinal motion will be restrained by both fenders and chains. Two or more fenders of appropriate performance are installed so as to absorb the kinetic energy of floating oil reservoirs calculated according to numerical simulations and/or model experiments. Figure 4 shows rubber fenders installed at the Kamigoto oil storage facility.

A different arrangement is adopted at the Shirashima Oil Storage Base with four mooring dolphins constructed at the side of each floating oil reservoir. The dolphins restrain lateral motions by side fenders and longitudinal motions by end fenders installed at the wing arms. Two or three fenders of appropriate performance have been installed so as to absorb the kinetic energy of floating oil reservoirs calculated according to numerical simulations and/or model experiments.

Figure 5 shows the dolphins for loading, unloading and mooring, as well as the rubber fenders that were installed on the dolphin for the Shirashima Oil Storage Base.

Fig. 3 Configuration of mooring dolphins and fenders (Kamigoto)

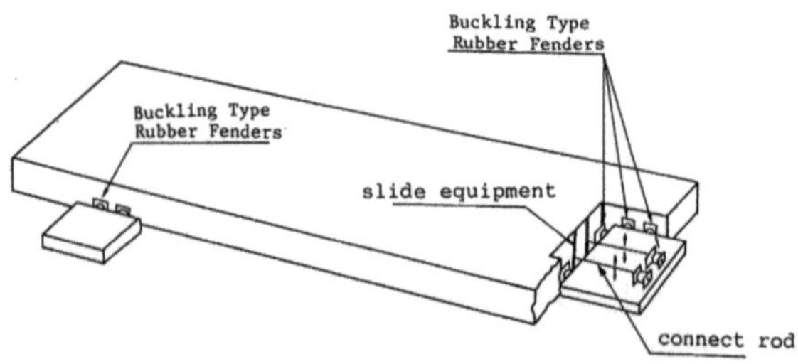

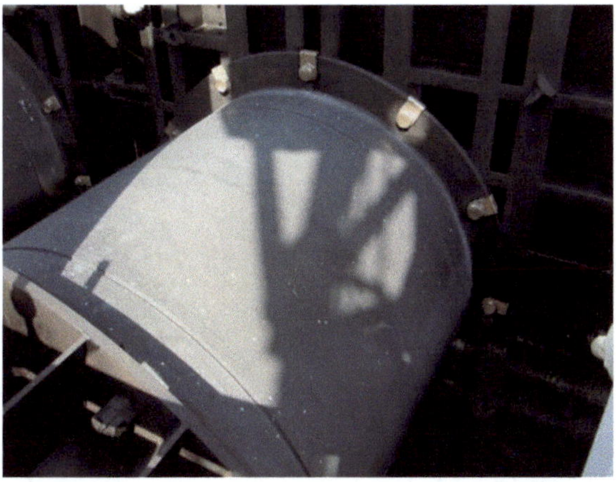

Fig. 4 Rubber fenders installed (Kamigoto)

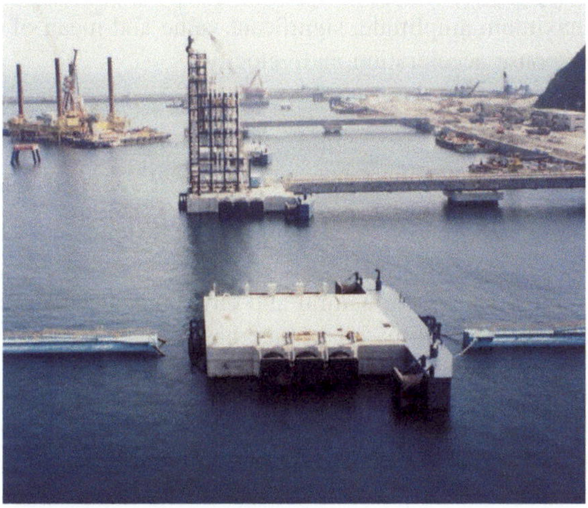

Fig. 5 Dolphins and rubber fenders (Shirashima)

4 Estimation of Mooring Forces and Design of Mooring Dolphins

4.1 General

The mooring force of a floating structure subjected to waves or seismic motion is usually determined by means of a numerical simulation. The equation of motions of a moored floating structure takes on a second order differential equation with six degrees of freedom. The time domain numerical simulation is carried out in consideration of forces from irregular waves, the second order wave force or the so-called steady or fluctuating drift force, current force, fluctuating wind force, seismic force, and the non-linear load-deflection characteristics including hysteresis of mooring facilities (refer to Fig. 6).

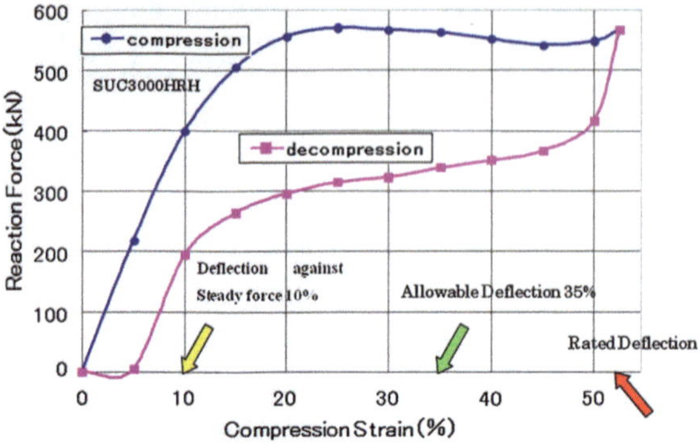

Fig. 6 Load deflection characteristics of rubber fender (SUC3000H)

The results of the time-domain numerical simulation of a floating structure subjected to wind and waves or seismic force, shall be processed by statistical analysis in order to obtain those quantities such as the maximum value, the maximum amplitude, significant value and mean of motions, mooring forces, the response acceleration and velocity.

4.2 Natural and Environmental Design Conditions for Design of Mooring Dolphins

Table 2 shows the natural and environmental design conditions of the Kamigoto and the Shirashima oil storage base.

4.3 Equation of Motions

The second order differential equation of motions (involving six degrees of freedom, namely surge, sway, heave, roll, pitch and yaw) is given by

Table 2 Design Conditions of Floating Oil Storage Bases

			Kamigoto	Shirashima
Water depth of base			−27.0 m	−25 m
Loading condition (Ballasted or half)	Return period 20 years	Wind speed	44 m/s	44 m/s
		Waves	$H_{1/3} = 0.32–0.70$ m $T_{1/3} = 10–13$ s	$H_{1/3} = 0.20–0.25$ m $T_{1/3} = 8–17.5$ s
Loading condition (Half or full laden)	Return period 100 years	Wind speed	51 m/s	51 m/s
		Waves	$H_{1/3} = 0.38–0.95$ m $T_{1/3} = 10–13$ s	$H_{1/3} = 0.25–0.30$ m $T_{1/3} = 8–17.5$ s

$$\sum_{i=1}^{6}\left(M_{ij}+m_{ij}(\infty)\right)\ddot{X}_i(t)+\sum_{i=1}^{6}\left\{\int_{-\infty}^{t}L_{ij}(t-\tau)\dot{X}_i(\tau)d\tau+D_i(t)\right\}$$
$$+\sum_{i=1}^{6}\left(C_{ij}+G_{ij}\right)X_i(t)$$
$$=F_j(t) \tag{1}$$

where M_{ij} is the mass of floating structure, $m_{ij}(\infty)$ the added mass, $X_i(t)$ the motion of floating structure at time t, $L_{ij}(t)$ the retardation function at time t, the damping force due to mooring lines and viscosity at time t, C_{ij} the restoring force coefficient, G_{ij} the mooring force coefficient, $F_j(t)$ the external forces at time t; i, j the mode of motions of floating structure ($i, j = 1$–6).

The retardation function $L_{ij}(t)$ and the added mass $m_{ij}(\infty)$ are calculated by

$$L_{ij}(t)=\frac{2}{\sigma}\int_{0}^{\infty}B_{ij}(\sigma)\cos\sigma t d\sigma \tag{2}$$

$$m_{ij}(\infty)=A_{ij}(\sigma)+\frac{1}{\sigma}\int_{0}^{\infty}L_{ij}(t)\sin\sigma t dt \tag{3}$$

where $A_{ij}(\sigma)$ is the added mass at angular frequency σ and $B_{ij}(\sigma)$ the damping coefficient at angular frequency σ.

The motion of a floating structure is dominated by the predominant angular frequency of external forces or at the natural angular frequency of the moored floating structure relating to the frequency characteristics and natural period of the moored floating structure. Therefore, radiation forces are usually represented by the value at the significant wave period or at the natural period of motions.

When the spectrum of external forces is not in a wide band, motions of floating structure are instead calculated by

$$\sum_{i=1}^{6}\left(M_{ij}+A_{ij}(\sigma_0)\right)\ddot{X}_i(t)+\sum_{i=1}^{6}\left\{\int_{-\infty}^{t}B_{ij}(\sigma_0)\dot{X}_i(\tau)d\tau+D_i(t)\right\}$$
$$+\sum_{i=1}^{6}\left(C_{ij}+G_{ij}\right)X_i(t)$$
$$=F_j(t) \tag{4}$$

where the added mass and the damping coefficient radiation forces are computed at the angular frequency σ_0 such as the predominant frequency of external forces or the natural frequency of motions.

4.4 Loads and Forces Act on a Floating Structure

The loads and external forces that act on a floating structure are the self-weight, buoyancy, and external forces such as wave force, wind force, current force, and seismic force. With the action of the loads and forces, motions of a floating structure are developed and the mooring system deformed and reaction forces generated. The load actions are given on the right side of Eqs. (1) and (4), and the motions of a floating structure and the mooring forces of the mooring system are determined from numerical simulations.

As loads such as wind force, wave force and seismic force are irregular and periodic, the frequency characteristics must be considered in the numerical simulations. An outline of the treatment of wind force, wave force and seismic force is given below.

4.4.1 Wind Force

Generally, wind speed is given as the average wind speed. Wind speed varies with respect to time and space, and the maximum instantaneous wind speed is usually larger than the average wind speed. The ratio of the maximum instantaneous wind speed and the average wind speed at a certain point is called the gust ratio.

Usually, the frequency spectrum of wind shall be determined according to the observed data at the construction site. However, if there is no observation data, the frequency spectrum of the wind speed by Davenport (1967), Hino (1976), or any other type may be used in the numerical simulations as an irregular and periodic wind speed. Equations (5) and (6) are the frequency spectrum of wind speed by Davenport and Hino, respectively,

$$fS_u(f) = 4K_r U_{10}^2 \frac{X^2}{(1+X^2)^{4/3}}, \quad X = 1200\frac{f}{U_{10}} \tag{5}$$

$$S_u(f) = 2.856\frac{K_r U_{10}}{\beta}\left\{1+\left(\frac{f}{\beta}\right)^2\right\}^{-5/6}, \quad \beta = 1.169\times 10^{-3}\left(\frac{U_{10}\alpha}{\sqrt{K_r}}\right)\left(\frac{z}{10}\right)^{2m\alpha-1} \tag{6}$$

Wind forces that are acting on a floating structure can be calculated using the following equations

$$R_X = \frac{1}{2}\rho_a U^2 A_T C_X \tag{7}$$

$$R_Y = \frac{1}{2}\rho_a U^2 A_L C_Y \tag{8}$$

$$R_M = \frac{1}{2}\rho_a U^2 A_L C_M \tag{9}$$

where C_X is the drag coefficient in the X direction (i.e. from the front of a floating structure), C_Y the drag coefficient in the Y direction (i.e. from the side of a floating structure), C_M the pressure moment coefficient about the center of gravity of a floating structure, R_X the component of resultant wind force (kN) in the X direction, R_Y the component of resultant wind force (kN) in the Y direction, R_M the moment (kJ) of resultant wind force about the center of gravity of a floating structure, ρ_a the density of air =1.23 × 10^{-3} (t/m^3),U the wind speed (m/s), A_T the front projected area above the water surface (m^2), $A_L$the side projected area above the water surface (m^2), and L the length of a floating structure (m). Note that the coefficients C_X, C_Y and C_M are to be determined from a wind tunnel test.

4.4.2 Wave Force

Wave force is the force exerted by the incident waves on a floating structure where the linear force component is proportional to the amplitude of the incident waves, and the nonlinear force component is proportional to the square of the amplitude of the incident waves. The radiation wave forces are exerted by motions of a floating body.

The linear force is the force that a floating structure receives from the incident waves as the reaction when the floating structure deforms the incident waves. It is expressed as the sum of the Froude-Krylov force and the diffracted wave force.

Radiation wave forces are induced according to developed motions of a floating structure. Radiation force is divided into two components that are proportional to the acceleration and velocity. They are treated as added mass and damping coefficient, respectively, in the equation of motions.

Both incident wave forces and radiation wave forces may be calculated analytically when the configuration of a floating body is simple. However numerical methods (such as the finite element method and/or boundary element method) are often used for any kind of floating bodies.

As sea waves are irregular and periodic, the wave force is given as a time series of irregular and periodic value. For the numerical simulations, one may use the Bretshneider (1968), Mitsuyasu (1970) frequency spectrum that is given by

$$S_u(f) = 0.257 H_{1/3}^2 T_{1/3} \left(T_{1/3} f\right)^{-5} \exp\left[-1.03\left(T_{1/3} f\right)^{-4}\right] \tag{10}$$

where $S(f)$ is the frequency spectrum, f the frequency, $H_{1/3}$the significant wave height and $T_{1/3}$ is the significant wave period. The equation is originally proposed by Bretshneider and is later modified by Mitsuyasu on the condition that there is the relation of $T_p \cong 1.0T_{1/3}$ between the peak period and significant wave period. In correcting the relation between the peak period and significant wave period as $T_p \cong 1.1T_{1/3}$ according to observed data, Goda (1987) proposed

$$S_u(f) = 0.205 H_{1/3}^2 T_{1/3}^{-4} f^{-5} \exp\left[-0.75\left(T_{1/3} f\right)^{-4}\right] \quad (11)$$

The wave spectrums of Eqs. (10) and (11) may be applied when the wind field is stationary. However, JONSWAP spectrum (1973) is applied to wind waves that grow in rather short fetch length under strong winds. Other types of frequency spectrum may also be used.

The wave drift force which is proportional to the square of the wave height must be considered when the length of a floating structure becomes equal or larger than the wavelength. By assuming that the floating structure is two-dimensional and the wave energy is not dissipated, the wave drift force is given by

$$F_d = \frac{1}{8} \rho_w g H_i^2 R \text{ and } R = K_R \left\{ 1 + \frac{\frac{4\pi h}{L}}{\sinh\left(\frac{4\pi h}{L}\right)} \right\} \quad (12)$$

where F_d is the wave drift force per unit length (kN), ρ_w the density of the seawater (kg/m^3), H_i the wave height of incident wave (m), K_R the ratio of reflection and R the coefficient of wave drift force.

4.4.3 Seismic Load

When a floating structure is located in a seismic region, the mooring system has to be designed in consideration of seismic forces. The floating structure is subjected to seismic forces that are proportional to its acceleration and mass as well as the interaction force between the floating structure and the mooring dolphins. The relative response acceleration of a floating structure is small because the natural period of a moored floating structure may be larger than the predominant period of the seismic motion. However, because the relative displacement with the mooring system is large, it is recommended that appropriate stiffness of mooring systems shall be selected according to the condition of use.

4.4.4 Load Deflection Characteristics of Rubber Fenders

Fenders used for mooring of floating oil reservoirs are termed ‘buckling-type’ rubber fenders. For buckling-type fenders, the reaction force increases rapidly with respect to small deflection and reaches the maximum reaction force at 20–25 % of strain. After this point, the reaction force drops a little but remains almost equal to the maximum reaction force until the strain reaches 50–60 %. Figure 6 shows the load-deflection characteristics of buckling-type rubber fenders. It resembles a buckling phenomenon, providing the inspiration behind its name. On the other hand, because of constant reaction at strains between 20 and 50 %, this kind of fender is also known as a constant reaction-type fenders. The rated deflection of

Table 3 Variation of load deflection characteristics of rubber fenders

Variation of reaction force to the same strain of nominal load deflection curve	
Manufacturing error	0.9–1.0
Aging	1.0–1.05
Velocity factor	1.0–1.10
Creep	The steady load or the mean load shall be less than the reaction force at 10 % strain, and to use the load deflection characteristics in consideration of creep
Repeated compression	0.8–0.9 (at 40 % deflection repeated ten cycles and more)
Inclination compression	Load deflection characteristics in consideration of the lateral force equivalent to 10 % of the axial force
Temperature factor	0.95–1.25 (at 0–50 °C)

this fender is 52.5 % strain. However, the allowable deflection for design was set at 35 % strain, with deflection caused by steady forces by wind and current set at less than 10 % strain in accordance of creep tests.

4.4.5 Variation of Load Deflection Characteristics of Rubber Fenders

Table 3 shows the variation of the reaction force of the rubber fender under environmental conditions, aging, and loading conditions. The determination of motions and mooring forces should be carried out using a numerical simulation method in consideration of these variations of fender characteristics. The upper and lower limits of the reaction force shall be determined from Table 3.

The upper limit of fender characteristics influences the maximum mooring force whereas the lower limit of fender characteristics influences the maximum motions and deflections of rubber fenders. A detailed explanation on major particulars can be referred to the paper by Ueda et al. (1998).

Aging tests, repeated compression tests, tests for temperature factors, tests for velocity factors, inclination compression tests, and creep tests should be executed by fender manufactures to determine the variation factors.

4.4.6 Statistical Analysis of Results

Motions and mooring forces calculated by means of numerical simulations and results of the hydraulic experiment are irregular and periodic. Generally, as the elapsed time of the numerical simulation or the hydraulic experiment is limited, the expected value of the variables is to be estimated in the duration of a rough weather condition. Results from computational analysis and from a hydraulic experiment shall be analyzed and presented in the form of time histories, with the maximum values, the significant values, the average values, the frequency spectrum, and the expected values.

It is recommended that the time step of the numerical simulation should be set at about one eighth or smaller of the minimum period of the external forces, in consideration of both frequency characteristics of external forces and the response characteristics of the moored floating structure. It is also recommended that the numerical simulation be performed as long as possible when more than one hundred effective amplitudes of motions and mooring forces are to be obtained in order to calculate the expected values accurately.

4.5 Design of Gravity Type Mooring Dolphins

The stability of the gravity type dolphin shall be examined with respect to sliding, overturning, bearing capacity of the foundation, and settlement. The safety factor F_s against sliding of a gravity type dolphin may be calculated by

$$F_s \leq \frac{fW}{P} \tag{13}$$

where W is the resultant vertical force acting on the dolphin (kN), P the resultant horizontal force acting on the dolphin (kN) obtained as fender reaction force from the numerical simulation, and f the coefficient of friction between the bottom of the dolphin and the foundation. In this case, the resultant horizontal resistance is given as the product of resultant vertical force and the coefficient of friction between bottom of the dolphin and the foundation. The resultant vertical force acting on the dolphin is calculated in consideration of the surcharge, self-weight of the dolphin, buoyancy, seismic force, dynamic water pressure during earthquake, and interaction force (i.e. mooring force from the mooring dolphins).

On the other hand, the safety factor F_s against the overturning of the gravity type dolphin may be calculated by

$$F_s \leq \frac{Wt}{Ph} \tag{14}$$

where t is the distance between the line of application of the resultant vertical forces acting on the dolphin and the front toe of the dolphin (m), h the height of the application of the resultant vertical forces acting on the dolphin, above the bottom of the dolphin (m). Note that the safety factor in Eq. (13) is given as a ratio of the resultant resistant moment and the resultant overturning moment.

Bearing capacity is assessed based on the foundation materials, size of structures and so on. The circular arc analysis is used when examining the bearing capacity for eccentric and inclined loads acting on a foundation of gravity type structures. Details on the formulae of bearing capacities are available in the codes, the design standards or the recommendations for port and harbour facilities and offshore structures.

5 Protective Facilities for Basins

Protective facilities to protect the basin for floating oil reservoirs include breakwaters, fender dikes for drifting ships and/or drifting matters, and oil fences. Breakwaters are the facilities which hold the function of keeping the adequate calmness of basins. Breakwaters shall be arranged at an appropriate position at the outskirts of basins and/or outer basins so as to shelter basins from waves effectively.

Fender dikes have the function of fending the floating structure from impact of drifting objects, working in tandem with the function of breakwaters. Fender dikes shall be arranged appropriately at the boundaries of basin neighboring to channels of general vessels and oil tankers loading and unloading for the base. Fender dikes should be stable and have sufficient strength against impact forces by drifting objects, as well as provide safety in a storm. Fender dikes shall also have the function of preventing diffusion of spilled oil.

Oil fences have the function of preventing oil spills while supporting the function of breakwaters. The outskirts of the basin shall be appropriately surrounded by oil fences. Oil fences shall be impermeable structures even in stormy conditions. It shall be examined by appropriate model experience to confirm that spilled oil would not leak according to occasional wave disturbance at the design high water level and the mostly low tide. Oil fences shall have the function of preventing diffusion of spilled oil if necessary.

6 Ancillary Facilities

Ancillary facilities of floating oil reservoirs and mooring facilities shall be constructed in order to stockpile oil safely and smoothly by means of floating oil storage facilities.

6.1 Security Measures and Disaster Prevention Equipments

Fire extinguishing equipments such as foam extinguishing facilities, water spray equipments, discharge oil prevention materials such as oil fences, oil treatment agent and oil adsorption materials, fire boats, oil recovery ships and work crafts shall be equipped as necessary. In order to prevent the occurrence of a disaster, inert gas systems and other facilities used for prevention of explosion have to be installed.

Equipments for navigation support and monitoring shall be appropriately installed to prevent ships navigating inside and in the vicinity of base from accidents of touching and collision. Marker lamp, buoy, acoustic signal devices, radar reflectors to indicate location of floating oil reservoirs shall be installed.

6.2 Equipment for Detection, Monitoring and Warning

Equipments for detection, monitoring and warning shall be installed to prevent base and floating oil reservoirs from the occurrence of disaster. Detection and monitoring shall be operated by a concentrated control system, and double monitoring systems shall be used as necessary.

6.3 Lighting Facilities

Lighting facilities shall be installed for convenience of cargo unloading, emergency measures and working at night.

6.4 Maintenance and Management of Facilities

The provider of floating oil storage bases shall be inspected under the respective ordinances on maintenance and management of the facilities. In addition, provider of floating oil storage bases shall establish voluntary inspection standard on the facilities and shall inspect facilities under the inspection standard. The effect of anticorrosion shall be confirmed for facilities on which anticorrosion measures were taken.

Maintenance and management of rubber fender shall be taken with understanding load deflection characteristics concerning repeated loading, deterioration by aging, and creep effect. Inspection of rubber fenders shall be carried out at least on those items such as appearance, compression performance and deflection.

7 Concluding Remarks

In this chapter, floating oil storage bases are introduced. The floating oil storage base requires a special construction methodology for mooring huge steel oil reservoirs to concrete caissons. The mooring system requires the use of rubber fenders on dolphins. These rubber fenders adopted are so called buckling type rubber fenders of a cylindrical shape. This kind of rubber fender is commonly used in oil terminals and/or container berths for very large ships. The unique load deflection characteristic of the rubber fender is suitable to absorb both the steady force and periodic motions simultaneously. The rubber fender is the key component in the successful functioning of the floating oil storage base. The floating structure-mooring system is the world's first completed large floating oil storage bases after which many large floating structures in Japan copied this system such as in the Kan-on Breakwaters (Hiroshma), the Yumemai Bridge (Osaka), and the Pukari-Sanbashi (Yokohama, passenger terminal).

References

Bretshneider, C. L. (1968). Significant waves and wave spectrum. *(Fundamental of Ocean engineering-Part 7), Ocean Industry*, pp.40–46.

Davenport, A. G. (1967). Gust loading factors. In *Proceedings of the ASCE*, ET3 (pp. 11–34).

Goda, Y. (1987). Statistical variability of sea state parameters as a function of wave spectrum. *Coastal Engineering in Japan, JSCE, 34*, pp.39–52.

Hino, M. (1976). Relationship between the instantaneous peak values and the evaluation time—A theory on the gust factor. In *Proceedings of the JSCE*, No.177, (pp. 23–33).

The Japan Port & Harbour Association. (1999). *Technical Standards and Commentaries for Port and Harbour Facilities*, Japan.

Mitsuyasu, H. (1970). Wind wave generation and their spectrum-wind wave spectrum in finite fetch length. *Coastal Engineering in Japan, JSCE, 17*, 39–52.

Ueda, S. et al. (1998). Properties of rubber fender in use of mooring for floating bridge. In: *14th Ocean Engineering Symposium*, The Society of Naval Architect of Japan, pp. 359–364.

Ujina Floating Ferry Pier and Kan-On Floating Breakwater, Japan

Tadasu Kusaka and Shigeru Ueda

Abstract This chapter describes the construction of the Ujina Floating Ferry Pier and Kan-on floating breakwater. This Ferry Pier was set up at Hiroshima Bay in Japan in 1993. At 150 m in length, 30 m in width, and 4 m in height, it is the largest in Japan. Five prestressed concrete pontoons (30 m in length, 30 m in width, and 4 m in height) were fabricated on land and towed to the pier site for assembly to form the floating pier with a total length of 150 m. The Kan-on Floating Breakwater was set up at Hiroshima Bay in 1994. The breakwater is one of the world's largest classes of floating breakwaters made from reinforced concrete and steel plates. Three units of the breakwaters measure 97.8 m in length, 20.0 m in width, and 2.5 m in height while another two units are 70.75 m long, 21.0 m wide, and 2.5 m high.

1 Preamble

This chapter describes the construction of the Ujina Floating Ferry Pier and the Kan-on floating breakwater. The Ujina Floating Ferry Pier was set up in Hiroshima Bay in Japan as a base for international tourist ships and Seto Inland Sea cruising, in 1993. At 150 m in length, 30 m in width, and 4 m in height, it is the largest floating concrete ferry pier in Japan. Conventional floating ferry piers require labor-intensive maintenance for checking leakage caused by concrete cracks. However, the Ujina Pier prevents cracks by adopting a prestressed concrete (PC)

T. Kusaka (✉)
Former Chuden Engineering Consultants Co. Ltd, 1704 Iida Hachihonmatsu-chou, Higashi-Hiroshima 739-0141, Japan
e-mail: kusaka22@bronze.ocn.ne.jp

S. Ueda (✉)
Institute of Environment Information, 2-2-2 Hayafuchi Tsuduki, Yokohama 224-0025, Japan
e-mail: ued21010@ideacon.co.jp

C.M. Wang and B.T. Wang (eds.), *Large Floating Structures*,
Ocean Engineering & Oceanography 3,
DOI 10.1007/978-981-287-137-4_5

pontoon. Furthermore, the immense size of the floating ferry pier was needed for large international cruise ships. Five pontoons (30 m in length, 30 m in width, and 4 m in height) were fabricated on land and towed to the pier site to assemble the 150 m long floating pier.

The Kan-on Floating Breakwater was set up in Hiroshima Bay as a protective marina barrier in 1994. The breakwater is one of the world's largest classes of floating breakwaters, at 97.8 m in length, 20.0 m in width, and 2.5 m in height, for three units. With two units, the Kan-on Floating Breakwater is 70.75 m long, 21.0 m wide, and 2.5 m high. As it was decided hurriedly that this floating breakwater was to be used as a yacht racing site for the XII Asian Games in 1994 in Hiroshima, the early completion of a floating breakwater was necessary. The construction period was remarkably shortened by using reinforced concrete and steel plate (RC-hybrid) as structural materials for the pontoons.

2 Introduction to Ujina Floating Ferry Pier

The Hiroshima Port prospered for a long time as a base for traffic and distribution, but continued enhancements as a traffic base for persons, goods, and information were planned as a node of the Hiroshima District. On the other hand, the formation of the wide area sightseeing network, which keeps the Seto Inland Sea alive, is beneficial because there are few long term resorts near Hiroshima. To match the needs of the recent cruising boom, a sightseeing ship base was needed. As a base for international tourist ships and the Seto Inland Sea cruising, a large-sized floating pier was commissioned.

This Ujina Floating Ferry Pier project (Sakurai 1994) was promoted by The Third District Port Construction Bureau Hiroshima Office, Ministry of Transport (the name of the organization at the time of construction), and was constructed in Hiroshima Bay in 1993. Resolving technical and social difficulties, all construction was completed within a period of 19 months. This floating pier serves as a main port facility for regular international passenger ships and cruisers in the Inland Sea in Japan. Additionally, it has created a waterfront in the Ujina district, as the gate of Hiroshima Port, where people gather around and enjoy the coastal area of Hiroshima Port. Because of the large tidal difference of 3.8 m in Hiroshima Bay, the large floating pier consists of five pontoons, joined by prestressing, to prevent problems with passengers being able to get on and off. The floating pier is 150 m in length, 30 m in width, and 4 m in height, and is the largest in Japan among floating piers for public use. The Ujina Pier project was the first in Japan to join such large-scale pontoons together while at sea. Figure 1 shows the location of the Ujina Floating Ferry Pier after the completion.

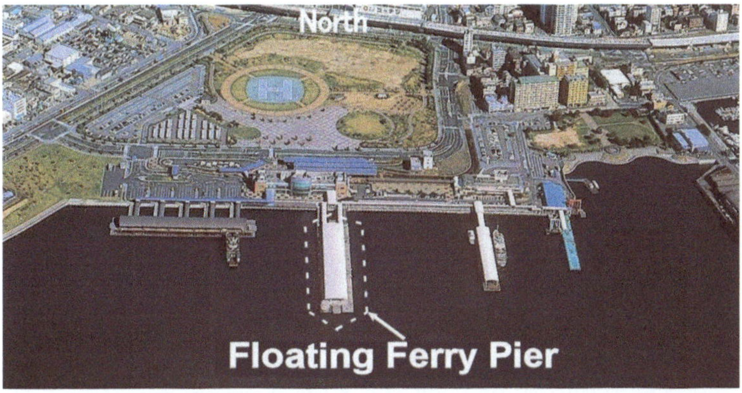

Fig. 1 Location of Ujina floating Ferry Pier

3 Technical Features of Floating Pier

Because the Ujina district is in an area surrounded by mountains on an island in the Seto Inland Sea, the climate and oceanographic conditions are relatively calm. However, the tidal difference is large at 3.8 m, which is more severe than other surrounding areas. Therefore a floating-type ferry pier is chosen instead of a fixed-type ferry pier. The floating concept also made it easier to get on and off of the International Tourist Ship, the Seto Inland Sea Cruising Ship and the ferry.

3.1 Dimensions and Materials

For a conventional floating pier design, an approximate total length of 60 m was the maximum possible because the hanging capability of the crane ship was limited at launching. However, a floating pier of 150 m in total length was necessary because of the upsizing of the ship. Construction using a dry dock is not economical, because dock charges and towing charges increase. Therefore, five pontoons of PC concrete (prestressed concrete), each 30 m long, 30 m wide and 4 m high, were manufactured on land and were sequentially joined at sea, creating the largest floating pier in Japan—150 m in total length. For the first time ever, the joining of pontoons at sea was accomplished using temporary drawing jacks and a guide material.

In conventional floating ferry piers made from reinforced concrete, a significant time was spent fixing water leaks caused by concrete cracks. Thus, for the Ujima Pier pontoon construction, prestressed concrete (PC) was selected to remedy these problems. However, it is difficult to apply a uniform tension to a PC structure and to control concrete cracks. Therefore, the maximum size of a single PC structure is

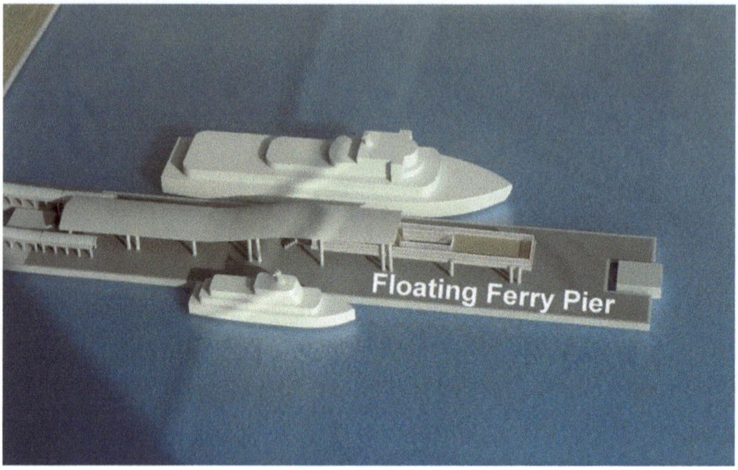

Fig. 2 Bird's eye view of Ujina Floating Ferry Pier

Table 1 Specifications

Pontoon	Material:	Prestressed concrete (PC)
	Dimension:	150 m (30 m × five blocks) long, 30 m wide, 4 m high
	Draft:	2.57 m
	Displacement:	110,800 kN
Dolphin	Material:	Reinforced concrete
	Dimension:	10 m long, 7.4 m wide, 8 m high

approximately 100 m. The floating ferry pier was constructed by dividing the total length into sections. Five blocks (30 m long and 30 m wide) were produced on land, after which the sections were joined at sea to create a pier that is 150 m in total length. Because of its sheer size, the floating ferry pier exhibits almost no sway from the waves and wind. In addition, the large floating ferry pier is much easier to use and much safer than smaller, less stable piers. Figure 2 shows a Bird's eye view of the Ujina Floating Ferry Pier.

In summary, the specifications of the Ujina Floating Ferry Pier are as follows: (Table 1).

3.2 Design Conditions

The design conditions are indicated below (Table 2).

Table 2 Design conditions

Objective ships	East Side: 5,000 G.T.	West Side: 600 G.T.
Depth of berth	East Side: −6.0 m	West Side: −4.0 m
Berthing velocity		0.15 m/s
Uniform load	4.9 kN/m^2	
Tidal level	Mx.H.H.W.L.: C.D.L. +4.61 m,	H.W.L.: C.D.L. +3.80 m,
	L.W.L.: C.D.L. +0.10 m,	Mi.L.L.W.L.: C.D.L. −0.43 m
Wind	Normal condition:	$V_{10} = 10$ m/s
	Extreme condition:	$V_{10} = 42$ m/s
Wave	Direction:	SSE—SSW
	Normal condition:	$H_{1/3} = 0.5$ m, $T_{1/3} = 2.5$ s
	Extreme condition:	$H_{1/3} = 1.1$ m, $T_{1/3} = 4.2$ s
Current	Normal condition:	0.2 m/s,
	Extreme condition:	0.5 m/s
Soil	Cohesive soil	

3.3 Joining of Pontoon

One of the main design features is the integration of five blocks into the full-length pier at sea. Each block was produced beforehand on land. The procedures to introduce prestress into longitudinal concrete sections, using PC wire, are shown in Fig. 3. The loading in the fully prestressed pier is analogous to a bending moment, and concrete cracks do not occur because the compression force is effective in all sections of the longitudinal hull concrete. On the other hand, the stress of reinforced concrete controls the occurrence of lateral concrete cracks. The first PC bar tensioned and inserted in the hull concrete, resists bending and shear in the frame of the joint part. Furthermore, the joining part is sealed by seal gum, and the secondary tension applied to the PC bar. Through the aforementioned procedure, the structure becomes entirely waterproof when completed.

3.4 Dolphin Mooring

In the Ujina Port area, the depth of the water is 4.0 m at low tide. The draft of Floating Ferry Pier is 2.5 m and thus, the effective depth of the water is only 4.0 m − 2.5 m = 1.5 m. A catenary curve is not sufficiently formed by the conventional chain mooring system when water depths are shallow. Because absorption of berthing and wave energies in chain mooring systems is difficult, the dolphin-type mooring system, as shown in Fig. 4, was proposed. This system has the following characteristics.

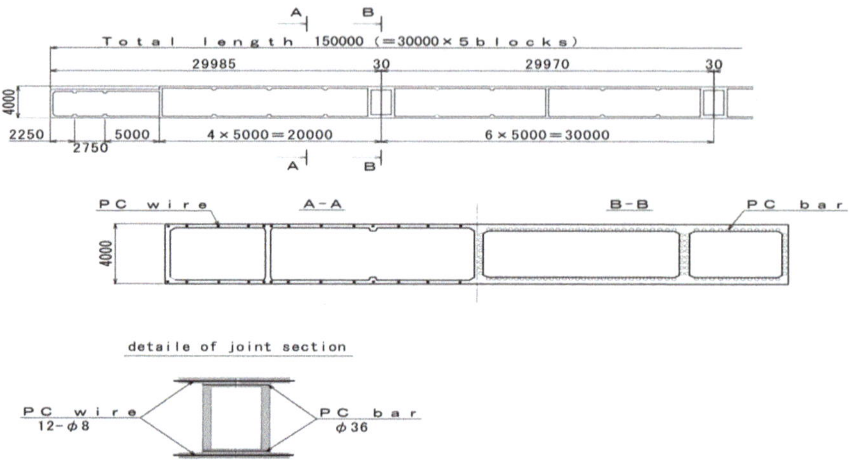

Fig. 3 Structure of pontoon

Fig. 4 Sketch of dolphin

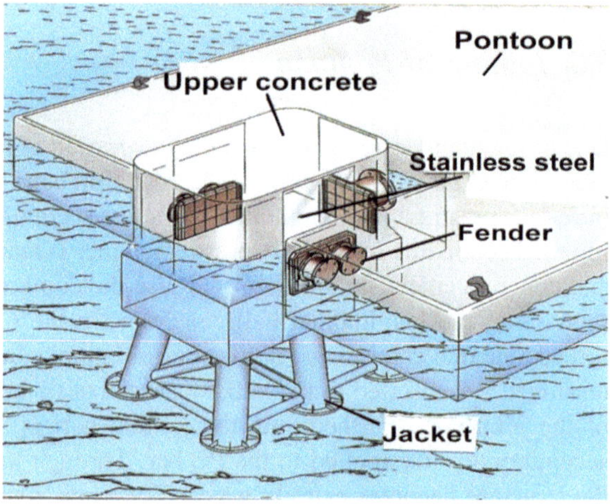

- The maintenance is easy because the main part of the dolphin (a fender and upper concrete) is either in the air or in the vicinity of the sea surface.
- The motions of a floating ferry pier at berthing or in waves are large in conventional chain-type mooring systems. However, the motion of a floating ferry pier is restricted by a fender with the dolphin-type mooring system. Therefore, usage quality and safety are improved at the time of getting on and off the ship, boat and ferry.

4 Simulation of Floating Pier's Motion

4.1 Explanation of Analysis

The motion of the floating pier and the fender's reaction force were examined using the time history analysis method because the reaction curve of the fender is nonlinear. The behavior of the floating pier is governed by

$$[A]\ddot{\eta} + [B]\dot{\eta} + [C](\dot{\eta} - u)|\dot{\eta} - u| + [K]\mu = E \tag{1}$$

where $[A]$, $[B]$, $[C]$, $[K]$ are the coefficient matrices, E the vector of the external wave force and wind load, the superdot represents differentiation with respect to time, Mitsuyasu's spectrum was used for a wave, and Davenport's spectrum for wind, η the vector of the quantity to be demanded and u the velocity vector. Because the coefficient matrix $[K]$ is a nonlinear spring of the fender and the external force vector E is an irregular load, Eq. (1) is calculated by numerical integration about the time series.

4.2 Simulation Results

Consider a tidal level of H.W.L., $V_{10} = 42$ m/s, $H_{1/3} = 1.1$ m and $T_{1/3} = 4.2$ s. A part of the numerical results, for the aforementioned conditions, is shown in Fig. 5.

The simulation results of the floating pier's motion are summarized as follows:

- the design limit criteria of the floating pier's motion are not described sufficiently, but there are no problems with the usage of the pier because the perturbations are small.
- The fender strain was less than the design criteria limit, and therefore, the safety of the design was confirmed.

5 Construction of Floating Pier

5.1 Procedure of Pontoon Assembling

The procedure for constructing the pontoon is shown in Fig. 6. In step 1, five blocks (30 m in length and 30 m in width) are manufactured on land. In step 2, the five blocks are relocated by a crane ship. In step 3, two blocks are joined at sea, and the first 60 m of the pontoon is completed. In step 4, the 60 m pontoon is towed to the site. In step 5, the second joining is conducted to assemble the pontoons to create its final length of 150 m (30 m × five blocks).

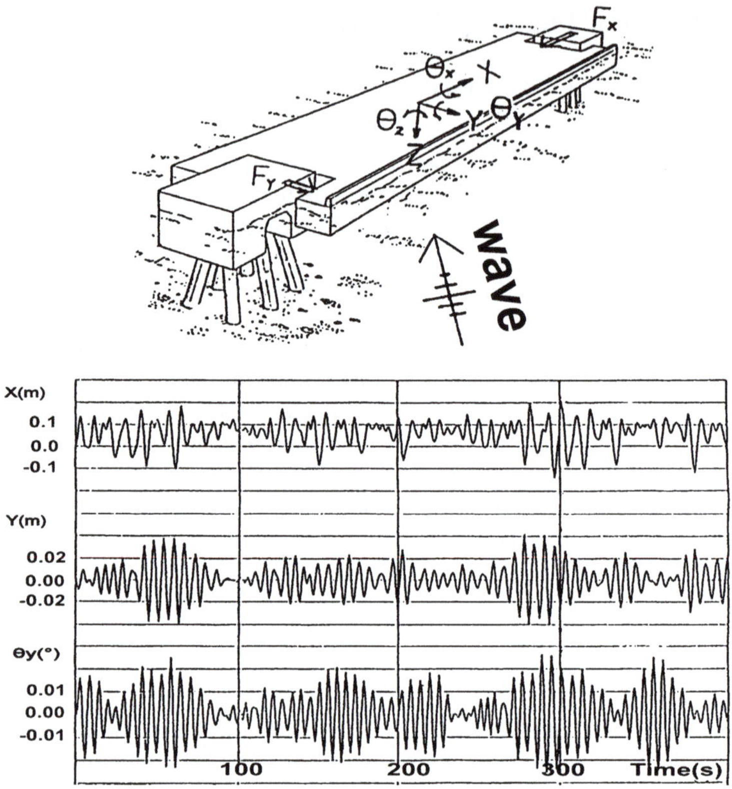

Fig. 5 Example of simulation

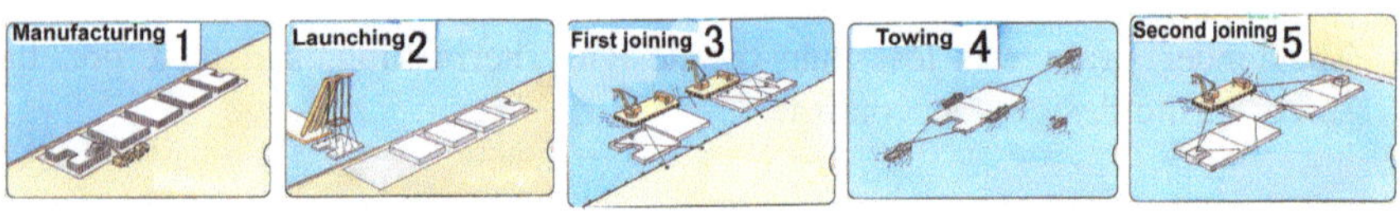

Fig. 6 Sketch of pontoon assembling

5.2 Counter Measures for Concrete Crack

Because the pontoon is used as a buoyant body, cracks in the concrete during construction need to be avoided. Therefore, the following measures were taken:

- Portland blast-furnace slag cement B class was used.
- High-range air-entraining and water-reducing agents were used.
- Concrete was mixed with cold water.

- Pipe cooling was carried out in curing concrete.
- The spray curing for concrete was provided.
- A shading net and curing mat were used.

5.3 Joining of Pontoon

For joining the five pontoon sections, a total of 52 PC bars (36 mm in diameter) were used per section, as shown in Fig. 7a. Figure 7b shows a sketch of the first joining at sea, made by a diver. The second joining, using the PC bars, was carried out after water stopping was performed. The P-shaped seal gum used during construction is shown in Fig. 7c.

6 Concluding Remarks on Floating Pier

The Ujina Floating Ferry Pier has encountered typhoons and earthquakes after its completion in 1993, but it is still in excellent condition. At present, this floating ferry pier serves as the main port facility for the Hiroshima district. The joining technology used in assembling the pontoons at sea allows for future expansion of the floating structure. The floating structure shown on the left in Fig. 1 is the second Ujina floating ferry pier, that was completed in 1998, by applying this joining technology at sea. After three blocks (55 m in length, 30 m in length, and 55 m in length) were joined, the second floating ferry pier has a total length of 140 m.

7 Introduction to Kan-on Floating Breakwater

After the 1980s in Japan, the demand for recreation was remarkably improved with a diversification of the sense of values and the increase in leisure time. Above all, oceanic recreation using pleasure boats steadily increases. Therefore, the Hiroshima Prefectural Government decided to construct the first public marina in the Kan-on district in Hiroshima Port (Fukuda et al. 1996). Both large and small pleasure boats are anchored in the marina. While large boats are not greatly affected by the motion of the waves, small boats are significantly affected. Thus, there is a need to protect small boats anchored in the marina from the waves of the open sea because the limiting wave height for a port where small boats can anchor safely is 0.5 m.

The Kan-on Floating Breakwater was constructed in 1994 at a location 5 km west of the Ujina Floating Ferry Pier. Because of the small influence of the river flow at the time of flooding, the high quality of water in the marina, and the view at low tide, the floating breakwater was selected as a protective facility in the marina.

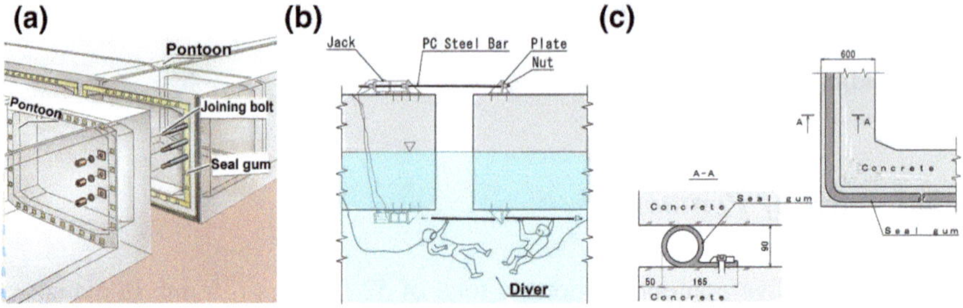

Fig. 7 Joining of pontoon. **a** Joining bolt. **b** Work of diver. **c** Seal gum

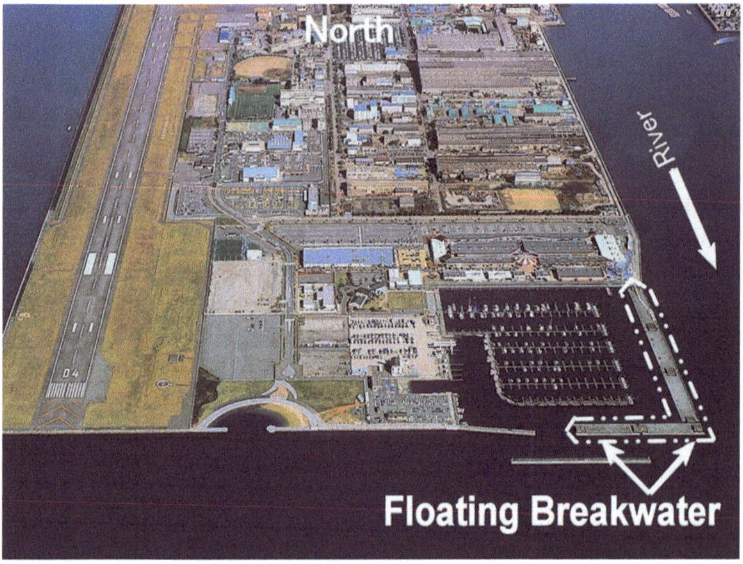

Fig. 8 Location of Kan-on floating breakwater

Figure 8 shows the location of the Kan-on Floating Breakwater after it was installed. The Kan-on Marina was also used as the yacht racing site of the XII Asian Games in Hiroshima in 1994.

8 Technical Features of Floating Breakwater

The OHTA River in Japan flows through the right side (the east side) of the Kan-on Floating Breakwater as shown in Fig. 8. The influence on the flow of this river by the floating breakwater is a technical subject. There are various types of breakwaters, and the floating-type breakwater, as a protective facility, was adopted for the following reasons:

- Because the lower part of the floating breakwater becomes the water passage mechanism, the influence of the backwater of the river is small in times of flooding.
- Because the draft of the floating breakwater is not large, the shielding rate (draft/water = 2.5 m/8.5 m = 0.29) is smaller than other breakwater types, such as the curtain wall-type. Thus, this provides sufficient seawater exchange below the floating breakwater, and the quality of the water in the marina is kept well.
- This area serves as the main marina of the Hiroshima Port. Because many people gather for leisure, the view quality and navigation safety of the boats were taken into account. The tidal range of approximately 3.8 m in Hiroshima Port is large in Japan, but the floating breakwater follows the sea level for low and high tides. Accordingly, the view quality is superior because there are few obstructions above the water surface during the low tide.

8.1 Dimensions and Materials

Floating structures produced using conventional RC concrete have problems with water leakage owing to a time dependent behavior. In order to prevent water leakage, reinforced concrete and steel plates (RC-hybrid) were used as design materials for the floating breakwater hulls. This system has an 8 mm thick steel plate inside the pontoon and makes certain that the damage due to water leakage is prevented. Furthermore, the 160–250 mm thick concrete surrounding the steel plate protects against rusting. This floating breakwater design is more likely to be damaged in collision with driftwood during flooding, or in a crash with an uncontrolled boat, and the RC-hybrid construction is sufficient for withstanding such extreme states. Furthermore, a rapid construction was required because the Kan-on floating structure was used as a yacht racing site for the XII Asian Games in Hiroshima in 1994. The large scale of the pontoon was made possible by using the RC-hybrid construction method, and the development cycle was remarkably shortened. Figure 9 shows a Bird's eye view of the Kan-on Floating Breakwater.

In summary, the specifications of the Kan-on Floating Breakwater are: (Table 3)

8.2 Design Conditions

The design conditions are indicated below (Table 4).

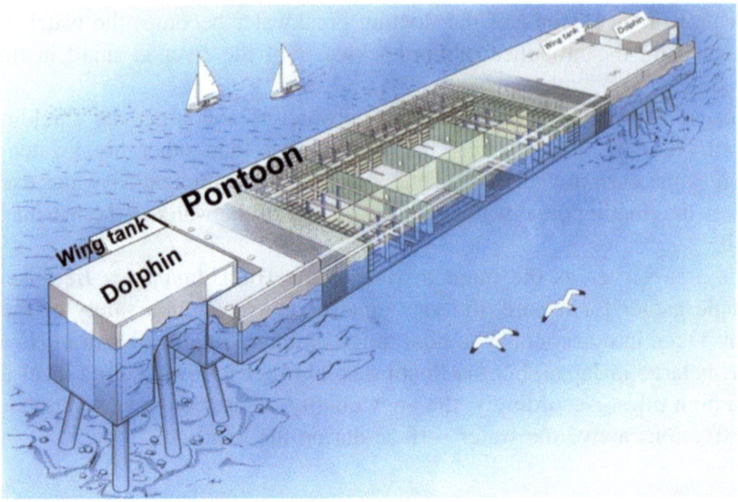

Fig. 9 Bird's eye view of Kan-on floating breakwater

Table 3 Specifications

Pontoon	Material:	Reinforced concrete and steel plate (RC-hybrid)
	Dimension:	97.8 m long, 20.0 m wide, 3.5 m high (3 units)
		70.75 m long, 21.0 m wide, 3.5 m high (2 units)
	Draft:	2.5 m
	Displacement:	45,000 and 30,900 kN
Dolphin	Material:	Reinforced concrete
	Dimension:	9.0 m long, 9.0 m wide, 7.8 m high (4 units)
		14.0 m long, 9.0 m wide, 7.8 m high (3 units)

9 Experimental Studies

9.1 Experiment of Backwater in River Flow

When a structure is set up in a river mouth, the river flow in the event of flooding becomes a critical design parameter. Because the pontoon is a floating type, the flow of the water under the lower section is permissible. A three-dimensional flow experiment was carried out to quantify the effects of the breakwater and flow interaction.

The river flow after the floating breakwater's setting was compared with the river flow before setting, and an influence on the river flow after the floating breakwater set up at the river mouth was evaluated. Because an instrument accuracy of the water level in the experiment is 0.1 mm at the minimum (as 5 mm

Table 4 Design conditions

Water depth	8.4 m	
Tidal level	Max. H.H.W.L.: C.D.L. +4.61 m,	H.W.L.: C.D.L. +3.80 m
	L.W.L.: C.D.L. +0.10 m,	Min. L.L.W.L.: C.D.L. −0.43 m
Wind	Normal condition:	$V_{10} = 10$ m/s
	Extreme condition:	$V_{10} = 33$ m/s
Wave	Direction: SSE-S	
	Normal condition:	$H_{1/3} = 0.5$ m, $T_{1/3} = 2.7$ s
	Extreme condition:	$H_{1/3} = 2.2$ m, $T_{1/3} = 4.5$ s
Current	Normal condition:	0.2 m/s,
	Extreme condition:	0.3 m/s
Soil	Cohesive soil	

for the value of the on-site conversion), an experimental apparatus cannot detect a water level difference of both of backwater. Therefore, the river flow before and after the floating breakwater's setting was examined by an experiment study, and the quantitative change of water level was supplemented by a non-uniform flow analysis.

Condition of Experiment:

Scale	Horizontal direction 1/100, vertical direction 1/50
Tide level	C.D.L. + 4.520 m (High Water Level + Flood Level)
Flow quantity	930 m^3/s (flood discharge)
Experiment case	Flow without floating breakwater,Flow with floating breakwater
Coefficient of roughness	0.028

Experiment of Flow:
Figure 10 shows a flow before the floating breakwater setting, and Fig. 11 shows the appearance of diffusion after the setting. When the spread of the dye is defined as a diffusion angle from a levee normal line of Breakwater, the comparisons of an observed diffusion angles in experiment are as follows (Table 5).

From visual observations, the river flows rapidly through the lower sections of the floating breakwater. Thus, the breakwater did not have a significant impact in the overall flow of the river. In addition, from the measured flow velocity, a notable difference in speed was not observed in the flow before and after the breakwater was established. It was confirmed that both the flow speed and the backwater flow were not significantly influenced after the floating breakwater was installed. This confirmation was established by performing a nonuniform analysis.

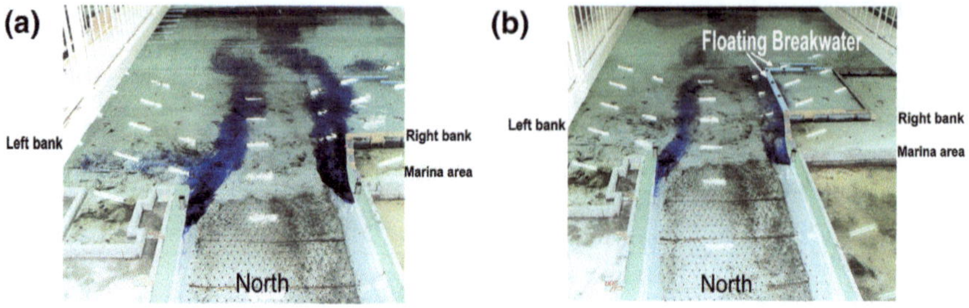

Fig. 10 Photos of flow. **a** Flow without floating breakwater. **b** Flow with floating breakwater

Fig. 11 Result of diffusion

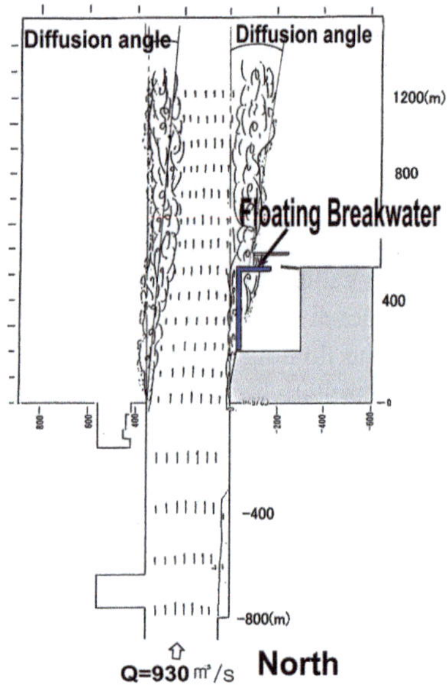

Table 5 Diffusion angles

Items	Diffusion angle of case without floating breakwater	Diffusion angle of case with floating breakwater
Left bank	−1° 18′	−6° 50′
Right bank	+11° 00′	+11° 00′

9.2 Wave Decrease Experiment of Floating Breakwater

As there was a request to safely anchor a boat on the inside of the floating breakwater, a 2.2 m high wave from the open sea has to be decreased to a height of 0.5 m or less in the marina. A two-dimensional wave-breaking experiment was carried out to examine the possible shapes of the floating breakwater that satisfy

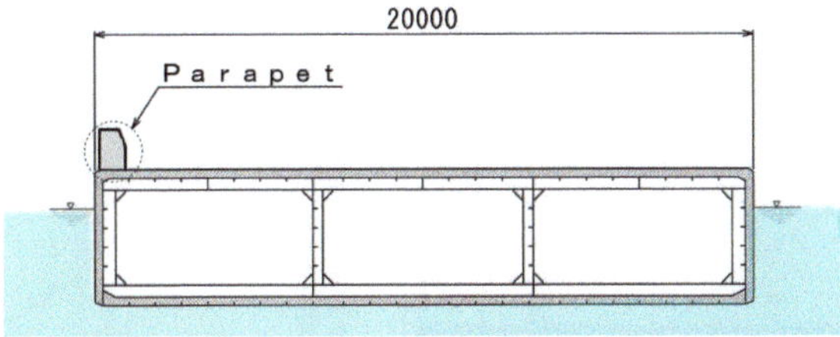

Fig. 12 Cross section of pontoon

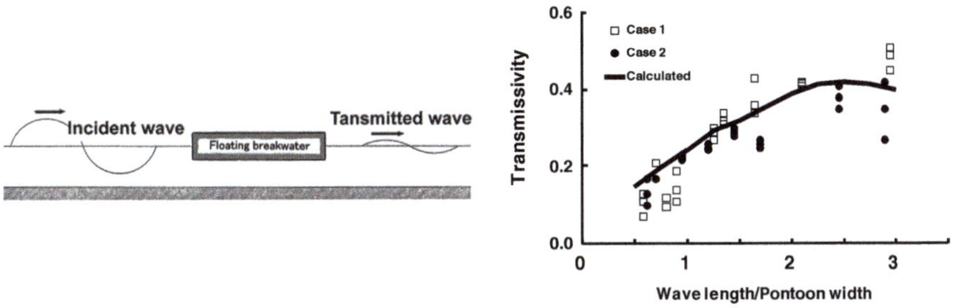

Fig. 13 Result of wave decrease experiment

this requirement. Figure 12 shows the cross section of the Kan-on Floating Breakwater.

A 1.0 m high parapet is placed on the weather side of the rectangular pontoon section as shown in Fig. 12. The wave energy is reduced by a wave overtopping, and accordingly, this parapet improves the wave reduction efficiency for the floating breakwater. Because such a parapet incurs wave breaking phenomena, it is difficult to predict the wave reduction efficiency by numerical simulation only. Therefore, a wave tank experiment at a 1:20 scale was carried out to examine the basic characteristics of wave reduction efficiency for the floating breakwater. Figure 13 shows the relationships for wavelength and transmissivity (transmitted wave height/incident wave height).

Case 1 parapet height: 0.7 m, draft: 3.5 m,
Case 2 parapet height: 1.0 m, draft: 2.5 m

As a result of the comparison of both cases, the specifications of the floating breakwater were selected in reference to Fig. 13. Case 2 reduces material costs and has good wave reduction efficiency for long waves. In addition, the calculated value from Fig. 13 was obtained by an energy loss coefficient of the wave breaking phenomenon, and a simulation using the region division method (Fujita and Kusaka 1977).

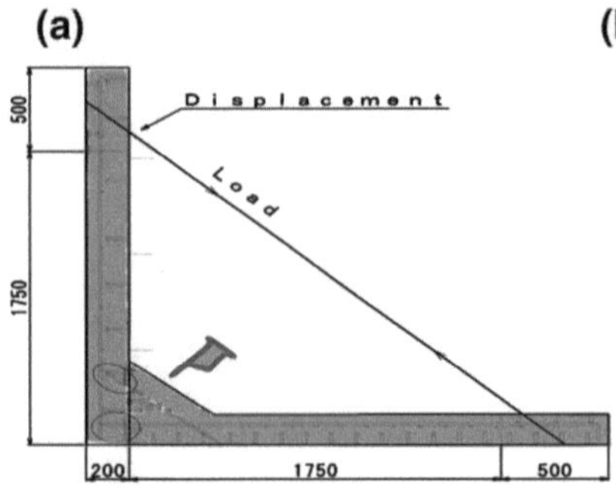

Fig. 14 Details of pontoon corner

9.3 Strength Test of Pontoon Corner

The hybrid structure increases rigidity by using a composite design of steel plates and reinforced concrete. However, there are many unknown properties in the corner strength, which is congested with steel plate, concrete and reinforced bars. Therefore, the characteristics of the hybrid structure at a lower corner, the part shown in Fig. 14, were examined using a structural experiment.

Case 1 This is the model where the main reinforcement bars and a haunch reinforcement, in the corners, penetrated a steel plate. Much labor is needed in steel plate processing and reinforcing bar arrangement.

Case 2 This is the model where the main reinforcement bars and a haunch reinforcement, welded to a steel plate, did not penetrate a steel plate. This was adopted because the methods of the construction were preferable.

Figure 15 shows the relationship of the displacement and the loading, which was obtained in the experiment (Table 6).

Fig. 15 Results of strength test

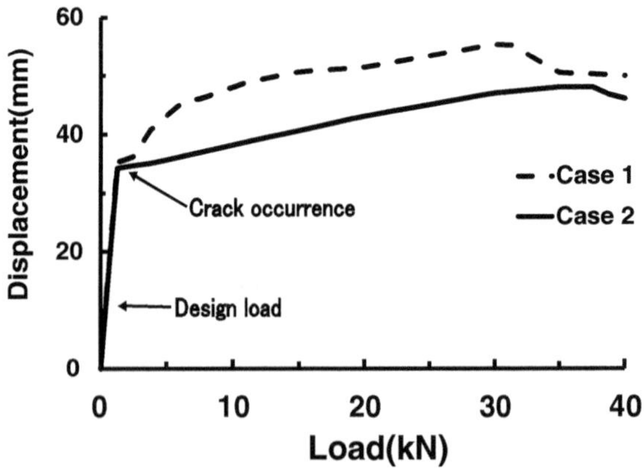

Table 6 Load of crack occurrence

Items	① Load of crack occurrence (kN)	② Design load (kN)	①/②
Case 1	35.3	11.8	3.0
Case 2	34.3	11.8	2.9

The crack occurrence load/design load is approximately 3.0 as measured in the experiment. Thus, it was confirmed that the corner structure was strong enough to endure the external forces. Furthermore, the crack occurrence load showed little difference for Case 1 and Case 2, but the shape of Case 2 was adopted because of the more favorable construction properties.

In addition, because the motion simulation method for the Kan-on Floating Breakwater is the same as the case for the Ujina Floating Ferry Pier discussed previously, a detailed description is not included here.

10 Construction of Floating Breakwater

10.1 Hybrid Structure

The maintenance in the case of the floating breakwater is not as easy as for the floating ferry pier. The hybrid construction shown in Fig. 16 was adopted for the floating breakwater, because it was necessary to pay particular attention to the water sealing and ease of the repair. Optimal construction is accomplished using an inward steel plate for concrete placing. In order to reduce the construction cycle as much as possible, this fabrication method was used. Furthermore, with this

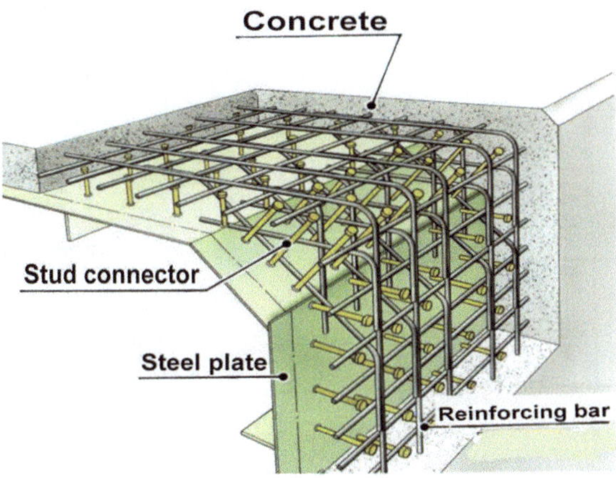

Fig. 16 Sketch of hybrid structure

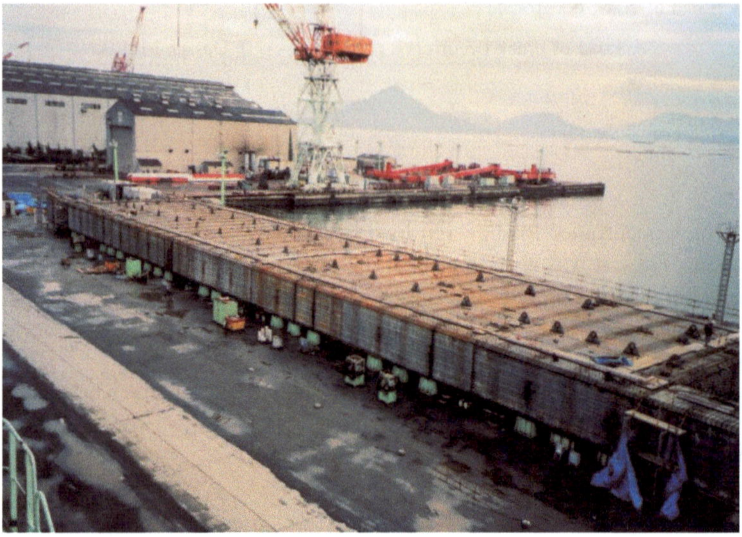

Fig. 17 Construction of pontoon

method, repair is easy because prestress was not induced during unexpected incidents, such as in a collision with driftwood of flooding, or in a collision with a recreational boat.

Figure 17 shows the construction of the pontoon breakwater. Five units of the pontoon were produced on land and were subsequently launched at sea. Figure 18 shows the launching of the pontoon from land to the coastal water by a 36,000 kN

Fig. 18 Launching of pontoon

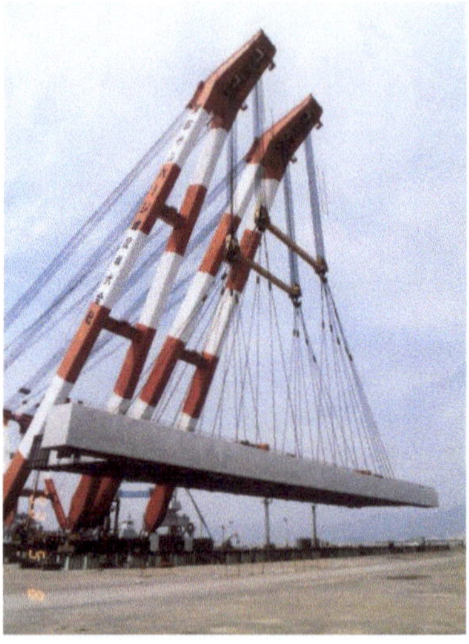

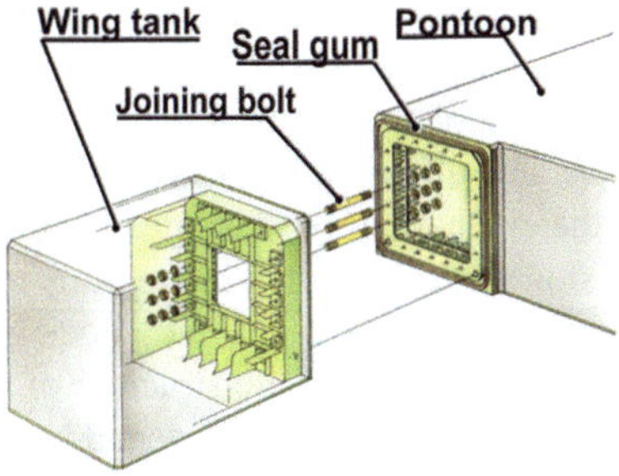

Fig. 19 Sketch of wing tank

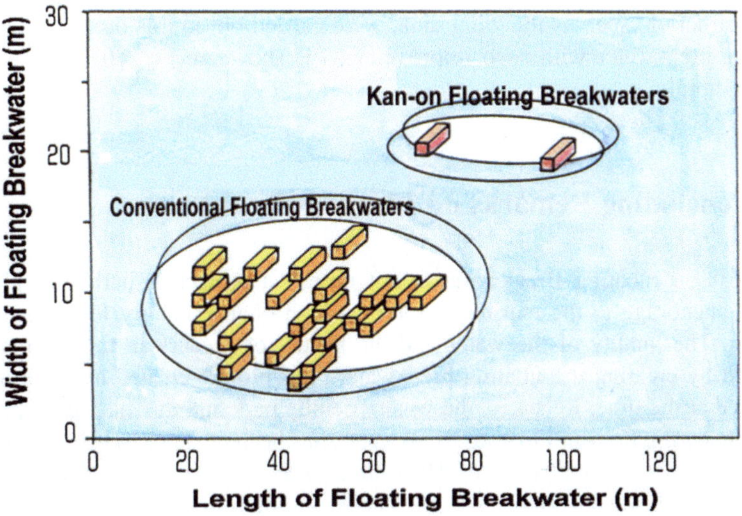

Fig. 20 Comparison of floating breakwater size

lifting capacity crane ship. The weight of hybrid pontoon structure without the ballast is relatively light. Therefore, it was possible to have a rapid launching operation.

10.2 Installation of Pontoon

Moving the pontoon on and off the dolphin is easy when using a wing tank with the floating breakwater. A wing tank system is a device that joins the wing tank to the main body of the pontoon with a joining bolt, as shown in Fig. 19. The joining technology was also used in the construction of the Ujina Floating Ferry Pier as mentioned previously. This technology was applied to the joining system for the floating breakwater. However, this system was used to enable an improved work efficiency onsite; the taking on and off was possible and the joining materials were downsized (the joining system of the Ujina Floating Ferry Pier was a one-time occurrence, and taking on and off several times was not possible). The floating breakwater, adjacent to the river, may be damaged by driftwood during a flood, and the risk of the collision with a boat by a steering mistake is larger than a general breakwater or floating pier. The repairs at a dock are relatively easy at the time of such an accident, using the removable structure of the wing tank. Also, the setting period of the construction onsite was shortened because of the mounting of the wing tank. The effect of the simplification is useful because the revetment and reclamation works were carried out at the same time as the construction of the floating breakwater.

Structural improvements, such as the RC-hybrid method, and the improvements in construction, such as the wing tank, were carried out in this project. The large floating breakwater with such improvements is shown in Fig. 20 and was completed rapidly.

11 Concluding Remarks on Floating Breakwater

The Kan-on Floating Breakwater is not merely a protective harbor facility for boats' anchorage in the marina. It serves a high quality facility for oceanic recreation. The quality of the water and the beauty of scenery in the marina were secured by creating the floating breakwater. The project created an environment required at the time. Recently, the water-friendly park and the marine shop, made possible by the Kan-on Floating Breakwater, has become a place of recreation and relaxation, for the megacity of Hiroshima. The spacious deck of the Kan-on Floating Breakwater is opened to the public on holidays as a fishing park, and is used by many citizens.

Acknowledgments A part of the document of the Project of Ujina Floating Ferry Pier was provided by the Hiroshima Port and Airport Office, Ministry of Land, Infrastructure, Transport and Tourism, and a part of the document regarding the Project of the Kan-on Floating Breakwater was provided by the Hiroshima Port and Harbor Promotion Office, Hiroshima Prefectural Government. The authors wish to express their sincere appreciation to Mr. Hideyuki Hokimoto (from Chuden Engineering Consultants Co., Ltd.) who provided much advice on both projects.

References

Fujita, K. & Kusaka, T. (1977). Two dimensional floating breakwater with energy dispersion. *Proceedings of 24th Coastal Engineering in Japan*, JSCE (pp. 285–289) (in Japanese).

Fukuda, K., Takamaru, T., Yoshida, H., Kusaka, T., et al. (1996). Breakwater efficiency and influence of river current due to floating breakwater settled near river mouth. *Annual Journal of Civil Engineering in the Ocean, JCSE, 12*, 195–200. (in Japanese).
Sakurai, T. (1994). Construction of floating Pier at Ujina Region in Hiroshima Port. *Proceedings of the International Workshop on Floating Structures in Coastal Zone* (pp. 22.1–22.11).

Floating Offshore Wind Turbine, Nagasaki, Japan

T. Utsunomiya, I. Sato, T. Shiraishi, E. Inui and S. Ishida

Abstract Offshore wind energy resources in Japanese EEZ (Exclusive Economic Zone) are now considered to be huge. In order to utilize the huge amount of energy located in relatively deep water areas (water depth range: 50 – 300m), Ministry of the Environment, Japan has kicked-off the demonstration project on floating offshore wind turbine (FOWT). The project will continue for six years beginning from 2010fy to 2015fy. In the project, two FOWTs have been installed. The first FOWT mounts a 100kW wind turbine of downwind type, and the length dimensions are almost half of the second FOWT (so called as half scale model). The second FOWT mounts a 2MW wind turbine of downwind type, and called as full scale model. The FOWTs consist of PC-steel hybrid spar (which is cost-effective) and are moored by three mooring chains. The half scale model was installed at the site on 11th June, 2012 as the first grid-connected FOWT in Japan. The half scale

T. Utsunomiya (✉)
Department of Ocean Energy Resources, Kyushu University, 744 Motooka Nishi-Ku, Fukuoka 819-0395, Japan
e-mail: utsunomiya@nams.kyushu-u.ac.jp

I. Sato
Toda Corporation, Kyobashi 1-7-1, Chuo-Ku, Tokyo 104-8388, Japan
e-mail: iku.sato@toda.co.jp

T. Shiraishi
Hitachi Ltd, Shiroganecho 1-1-1, Hitachi, Ibaragi 317-0056, Japan
e-mail: takashi.shiraishi.kx@hitachi.com

E. Inui
Fuyo Ocean Development and Engineering Co Ltd, Kuramae 3-15-7, Taito-Ku
Tokyo 111-0051, Japan
e-mail: inui@fuyokaiyo.co.jp

S. Ishida
National Maritime Research Institute, Shinkawa 6-38-1, Mitaka, Tokyo 181-0004, Japan
e-mail: ishida@nmri.go.jp

C.M. Wang and B.T. Wang (eds.), *Large Floating Structures*,
Ocean Engineering & Oceanography 3,
DOI 10.1007/978-981-287-137-4_6

model was attacked by very severe typhoon Sanba (1216), the greatest tropical typhoon in 2012 in the world. The behavior during the typhoon attack, including the measured environmental data and the FOWT responses, is described. The behavior during power production is also described. The installation of the full scale model has successfully been made; the opening ceremony was held on 28th October, 2013 as the first multi-megawatt FOWT in Japan. The installation procedures are briefly mentioned.

1 Introduction

In order to mitigate greenhouse gas emission, it is necessary to increase renewable energy production. In Japan, the production of renewable energy is still very limited. However, offshore wind energy resources in the Japanese Exclusive Economic Zone (EEZ) are now considered to be huge. Table 1 shows the potential of wind energy as estimated by the Ministry of the Environment, Japan. As it can be seen in the table, the technical potential at offshore locations is huge. In particular, the offshore wind energy resource in deeper waters (where the water depth is greater than 50 m) is remarkable. The total electrical power production facility in Japan as at 2010 is about 200 GW. Of course, if one accounts for the economic feasibility, the potential is reduced from the technical potential. However, the offshore wind energy resources are still maintaining a huge potential, particularly in deeper waters.

For developing deep water wind energy, the use of floating-type foundations is considered to be more economical than the use of bottom-fixed foundations (EWEA 2013). However, at this moment, the floating-type offshore wind turbine is not at the commercial stage but at the demonstration stage. There existed only one multi-megawatt floating wind turbine and that is Norway's Hywind, built in 2010 (Statoil 2012). The second multi-megawatt floating wind turbine, WindFloat—a semisubmersible-type floating wind turbine was installed in Portugal in June 2012 (Principle Power, Inc. 2012). Since then, Japan's Ministry of the Environment kicked-off a demonstration project on floating offshore wind turbine (FOWT). The project will take 6 years; beginning from September 2010 to March 2016. In this project, two floating offshore wind turbines have been installed. This chapter introduces the demonstration project.

Table 1 Wind energy potential in Japan (Ministry of the Environment 2012)

Location	Technical potential (GW)	Economical potential
On land	283	133 GW (20 JPY/kWh FIT)
Offshore (<50 m)	309	159 GW (30 JPY/kWh FIT)
Offshore (>50 m)	1,264	300 GW (30 JPY/kWh FIT)

FIT Feed-in tariff

2 Outline of Demonstration Project

2.1 Objectives of Project

The ultimate objective of the demonstration project is to reduce the greenhouse gas emission through commercialization of FOWT in the Japanese EEZ. Towards the commercialization of FOWT, a mandatory and important step is to demonstrate its technical feasibility. In particular, a demonstration of installation and operation of a multi-megawatt floating offshore wind turbine at sea is the primary objective of the project.

Although the installation/operation of a multi-megawatt FOWT is the main target of the demonstration project, a step-by-step approach is generally preferred, in order to reduce possible risks. Also, social acceptance may be gained by such a step-by-step approach. Thus, the half-scale model, of which length scale is almost half of the full-scale model, was planned to be installed before the installation of the full-scale model.

At the same time, the environmental impact assessment and the influences to local fishery community are considered to be most important matters for the demonstration project. Thus, the objectives of the project are related not only to technical and economic matters, but also to social and environmental matters.

2.2 Outline of Schedule

Figure 1 shows the master schedule of the demonstration project. The project will spread over 6 years. In the first year, the offshore site for the demonstration was selected. In the site selection, obtaining permissions/agreements from the local fishery cooperative and the local community is the most critical matter. After establishing the demonstration site, the meteo-ocean measurement and the survey for environmental impact assessment got started. The half-scale model was designed, constructed, installed, operated, and finally removed. Following the half-scale model test, the full-scale model was designed, constructed, installed, and is now in operation. The meteo-ocean measurement and the survey for the environmental impact assessment have been and will be made through the demonstration project.

2.3 Location

Figure 2 shows the location of the at-sea demonstration site. The site is about 1 km offshore of Kabashima Island, Goto city, Nagasaki prefecture, Japan. The mean water depth is 97.2 m (at mean sea level; MSL). The site is surrounded by the island on the north-western side, but it is opened in the south-eastern side. The marine cable has been installed for the grid-connection. The distance to the shore from the FOWT along the marine cable is about 1.8 km.

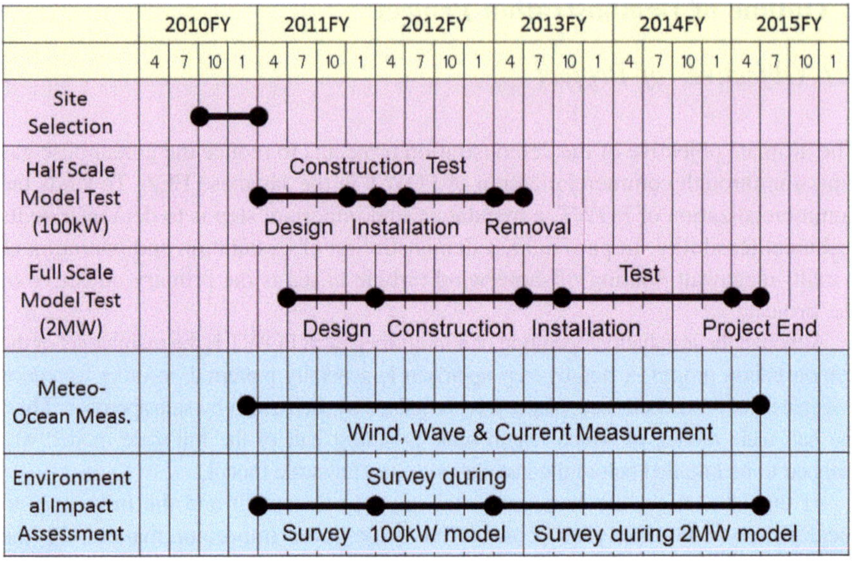

Fig. 1 Master schedule of demonstration project

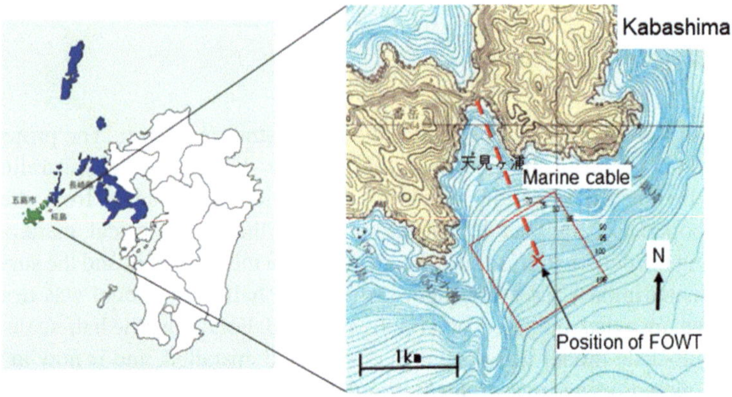

Fig. 2 Demonstration site—Kabashima, Goto city, Nagasaki prefecture

3 Meteo-Ocean Measurements

3.1 Wind Measurement and Simulations

As shown in Fig. 3, the wind speed measurements have been made at St. W (1) and St. W (2) of Kabashima Island. The altitude of the base of St. W (1) is about 40 m from the sea level, and the wind-mast height is 15 m. The annual average wind speed at St. W (1) (at 15 m height) was 6.2 m/s. The altitude of the base of

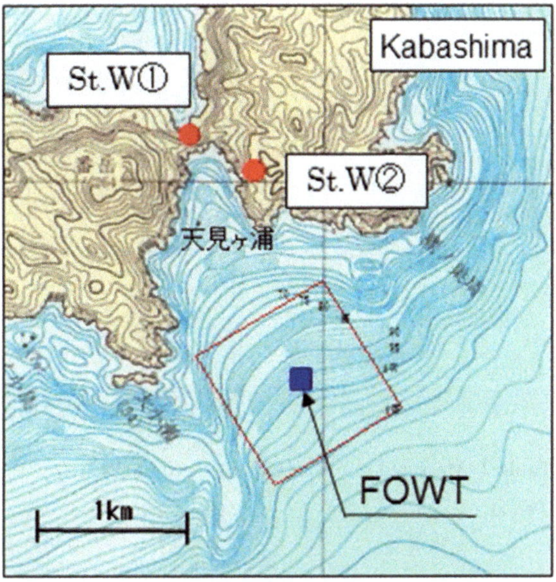

Fig. 3 Location of the wind masts

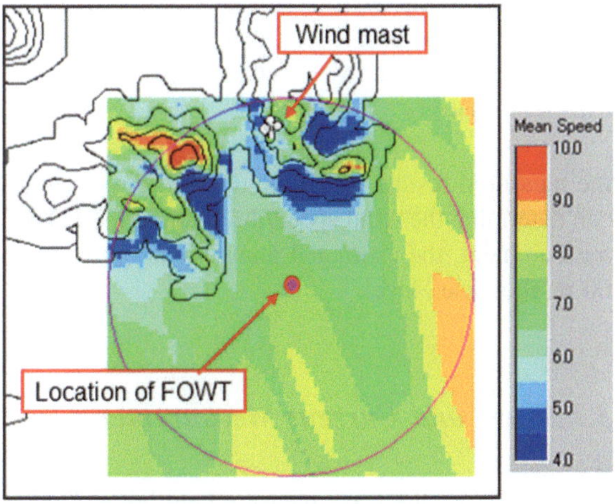

Fig. 4 Simulation of wind speed distribution

St. W (2) is about 100 m from the sea level, and the wind-mast height is 40 m. The annual average wind speed at St. W (2) (at 40 m height) was 5.2 m/s. The reason for a larger annual average wind speed at St. W (1) than at St. W (2) is due to topology effect around the wind mast.

Figure 4 shows the wind speed distribution by a numerical simulation. It corresponds to the wind speed distribution at the height of 57 m above the sea/land surface for wind direction of NNW (the wind incident from NNW direction). The wind speed at a very high position is 10 m/s. From the numerical results for 16 different wind directions, one can obtain a relationship between the wind speed/direction at the wind mast on land and those at the FOWT location. By combining

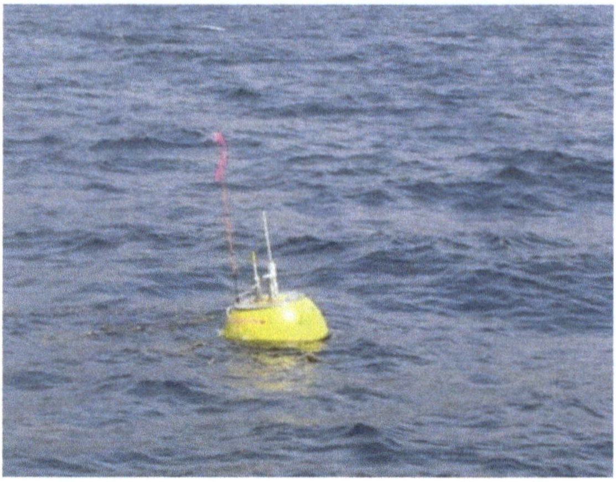

Fig. 5 Wave measurement buoy

Table 2 Measured results of current speed at site

Period	Mean current speed		Maximum current speed	
	5 m BSL (m/s)	75 m BSL (m/s)	5 m BSL (m/s)	75 m BSL (m/s)
23 Feb. 2011–23 Mar. 2011	0.12	0.07	0.40	0.32
12 Aug. 2011–10 Sep. 2011	0.09	0.07	0.47	0.29
29 Jun. 2012–30 Jul. 2012	0.10	0.08	0.44	0.30
24 Oct. 2012–27 Nov. 2012	0.10	0.07	0.35	0.33

BSL Below sea level

the measured results of the wind speed/direction at wind mast (at St. W (2)), one can determine the wind speed/direction at the FOWT location. After one-year field measurement, the estimated annual average wind speed is found to be 7.5 m/s at 60 m above sea level.

3.2 Wave and Current Measurements

Figure 5 shows the wave measurement buoy. The buoy has been installed near the FOWT location. The wave height, the wave period, and the wave direction have been obtained from the buoy records. The annual average wave height and period were $H_{1/3} = 0.58$ m and $T_{1/3} = 5.7$ s for 2011 fy, and $H_{1/3} = 0.64$ m and $T_{1/3} = 6.1$ s for 2012 fy, respectively. When wind blows from north–northwest directions, the wave height is relatively small because of the limited distance of fetch. On the other hand, the swells and the waves from east–south directions directly propagate to the FOWT location. Thus, higher waves were observed when the wave direction is in the east–south directions.

Table 2 summarizes the measured results of the current speed at the site. The current speed was measured at different depths for about 1 month duration in all

Table 3 Survey items for environmental impact assessment

Environmental factor	Items of environmental impact assessment
Atmospheric environment	Noise, low frequency noise
Water environment	Turbidity, bottom sediment
Topography	Sea bed topography
Sea plant	Seaweed species
Sea animal	Benthos, fish and shellfish, marine mammals
Animal	Avian species
Ecosystem	Relationship among important species
Other biological environment	Underwater noise
Scenery	Scenery resources
Fishery	Basic survey on fisheries environment (water temperature, salt content, pH, DO, COD, plankton, spawn, larva), fisheries activity
Others related to fishery	Fish gathering effect
Floating body related factor	Marine growth

four seasons. The mean current speed at the sea surface level is about 0.1 m/s, and the maximum current speed is about 0.5 m/s. There are no significant differences in the current speed for the four seasons.

4 Environmental Impact Assessment

Table 3 shows the survey items for the environmental impact assessment. According to the Japanese law, an environmental impact assessment is mandatory for wind power plants whose output is greater than 10 MW. Thus, an environmental impact assessment is not required for the demonstration power plant by the Japanese law. However, an environmental impact assessment was considered to be necessary for this research facility since this is the first floating wind turbine to be built in Japan. As it can be seen from Table 3, many items have been surveyed. At this moment, no major factors affecting the surrounding environment are observed. It is noted that fishery related environmental items are also examined, since the fishery activity is relatively high around the demonstration site.

5 Half-Scale Model

5.1 Description

Figures 6 and 7 show the outline and main dimensions of the half-scale model, respectively. In Table 4, the main specifications of the model are presented. The floating foundation has a slender cylindrical shape (spar), with a draft of 37.05 m, an outer diameter of 3.8 m at the bottom and an outer diameter of 2.375 m at the sea level. The bottom half of the floating foundation is made of precast PC (pre-stressed concrete) segments whereas the upper half is made of ring-stiffened steel. In order to mitigate the yaw motion, four straight fins are attached along the PC part. Conventional catenary chain system is used as the mooring system. Three stud-link chains with a nominal diameter of 56 mm are used. For the anchors, two concrete sinkers and one Danforth-type anchor were selected after considering the sea-bed condition.

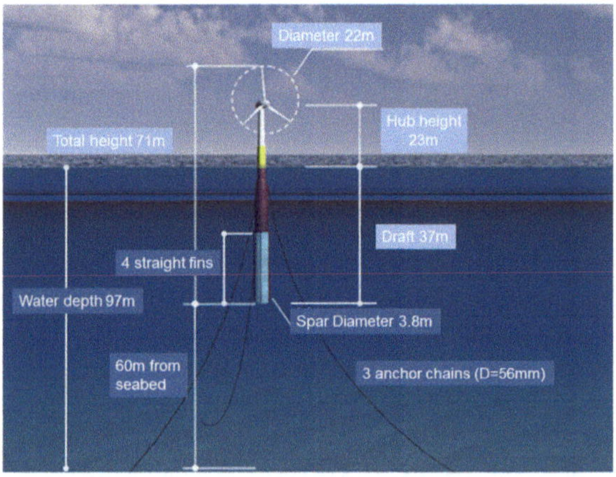

Fig. 6 Outline of half-scale model

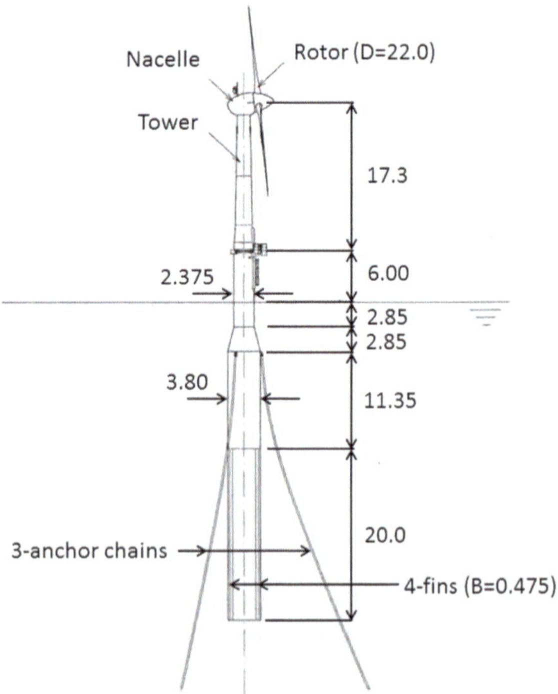

Fig. 7 Main dimensions of half-scale model

Table 4 Main specifications of half-scale model

Items	Specifications
Wind turbine	SUBARU22/100 (down-wind type)
Rated power (kW)	100
Rotor speed (rpm)	18–72
Cut-in wind speed (m/s)	3–4
Cut-out wind speed (m/s)	20
Rated wind speed	7.5 m/s (when power limited to 40 kW)
Dimensions of fins	0.475 m (width) × 20 m (height)
Number of fins	4
Mooring chains	Nominal diameter of 56 mm (Grade 3) × 3
Concrete sinker	Weight in air: 200 tf × 2
Danforth anchor	Weight in air: 10 tf × 1

The wind turbine has a rated output of 100 kW. However, during the demonstration test, the maximum power was limited to 40 kW so as to increase the possibility of occurrence of wind speed above rated wind speed, where pitch control of blades is made. The wind turbine was modified to a down-wind type from the original up-wind type design. The down-wind type turbine is considered to have advantages such as weather-vane effect in yaw direction and the rotor axis being horizontal when the tower is tilted due to wind action, and increasing of power efficiency when tilted main shaft is parallel to the water line.

5.2 Structural Design

Table 5 shows the design conditions used for the structural design of the half-scale model. The structural design was made for both power production situation and parked situation. The power production situation corresponds to DLC (design load case) 1.1 defined in IEC 61400-3 (IEC 2009), and the parked situation corresponds to DLC 6.2a. The wind speed of 45.9 m/s is determined from the design reference wind speed of 36 m/s defined as the 10 min average wind speed at 10 m above the ground surface of roughness category II (in the Japanese building code). The roughness category I is assumed for the sea surface, and the design wind speed of 45.9 m/s is defined at the hub height (23.3 m above sea level). The conversion factor of 0.95 from 10 min average wind speed to 1 h average wind speed is adopted (IEC 2009). The power law is used to define the vertical profile of the mean wind speed with the power law exponent of 0.10 for parked situation and 0.14 for power production situation.

The design wave height of 7.73 m was estimated by Mase et al. (2011) as the extreme wave height corresponding to a 50 year return period. Since the design wave height of 7.73 m is considered to be defined for a 3 h duration, it was multiplied by

Table 5 Design conditions

Design situation	V_{hub} (m/s)	I_1 (%)	$H_{1/3}$ (m)	$T_{1/3}$ (s)	MIS (deg)	U_{ss} (m/s)	U_w (m/s)	Water depth (m)	Duration (min.)	Seeds
Power production	4–20	IEC A	1.0–2.9	8.0	0, 90, 180	0.40	0.03–0.16	97.2 (MSL)	10	6
Parked	45.9	15.5	8.43	14.0	0, 90, 180	0.56	0.40	99.1 (HHWL)	60	6

V_{hub} Mean wind speed at hub-height
I_1 Turbulence intensity
$H_{1/3}$ Significant wave height
$T_{1/3}$ Significant wave period (for irregular waves)
MIS Misaligned angle between wind and wave directions
U_{ss} Surface speed of the subsurface current
U_w Surface speed of the wind-generated current

1.09 in order to convert to a 1 h duration, that is, $7.73 \times 1.09 = 8.43$ m (IEC 2009). The misalignment between wind and wave directions is considered for 0, 90, and 180°. For current, the subsurface current speed is assumed to take on a power law distribution with the power law exponent of 1/7. The wind-generated current speed is assumed to be linear up to 50 m from the sea surface.

Figure 8 shows the schematic representation of the aero-hydro-servo-mooring dynamics integrated dynamic analysis tool used for the structural design (Utsunomiya et al. 2012). The equations of motion for a whole system are generated and solved by a commercially-available multibody-dynamics solver (MBS), MSC. Adams. NREL/FAST (Jonkman and Buhl 2005) is utilized to generate the input data for Adams as a preprocessor, but some modifications have been made. The modifications include

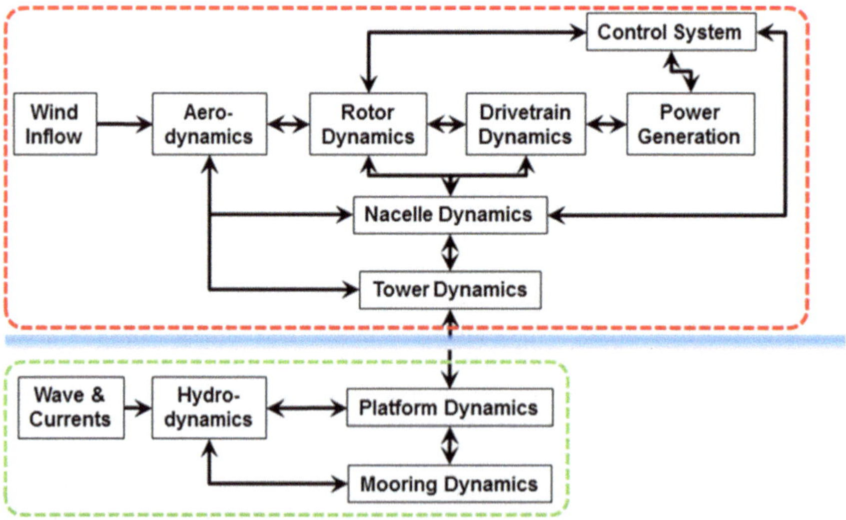

Fig. 8 Block diagram of simulation method

- the generation of the platform's data connected by linear springs/dampers,
- the application of wind loadings to the tower, the nacelle and the platform; and
- the application of wave and current loadings to the platform.

The aerodynamic forces acting on the rotor are evaluated by NREL/AeroDyn (Laino and Hansen 2002). The AeroDyn is based on the blade element momentum (BEM) theory, of which details may be obtained from Moriarty and Hansen (2005). The wind forces acting on the tower, the nacelle and the platform are evaluated as drag forces by using the relative wind speed.

The wave forces are evaluated by Morison's equation with a relative velocity formulation. The wave particle velocities and accelerations are evaluated along the instantaneous position of the platform axis. The current velocity is added to the horizontal wave particle velocities.

The Airy wave theory is used to generate irregular wave kinematics, but the Wheeler's stretching method is used to obtain the wave kinematics up to the finite water surface. The wave loadings to the platform are then evaluated in the body-fixed coordinates up to the instantaneous position of the water surface.

The mooring line forces are evaluated by the quasi-static catenary theory, and the routine is named Moorsys. Alternatively, dynamic MBS modeling of the mooring lines using mass-spring model is possible. However, the CPU time requirement is about twice of the quasi-static analysis, although the overall behavior is similar. So, the dynamic analysis and load estimation have been made by using the quasi-static routine.

The control system for the wind turbine including control of blade pitch angles, generator torque, and yaw angle is given by the Bladed-style DLL (Dynamic Link Library).

Figure 9 shows the numerical model used for Adams. In the numerical model, the blades are discretized into 11 elements for each, the tower into 14 elements, and the platform into 28 elements. They are connected by elastic springs that represent flexural, shear and axial rigidities. The mooring forces are evaluated at the mooring point as the 6 degrees-of-freedom restoring forces from the mooring lines by using the quasi-static catenary theory.

Figure 10 presents the dynamic responses of the floating body for parked situation at MIS $= 0°$ and the seed number 1. Figure 11 shows the bending moment at the tower base while Fig. 12 shows the tension in an anchor chain. From these time-series data, the maximum/minimum values are extracted for the structural design.

Table 6 shows the maximum tension in chains and at anchors. These values are obtained for MIS $= 0°$. Since the maximum values fluctuate for seed numbers, the mean value is used as the characteristic value for the design.

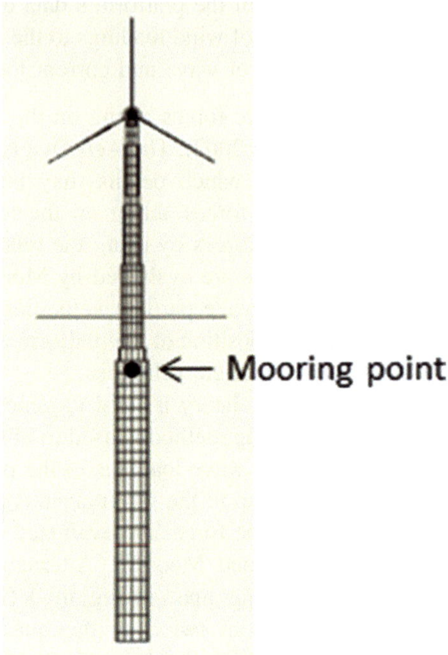

Fig. 9 Numerical model for ADAMS

5.3 Construction

The PC part and the steel part of the floating body were constructed separately by different manufacturers. Figure 13 shows the manufacturing of PC precast segments. The ring segments, with an outer diameter of 3.8 m and length 2.0 m, have been produced by a centrifuge manufacturing process. Figure 14 shows the manufacturing of the steel part. It was fabricated at a shipyard.

The ring segments and the steel part were then transported to a quay yard. Figure 15 shows the assembly of the floating body and the tower section at the quay yard. They were assembled in the horizontal position.

5.4 Installation

Figure 16 shows the installation procedure. The completed floating body with tower and the wind turbine was transported by a barge to the northern side of Kabashima Island. The northern side of the Island is rather calm with respect to wind/wave conditions when compared to the southern side. The entire floating body was upended by using a large floating crane. After the successful upending, the floating wind turbine was towed to the southern side of Kabashima Island. At the demonstration site, the mooring chains were hooked-up to the floating body.

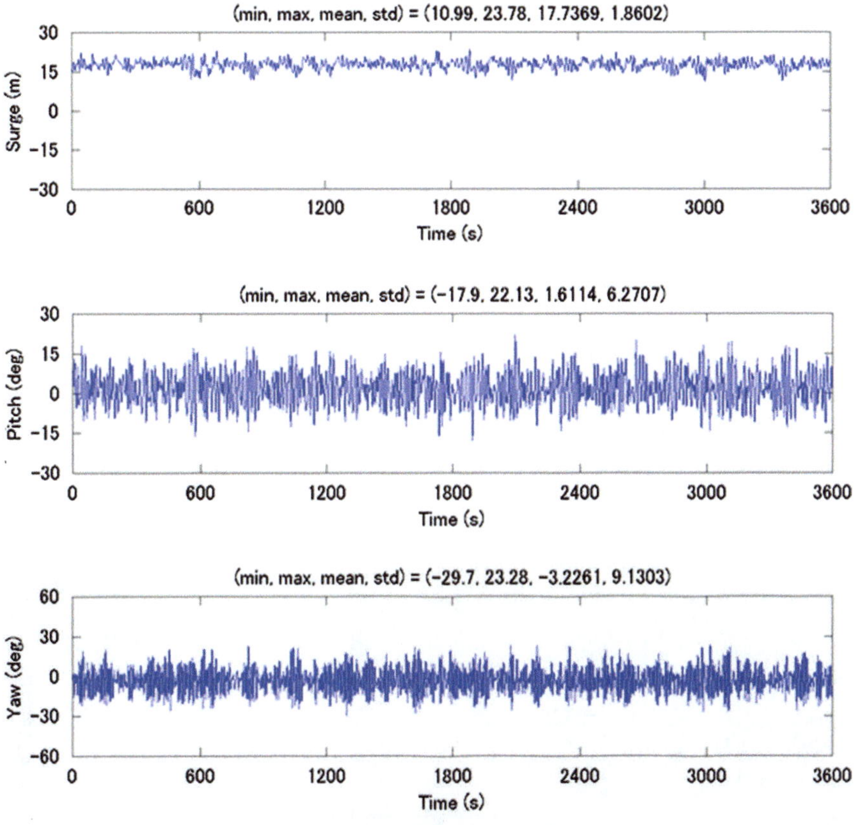

Fig. 10 Dynamic responses of floating body (at MSL-26.542 m)

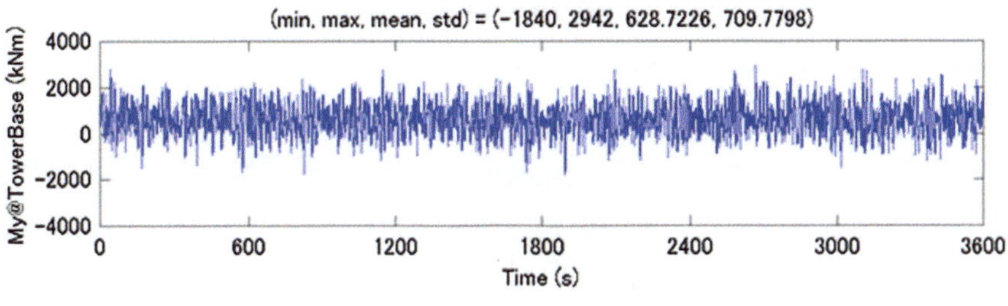

Fig. 11 Bending moment at tower base (at MSL + 6.0 m)

The mooring chains with the sinkers/anchor were pre-laid at the site. The hook-up of the mooring chains were completed on 11 June 2012. Figure 17 shows the general view of the half-scale model after completion with an access boat for maintenance.

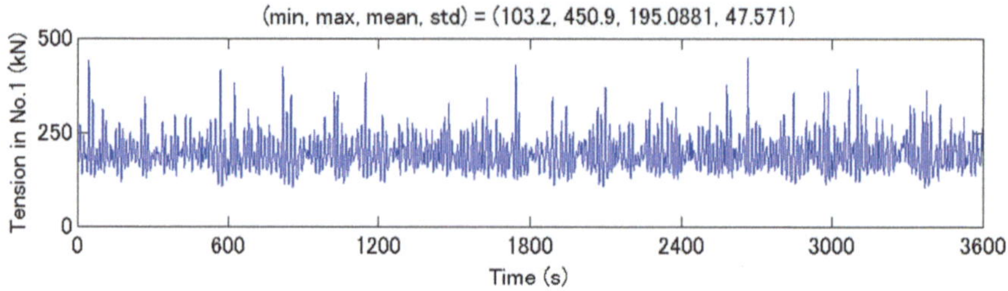

Fig. 12 Tension in anchor chain

Table 6 Maximum tension in chains and at anchors

Seed no.	1	2	3	4	5	6	Mean
Max. tension in chains	448	407	492	432	461	405	441
Max. tension at anchors	392	352	436	376	406	350	385

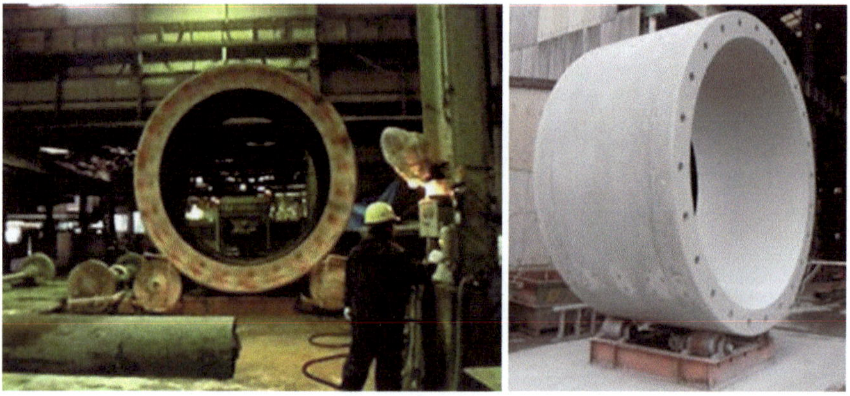

Fig. 13 Manufacturing of PC precast segment

Fig. 14 Manufacturing of steel part

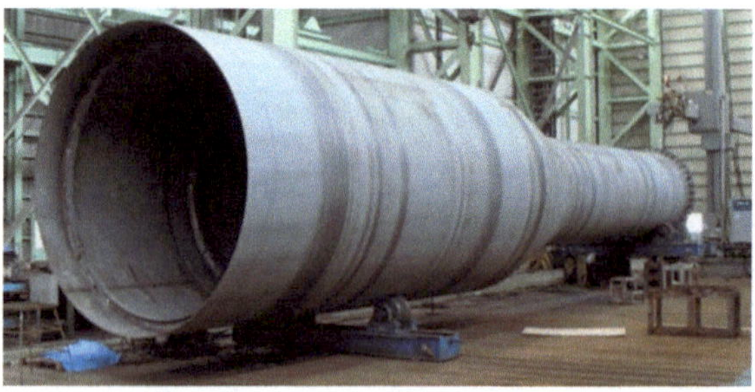

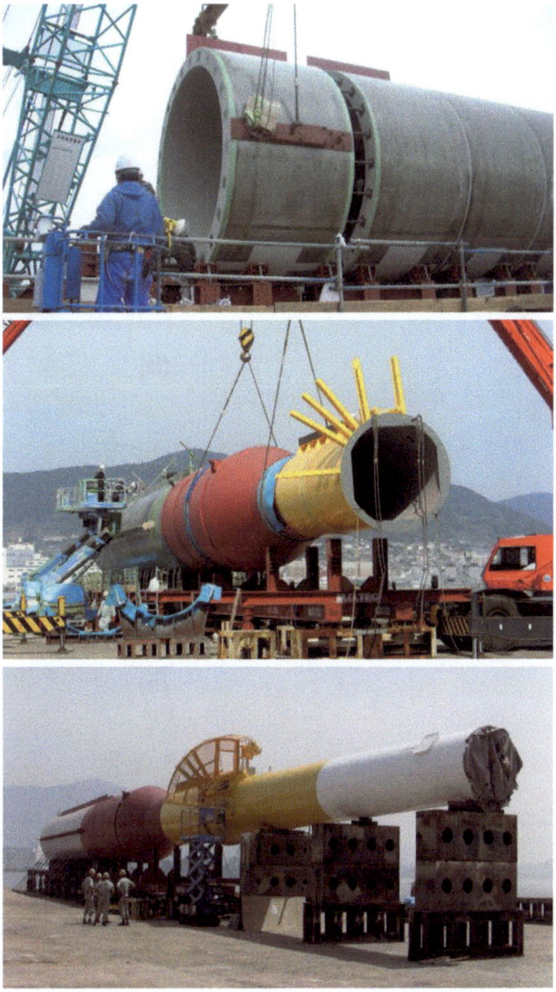

Fig. 15 Assembly at quay yard

5.5 *Response During Typhoon Attack*

During the demonstrative experiment of the half-scale model, the FOWT was attacked by two separate severe typhoons, that is, by Bolaven (international designation: 1215) and Sanba (international designation: 1216). Sanba (1216) was a record-making typhoon event. Figure 18 shows the track of the typhoon Sanba (1216). It was closest to the at-sea experiment site at around 5:00 a.m. on 17 September 2012. At that time, the central atmospheric pressure of the typhoon was 940 hPa. During the typhoon, several data were obtained, that include the wind speed, wave height, motion of the floating body, strains of the tower and the floating body, tension of a mooring line. The details of the measurement can be

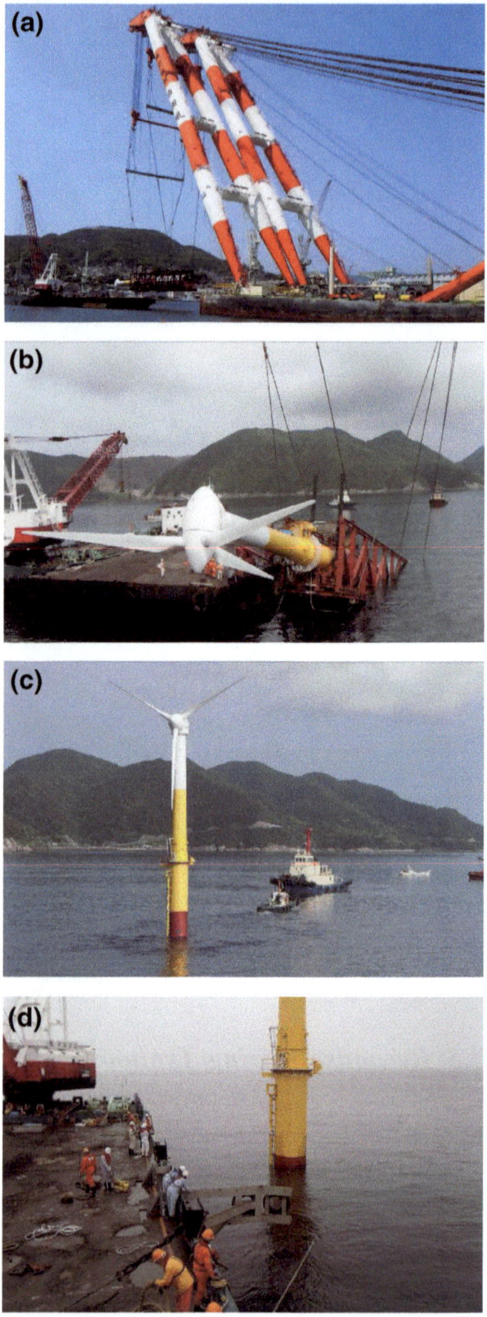

Fig. 16 Installation of half-scale model **a** Stowage to barge **b** Upending **c** Towing to the site **d** Hook-up of mooring chains

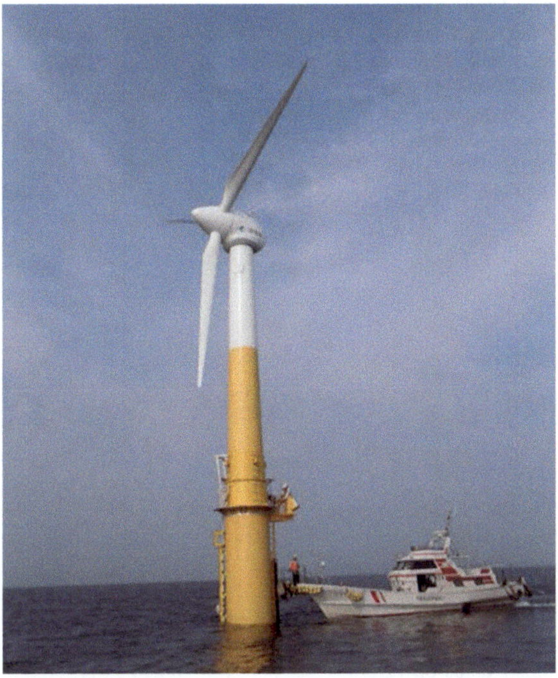

Fig. 17 General view of half-scale model

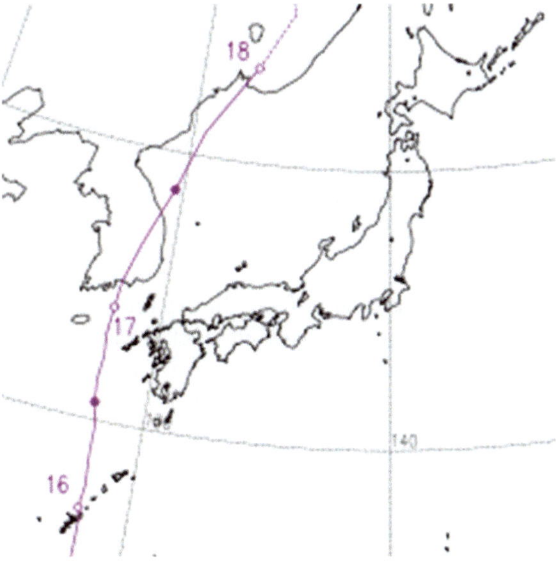

Fig. 18 Track of typhoon Sanba (1216)

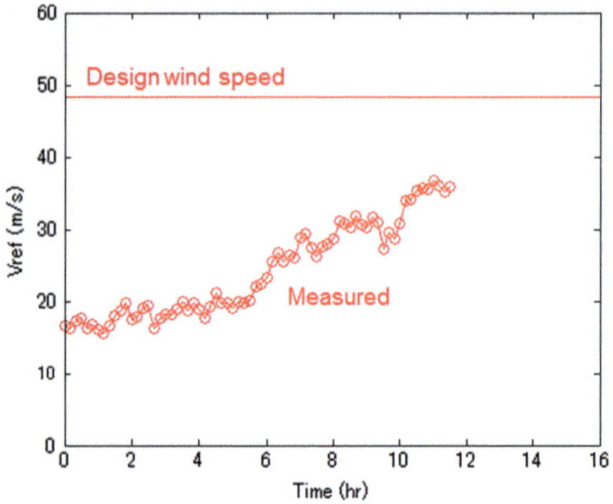

Fig. 19 10-min average wind speed measured on top of nacelle. The time 0 h corresponds to data for 18:00–18:10 on 16 September, and 11 h to 05:00–05:10 on 17 September

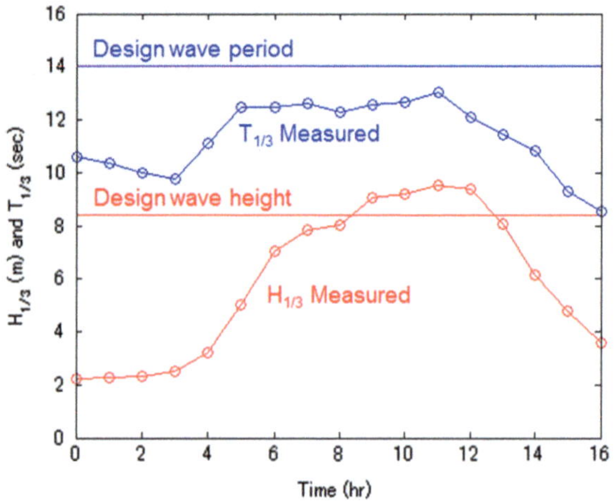

Fig. 20 Significant wave height $H_{1/3}$ (m) and significant wave period $T_{1/3}$ (s). The time 0 h corresponds to data for 18:00–19:00 on 16 September, and 11 h to 05:00–06:00 on 17 September

found in Ishida et al. (2013). In the following, some details during the typhoon attack are presented (Utsunomiya et al. 2013).

Figure 19 shows the 10-min average wind speed measured at the top of the nacelle by the cup-type anemometer. The maximum 10-min average wind speed was 36.8 m/s taken during 05:00 a.m–05:10 a.m. on 17 September 2012. The design wind speed corresponding to the return period of 50 years as the 10-min average wind speed for the FOWT is 48.3 m/s at the hub-height (23.3 m above sea level). Thus, the maximum wind speed during the typhoon event was 24 % lower than the design wind speed.

Figure 20 shows the significant wave height and the significant wave period measured by the wave measuring buoy. The maximum significant wave height was 9.5 m taken during 05:00–06:00 a.m. on 17 September 2012, and the maximum significant wave period was 13.0 s taken also at the same time. The maximum

Table 7 Numerical calculation cases

	C_D of fins	Mooring dynamics	Yaw damping (%)
Case 1	1.5	Quasi-static catenary theory	2
Case 2	10.0		
Case 3	1.5	Dynamic mass-spring model	0

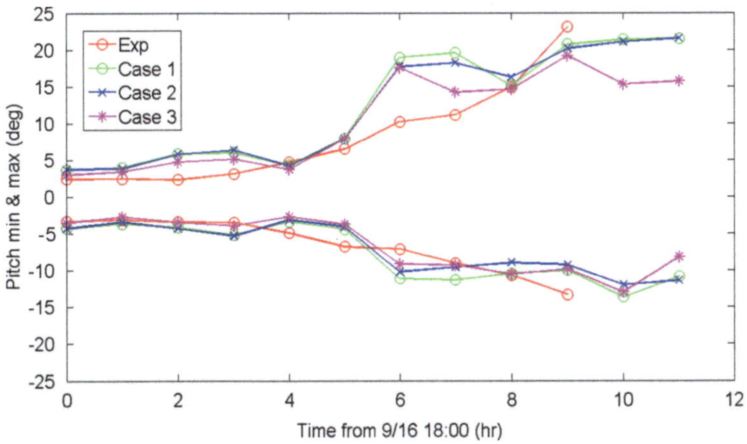

Fig. 21 Minimum and maximum values of the platform motion

wave height of 9.5 m exceeded the design wave height of 8.4 m (for 1-h reference period), but the maximum significant wave period was below the design value of 14.0 s.

In order to validate the numerical simulation method, the dynamic responses have been reproduced by using the measured wind and wave time series data. Table 7 shows the numerical calculation cases. Figure 21 shows the minimum and maximum values of the platform motion in pitch. The reference point of the motion is at 6 m above the sea level (at the tower base). Basically, good agreement can be seen between the experimental (Exp) and simulation results (Case 1–3).

Figure 22 shows the maximum values of the bending moments M_{xy} at the tower section #2. As it can be seen from Fig. 22, all cases can reproduce the experimental values rather well. Also the experimental values are well below the design load, for which the load factor of 1.35 is applied.

Figure 23 shows the maximum value of a chain tension. The basic agreement is good and it can also be observed that the tension values are smaller than the design value (= 441 kN; the maximum tension used in the design before applying load factors).

It should be emphasized that although the significant wave height of 9.5 m exceeded the design wave height of 8.4 m that corresponds to the return periods of 50 years, the FOWT experienced no damage by the typhoon attack.

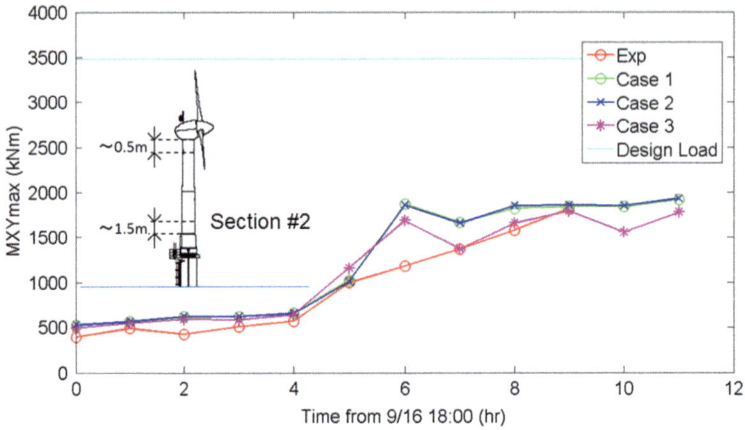

Fig. 22 Maximum values of the bending moments at tower section #2

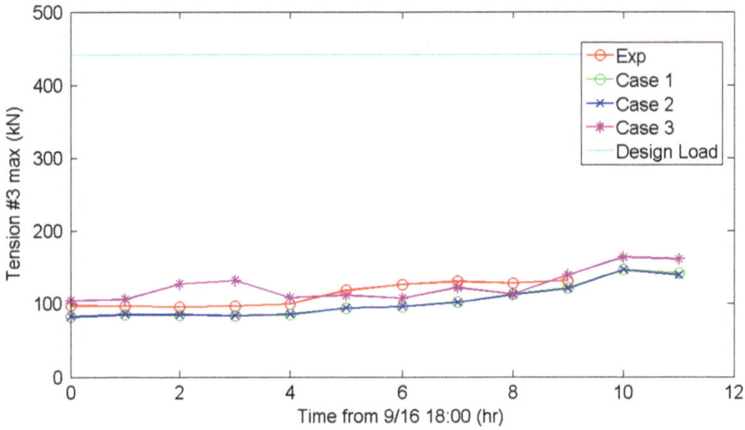

Fig. 23 Maximum values of chain tensions

5.6 Response During Power Production

The dynamic response of the FOWT at power generation is presented below. Nielsen et al. (2006) reported that for a spar-type FOWT, a conventional wind turbine control scheme may induce a magnification of the pitch motion. The effect may be referred to as “negative damping effect”. The wind turbine control scheme for the FOWT at Kabashima was modified to suit a floating wind turbine so as to avoid the negative damping effect. In order to confirm the effectiveness of the control scheme, the field measurement at power generation has been made, and the results are presented below (Utsunomiya et al. 2014).

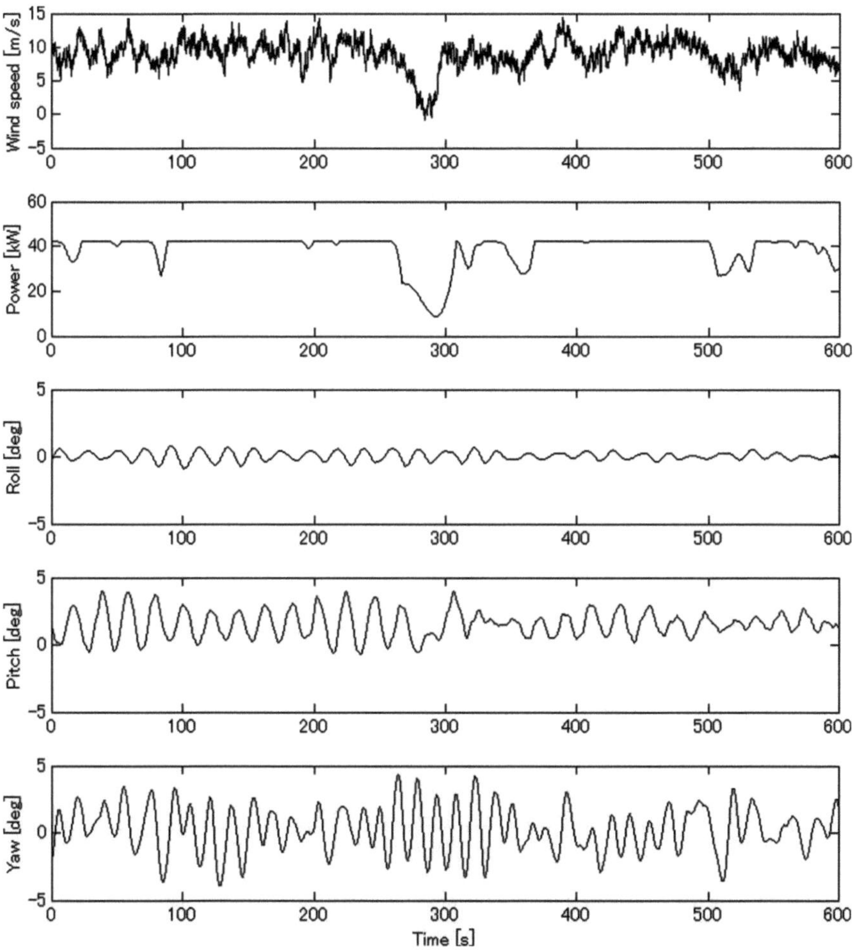

Fig. 24 Simulation time series for Case 1; $V_{hub} = 9$ m/s, T.I. = IEC Class A, MIS = 0°, $H_{1/3} = 0.5$ m, $T_{1/3} = 5.8$ s

Figure 24 shows an example of the numerical simulation results as time-series data. In the simulations, the duration of 660 s is analyzed, and then 60 s from the beginning are removed in order to eliminate the transient response part. As it can be seen from Fig. 24, the magnification of the pitch response due to negative damping effect is not significant.

Figure 25 shows the turbulence intensities of the wind data used for comparison herein. Figure 26 shows the mean values of the pitch response. A fairly good agreement between the simulation results and the measured values is observed. The effect of turbulence intensities is insignificant for the mean values.

Figure 27 shows the standard deviations of the pitch response (pitch SD). As it can be seen in Fig. 27, the numerical simulation results show that the turbulence

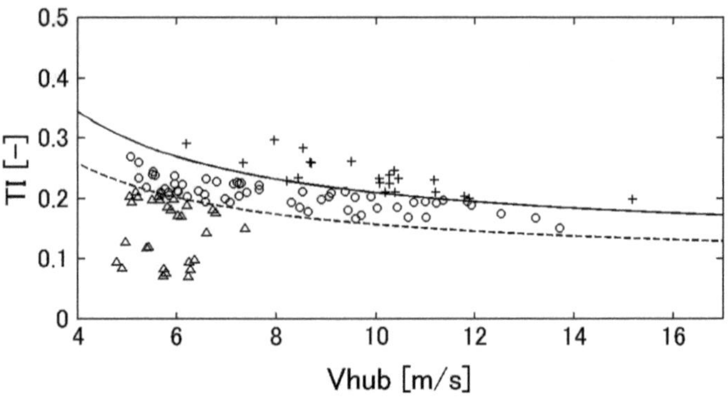

Fig. 25 Turbulence intensities (T.I.) (*Solid line* IEC Class A, *Broken line* IEC Class C, +: T.I. greater than IEC Class A, *circle* T.I. between IEC Class C and Class A, *triangle* T.I. smaller than IEC Class C)

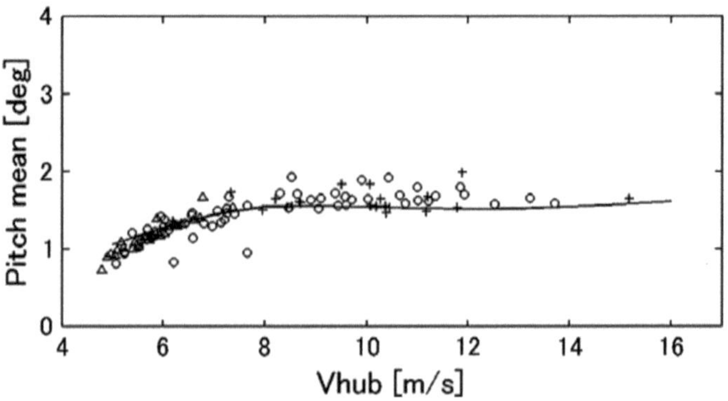

Fig. 26 Mean values of pitch response (*Solid line* calculation at T.I. equals to IEC Class A, +: T.I. greater than IEC Class A, *circle* T.I. between IEC Class C and Class A, *triangle* T.I. smaller than IEC Class C)

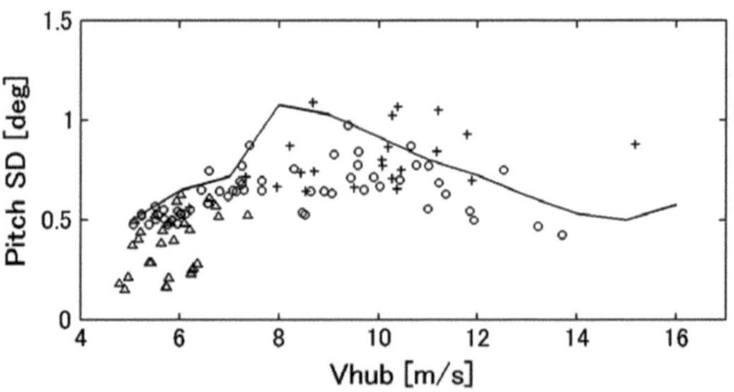

Fig. 27 Standard deviations of pitch response (*Solid line* calculation at T.I. equals to IEC Class A, +: T.I. greater than IEC Class A, *circle* T.I. between IEC Class C and Class A, *triangle* T.I. smaller than IEC Class C)

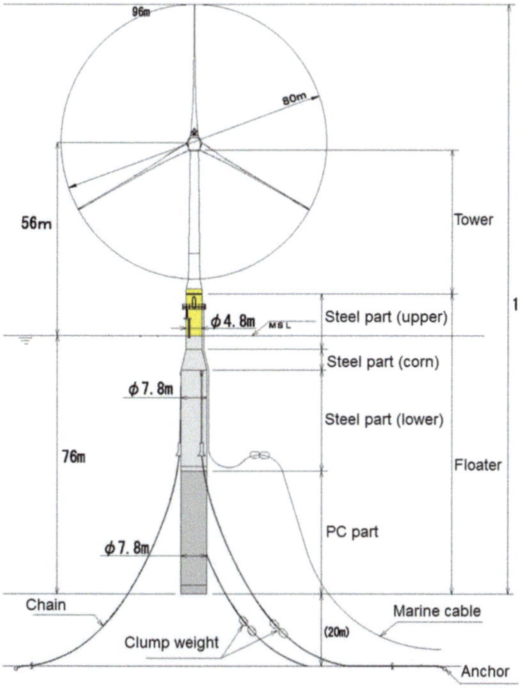

Fig. 28 Main dimensions of full-scale model

intensities affect the pitch SD considerably. Thus, a large scatter of the pitch SD in the measured values is due to the large scatter of the turbulence intensities themselves (see Fig. 25). The simulation results of pitch SD predict fairly well the upper bound of the measured data for the corresponding turbulence intensity class (IEC Class A).

6 Full-Scale Model

The half-scale model was removed from the demonstration site as scheduled. Subsequently, the full-scale model was installed at the same site. Figure 28 shows the main dimensions of the full-scale model. Basically, the design concept of the spar platform is the same as the half-scale model except for the scaling up to about twice in the length dimensions.

Figure 29 shows the installation procedure:

(a) The PC parts and the steel part were connected at a quay yard in the horizontal position
(b) The completed floater (hybrid spar) was towed to the northern part of the Kabashima Island by using a barge

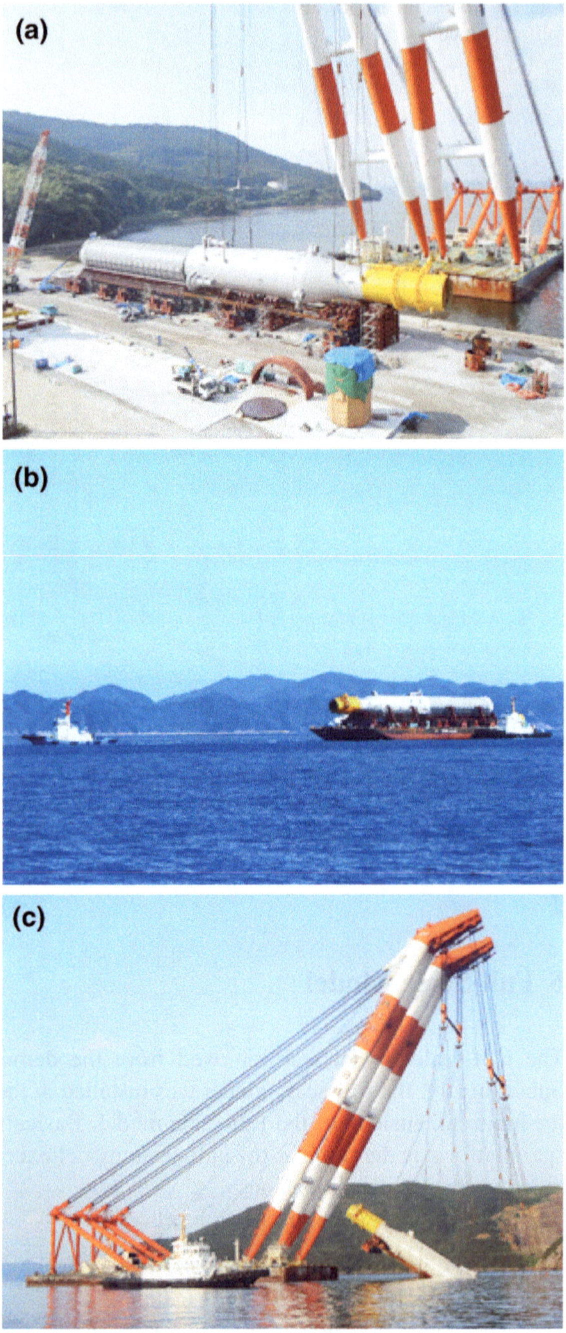

Fig. 29 Installation of the full-scale model
a Connection between PC part and steel part **b** Towing of the floater **c** Upending of the floater **d** Assembly of the rotor **e** Towing to the demonstration site **f** Hookup of anchor chains

Fig. 29 continued

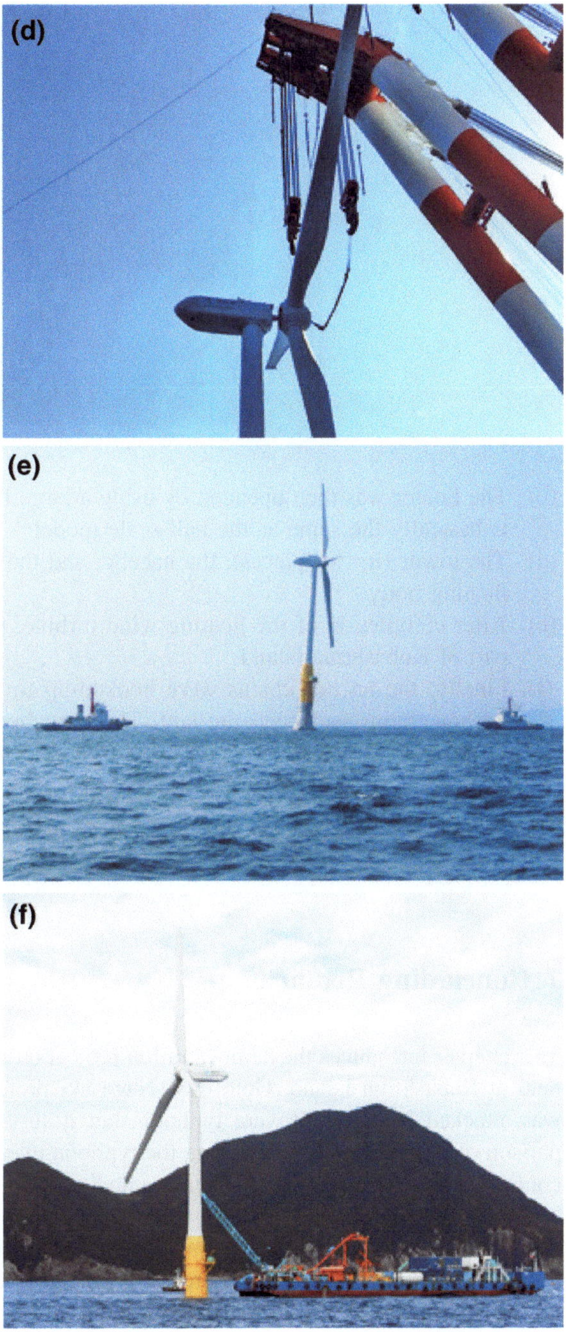

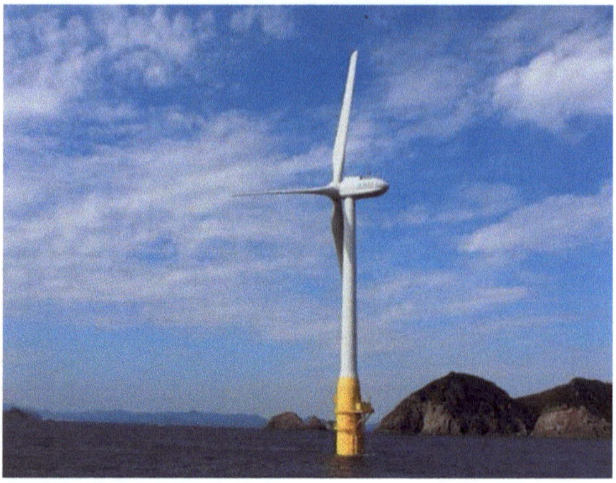

Fig. 30 General view of full-scale model

(c) The floater was then upended by using a large floating crane. The procedure is basically the same as the half-scale model
(d) The tower (in two pieces), the nacelle, and the rotor were assembled to the floating body
(e) After completion of the floating wind turbine, it was towed to the southern part of Kabashima Island
(f) Finally, the anchor chains were hooked-up to the floating body. Here, the anchor chains had been pre-laid. The final hook-up was completed on 18 October 2013.

Figure 30 shows the general view of the full-scale model. After the electric marine cable hook-up and the grid-connection, the opening ceremony, as the first multi-megawatt floating wind turbine in Japan, was held on 28 October 2013.

7 Concluding Remarks

This chapter introduces the demonstration project on floating offshore wind turbine held at Kabashima Island, Goto city, Nagasaki prefecture. The half-scale model was attacked by a very severe typhoon, but it survived with no damage. Comparison of the measured data during the typhoon event with the simulation results confirmed the validity of the design method.

The full-scale model has also been installed with a great success. It also demonstrates the feasibility of the spar-type FOWT in Japan. The success of the hybrid spar combining lower PC precast segments and upper steel part has opened up the realization of a low-cost floating wind turbine system in near future.

Acknowledgments This work is a part of the floating offshore wind turbine demonstration project funded by the Ministry of the Environment of Japan.

References

EWEA. (2013). Deep water—The next step for offshore wind energy. Report. European Wind Energy Association. Retrieved January 16, 2014, from http://www.ewea.org/publications/reports/deep-water/.

IEC. (2009). Wind turbines—Part 3: Design requirements for offshore wind turbines. IEC 61400-3, Edition 1.0, 2009-02.

Ishida, S., Kokubun, K., Nimura, T., Utsunomiya, T., Sato, I., & Yoshida, S. (2013). At-sea experiment of a hybrid SPAR type offshore wind turbine. In *ASME 2013 32nd International Conference on Ocean, Offshore and Arctic Engineering, Nantes, France*, June 9–14, 2013. Paper No. OMAE2013-10655. doi:10.1115/OMAE2013-10655.

Jonkman, J., & Buhl, M. L., Jr. (2005). FAST user's guide, Technical Report. NREL/EL-500-38230.

Laino, D. J., & Hansen, A. C. (2002). AeroDyn user's guide, version 12.50, NREL.

Mase, H., Mori, N., Nakajo, S., Yasuda, T., Dong, S., & Ikemoto, A. (2011). Evaluation of design wind and wave for floating type wind farm using meteorological re-analysis and prediction data. *Journal of Japan Society of Civil Engineers, 67*(2), I_1226–I_1230, Ser. B2 (Coastal Engineering). doi:10.2208/kaigan.67.I_1226.

Ministry of the Environment. (2012). Study of basic zoning information concerning renewable energies (FY2011). Report. Ministry of the Environment. Retrieved January 16, 2014, from http://www.env.go.jp/earth/report/h24-04/.

Moriarty, P. J., & Hansen, A. C. (2005). AeroDyn theory manual, NREL/EL-500-36881.

Nielsen, F. G., Hanson, T. D., & Skaare, B. (2006). Integrated dynamic analysis of floating offshore wind turbines. In *25th International Conference on Offshore Mechanics and Arctic Engineering, Hamburg, Germany*, June 4–9, 2006. Paper No. OMAE2006-92291. doi:10.1115/OMAE2006-92291.

Principle Power, Inc. (2012). Principle Power and Partners inaugurate the first Portuguese Offshore Wind Turbine. Press Release. Retrieved January 20, 2014, from http://www.principlepowerinc.com/news/press_PPI_WF_inauguration.html.

Statoil. (2012). Hywind—The world's first full-scale floating wind turbine. Retrieved January 20, 2014, from http://www.statoil.com/en/technologyinnovation/newenergy/renewablepowerproduction/offshore/hywind/pages/hywindputtingwindpowertothetest.aspx.

Utsunomiya, T., Yoshida, S., Ookubo, H., Sato, I., & Ishida, S. (2012). Dynamic analysis of a floating offshore wind turbine under extreme environmental conditions. In *ASME 2012 31st International Conference on Ocean, Offshore and Arctic Engineering, Rio de Janeiro, Brazil*, July 1–6, 2012. Paper No. OMAE2012-83985. doi:10.1115/OMAE2012-83985.

Utsunomiya, T., Sato, I., Yoshida, S., Ookubo, H., & Ishida, S. (2013). Dynamic response analysis of a floating offshore wind turbine during severe typhoon event. In *ASME 2013 32nd International Conference on Ocean, Offshore and Arctic Engineering, Nantes, France*, June 9–14, 2013. Paper No. OMAE2013-10618. doi:10.1115/OMAE2013-10618.

Utsunomiya, T., Yoshida, S., Kiyoki, S., Sato, I., & Ishida, S. (2014). Dynamic response of a spar-type floating wind turbine at power generation. In *ASME 2014 33rd International Conference on Ocean, Offshore and Arctic Engineering, San Francisco, California, USA*, June 8–13, 2014. Paper No. OMAE2014-24693. doi: 10.1115/OMAE2014-24693

Large Marine Concrete Structures: The Norwegian Design Experience

Tor Ole Olsen, Olav Weider and Anders Myhr

Abstract The beauty of buoyancy has fascinated man, and the effect has been utilized at all times. Floating structures as ships or floating bridges have been used for several thousand years, predominantly for transport purposes. Norway, with its long coast and dependency of the ocean, has always had a special relationship to the sea and the buoyancy force. Norwegian Vikings built wooden long ships and sailed from Europe to America some 500 years before the Italian explorer C. Columbus did. Their ships were slender with a low draft to enable them to easily manoeuvre up shallow rivers and are considered by many to be the most beautiful ship design ever built. In modern times, Norwegians utilize the oceans in many ways, and now the need for structures to be put in the oceans is large and increasing, as population grows and environmental concerns increase.

1 Introduction

Floating structures as ships or floating bridges have been used for several thousand years, but predominantly for transport purposes. Norwegian Vikings built wooden long ships and sailed from Europe to America some 500 years before the Italian explorer C. Columbus did. Their ships were slender with a low draft to enable them to easily manoeuvre up shallow rivers and are considered by many to be the most beautiful ship design ever built. The design allowed for both sailing and efficient propulsion by manual labour applied to as many as 16 pairs of oars. Figure 1 shows the Gokstad ship, built approximately year 890 AC, and buried some 10 years later. It was found in a grave, and dug out in 1880. The ship was possibly used for traversing the Atlantic Ocean.

T.O. Olsen (✉) · O. Weider (✉) · A. Myhr
Dr.techn.Olav Olsen, 1325 Lysaker, Norway
e-mail: too@olavolsen.no

O. Weider
e-mail: owe@olavolsen.no

C.M. Wang and B.T. Wang (eds.), *Large Floating Structures*,
Ocean Engineering & Oceanography 3,
DOI 10.1007/978-981-287-137-4_7

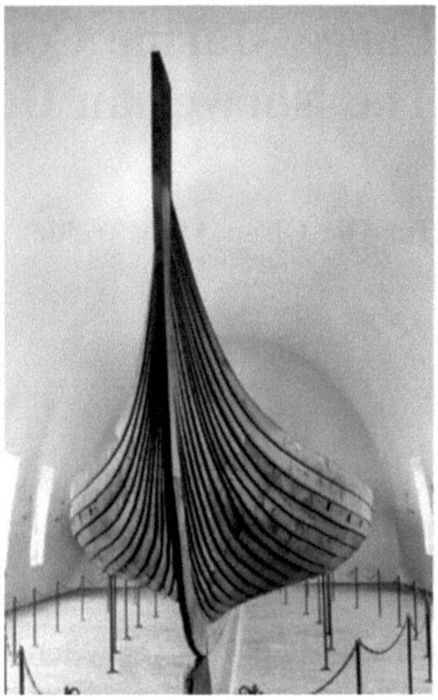

Fig. 1 The Gokstad ship from approximately year 890 AC. Designed for both rowing and sailing in open waters while maintaining a low draft for the ability to venture far up shallow rivers (KHM 2013)

It appears that the Vikings were able to utilise their wits and clever craftsmanship to construct such magnificent vessels. Where the Vikings addressed technical challenges such as hydrodynamic loading, floating stability, structural strength and complex construction methods through trial and error, we now use computers and numerical software to solve these problems. However, even with the technology we now have available, we struggle to predict the performance of such complex wooden structures.

Since the Vikings' conquest of the Atlantic Ocean, the structures built by man have increased in both size and complexity. Thus, full-scale trial and error is not a favourable design methodology for today's expensive floating superstructures. We need to base our work on experience and safety predictions, founded in sound principles and analytical tools.

2 Concrete Background and Basic Properties

There is a common understanding that structures should be constructed out of the most suitable material. The wooden ships of the Vikings were impressive in both function and make. Wood has been used for floating structures for thousands of years, and is still being used. For larger structures, steel and concrete have been shown to be more appropriate due to favourable fabrication aspects and service life. In the following section, interesting aspects in the development of floating concrete structures will be presented.

The history of floating concrete sea structures goes back to the 19th century. In 1848, Joseph L. Lambot made the first known reinforced concrete boat. However, many of the subsequent structures did not possess inherently favourable characteristics. This was mainly due to large mass, improper designed and that they were built on the principles of traditional ships. Thus, the material properties of the reinforced concrete were not put to optimal use and it resulted in flawed construction.

During World War I, 14 concrete ships were built due to steel shortage. This included the 130 m long U.S.S. Selma. At the time, reinforced concrete had already been used in shipbuilding in the Scandinavian countries, but were restricted mainly to smaller vessels like barges. However, due to what the authors claim are inappropriate design and the eventual weight of the structure, after they served during the war, typical wartime concrete ships were grounded due to fuel costs and found themselves used as silos or quays.

World War II concrete ships saw widespread service in battle zones. Around 24 of these ships were large sea-going vessels and about 80 of them were sea-going barges of large size. The cargo capacities ranged from 3,200 to 140,250 tonnes. Morgan has produced subsidiary information that gives a good description of the early development of the concrete hull (Morgan 1975, 1977).

In the late 1950s, a number of pre-stressed concrete ocean-going barges were constructed in the Philippines. This was followed by an additional 19 barges from 1964 to 1966. There were also several concrete lighthouses constructed as caissons in the 1960s, and installed in the Irish Sea; Eastern Canada; and in the Gulf of Bothnia. Most of them are still standing. Several pontoons, barges and other crafts have also been successfully built in the former USSR, Australia, New Zealand and the UK.

From 1950 to 1982, it was registered that approximately 1,130 concrete hulls had been built. Most of them were small with overall length less than 50 m. Among the bigger ones, two groups of sizes are dominant,—approximately 250 hulls with length ranging from 58 to 67 m, and 40 hulls with a length of 110 m.

The American Concrete Institute Committee 357 (ACI 357) has written several publications (ACI Committee 2010) on barges and floating structures. The properties of the barges are very good, although somewhat heavy. The weight issue derives in part from a design flaw: concrete ships should not be designed in the same manner as a steel ship. Use of recently developed materials and technology may somewhat remedy this. An example is the barges built by Densit in Asia, which may be argued to be as light as steel barges.

Figure 2 shows Sylvestre, a concrete ship that was equipped with its own engine, built just after World War I. It was grounded in 1927, in Norway, where she also was built, and notably, is not in shape to float again after years of pounding from waves and ice.

Over the years, starting back in the mid 1920s, some 100 immersed concrete tunnels have been built in the following countries: USA, Canada, Argentina, Cuba, UK, Denmark, Sweden, Holland, Belgium, Germany, France, Hong Kong, Taiwan, Japan, Norway and Australia.

At present, there is yet to be a submerged floating tunnel for traffic. However, a submerged tunnel, Statpipe, was built to protect gas pipes on the west coast of

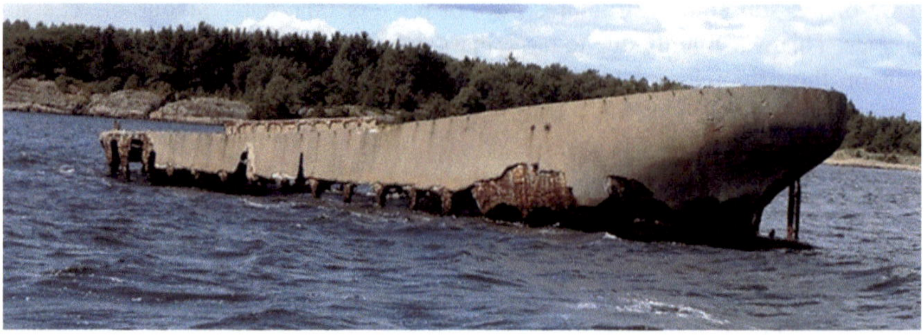

Fig. 2 Sylvestre, grounded in 1927 off the coast of Norway (OO 2012)

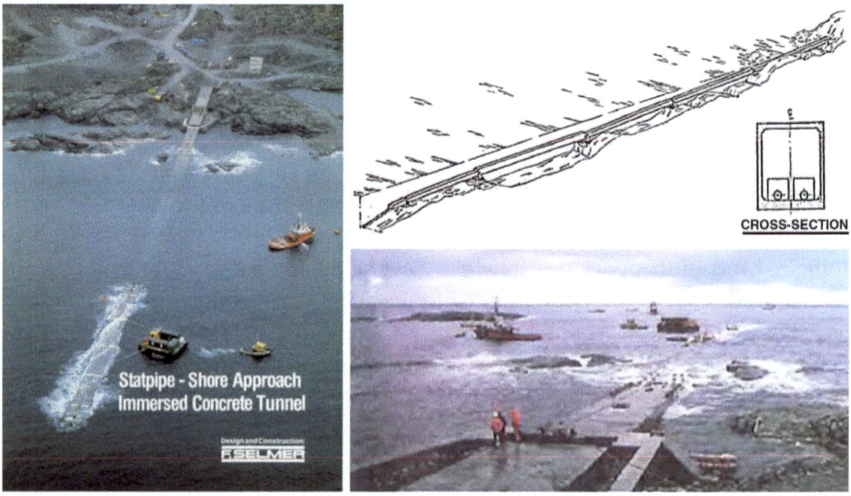

Fig. 3 Statpipe shore protection tunnel constructed in 1982 on west coast of Norway (Selmer 1988)

Norway in 1982. The project was constructed and designed by Selmer A/S (presently known as SKANSKA Norway AS), and Dr. techn. Olav Olsen. Statoil was operator of the 590 m long tunnel that was constructed in 5 individual segments (see Fig. 3).

A number of notable pontoon bridges have also been built of concrete. Gloyd (1988) has written an overview of the long traditions within this area, beginning with the first floating concrete bridge was built across Lake Washington in 1940 (Gloyd 1988).

Within the petroleum industry, permanent floating offshore concrete structures are now installed in the Java Sea (the Arco barge) (Anderson 1976), in the North Sea (Heidrun and Troll B) and outside the coast of Congo in West Africa (N'Kossa barge, 220 m long and 46 m wide) (Valenchon and Nagel 1995).

The 1970s and 1980s saw the spectacular development of offshore bottom fixed concrete structures, installed in water depths of up to 300 m in the middle of one of the world's stormiest oceans, the North Sea. It is remarkable how well these structures have performed in the hostile marine environment, while withstanding the extreme loads from waves approaching 30 m in height as well as the dynamic cyclic forces. Experience has shown that the offshore concrete structures currently in use are virtually maintenance-free. It is generally recognized that the first concrete platforms in the North Sea were over-inspected and that the need for extensive instrumentation of platforms should be reconsidered.

Offshore concrete platforms are now proven concepts with well-known characteristics and proven service records. National and international Codes and Recommendations for design and construction have been developed over the later years, and much progress has been made with respect to materials and construction methods to ensure satisfactory performance and durability in severe marine environments. The Norwegian Code NS3473 (NS 1998) has been widely used, as well as the rules provided by Det Norske Veritas, DNV (now DNV GL). Whereas the early North Sea platforms, installed in the mid 1970s, were designed for a service life of 25–30 years they are still in excellent condition and capable of significantly prolonged service. Adapting to large modifications to topside weight and layout has accommodated changing production requirements, which are common in the oil and gas industry.

Design codes give well-established rules also for assessing fire resistance. Two hydrocarbon fires inside North Sea concrete platform shafts in the late seventies have been reported. This resulted in surface scaling about 10–20 mm deep over a height of 5–10 m. This marginal impact is attributed to the large heat capacity and low thermal conductivity of concrete. No repair was found to be necessary—clearly demonstrating the excellent fire resistance of concrete. This highlights concrete as one of the best fire proofing materials available, a factor of unquestionable importance for an offshore oil or gas platform or storage unit. There are many instances, both ashore and afloat, of fire causing no more than localised or no damage to concrete structures. However, if a fire is allowed to continue, the steel reinforcement within the concrete may experience reduced capacity. As an example constituting the most impressive testimonial that could possibly be called for, Derrington (Derrington 1977) reported on how two concrete barges survived the Bikini Atoll nuclear bomb tests in good shape when their cargo of fuel oil was set alight— moored only 100 yards from the test center.

One of the argued downsides to concrete offshore structures has been the concern for deterioration arising from chloride-induced corrosion. This has been a focus when designing marine offshore concrete structures. Limiting values for concrete composition and cover to the reinforcement have been imposed based on experience with other marine structures as harbour works and bridges. In the 1990s initiatives were taken to collect the in-field performance from a number of North Sea structures with service record from two to 20 years (Fjeld et al. 1994; Helland et al. 2010). The chloride profiles were evaluated according to scientific analysis and probabilistic methods. This full scale field experience confirmed that all the

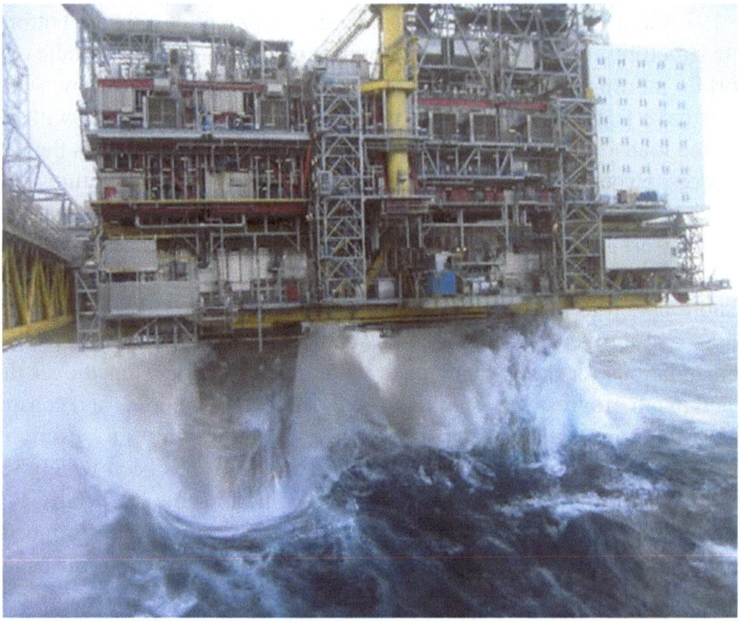

Fig. 4 Harsh environment; waves engulfing Oseberg A platform in the North Sea (Nilssen 2008)

structures will exceed, with a considerable margin, their respective design lives and that the specification requirements applied over the last 30 years has served the industry very well.

Concrete is made of various constituents, and the mix can be designed. A variation in the concrete mix may alter both mechanical and environmental properties significantly, and notably there has been a considerable development of concrete qualities over the last decade. Modern admixtures and material technology allows for manipulation of properties in all phases of the material life, from pre-casting to decommissioning. The versatility and flexibility of the material has made concrete a desired material for large, inaccessible structures in harsh environments. Figure 4 illustrates the environmental loading on an offshore structure, the Oseberg A platform.

The North Sea is a harsh environment with respect to waves and wind, but it does not feature ice. Where ice is present, it also presents itself as a source of harsh environmental conditions. This is the dominant environmental design requirement in the Arctic. Platforms designed for operation in the Artic have been built and operated for many years. Conservative loads have been assumed, and the platforms behaved within design parameters. Figure 5 shows the Sakhalin II platform in sheet ice.

Yet another arctic challenge is the icebergs, the picture in Fig. 6 was taken at the site where the Hibernia Platform sits on the seabed, and where the Hebron Platform will soon arrive. Active ice management is often employed to redirect

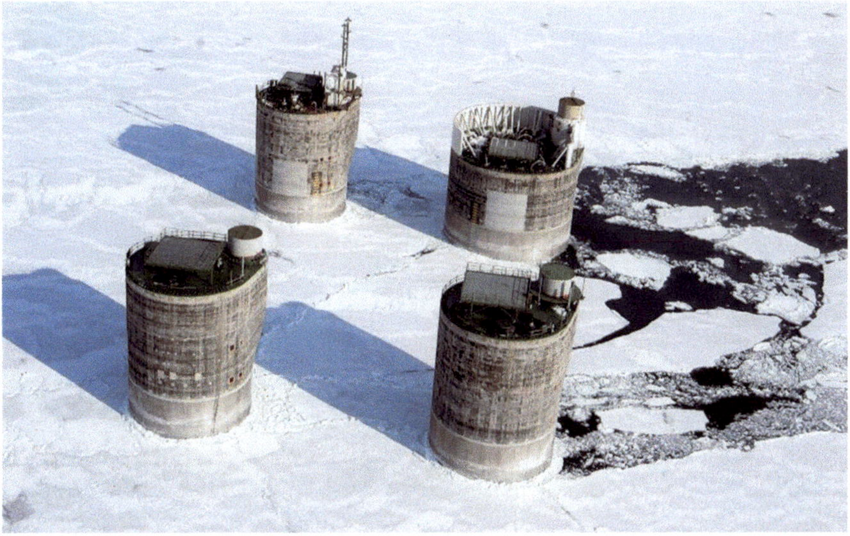

Fig. 5 Harsh environment (the Sakhalin II platform) (Gazprom 2013)

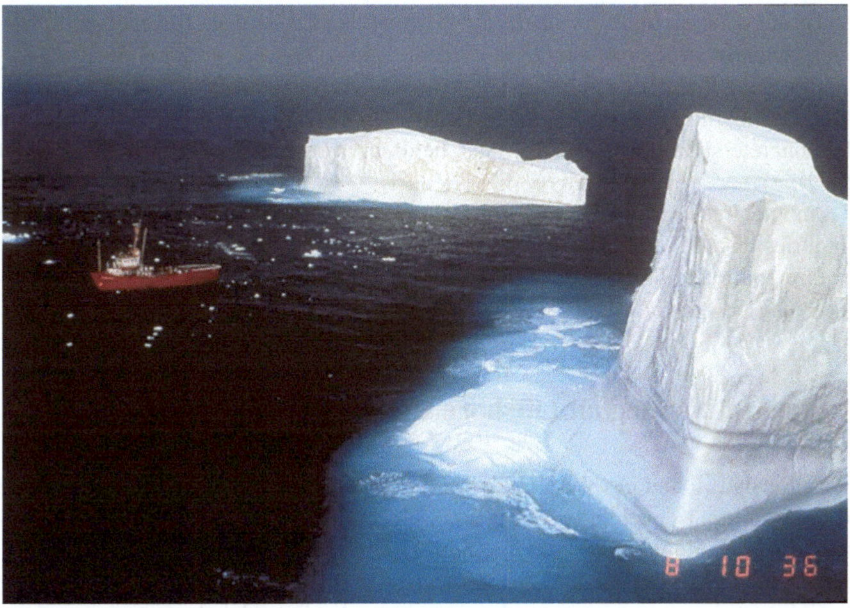

Fig. 6 Harsh environment (Iceberg, outside Newfounland) (ESA 2012)

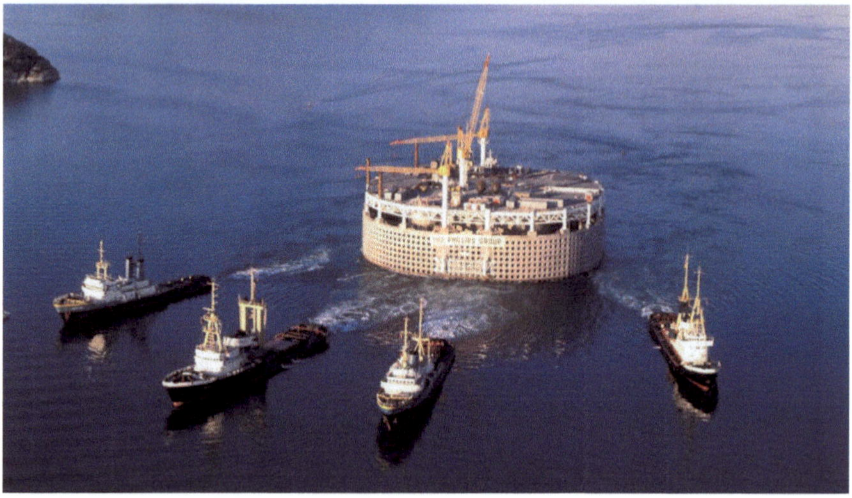

Fig. 7 Ekofisk tank during tow out

icebergs so as to avoid collision with a platform. The Hibernia platform is designed for iceberg impact, but has never experienced a collision.

To the concrete industry itself, the positive aspects of well-designed and well-built concrete structures should be evident. The robustness of concrete structures should make them even more competitive in very tough environments such as the Arctic. The International Concrete Federation (FIB) has published a report on the concrete structures for oil and gas fields in hostile marine environments (FIB 2009).

Another reason for selecting concrete as a construction material is that most countries have its own concrete construction industry. Concrete marine structures require some additional special skills, but these skills may be taught and experience may be bought.

3 Start of Norwegian Offshore Concrete Experience

Fortuitously, oil and gas were found on the Norwegian Continental Shelf; leading to the important development of a domestic industry geared towards oil and gas production. Foreign companies brought experience to Norway, and the Norwegians gradually built up their own competence. Within the structural concrete discipline, this competence was developed to a leading global position.

The American oil company Phillips Petroleum found oil on the Norwegian Continental Shelf in 1969, and began with the Ekofisk Platform as part of the offshore Field Development. The pioneer of offshore concrete structures for the North Sea was the French company Doris, who designed and built the Ekofisk Tank, and installed in 1973. Figure 7 shows the Ekofish Tank being towed to site.

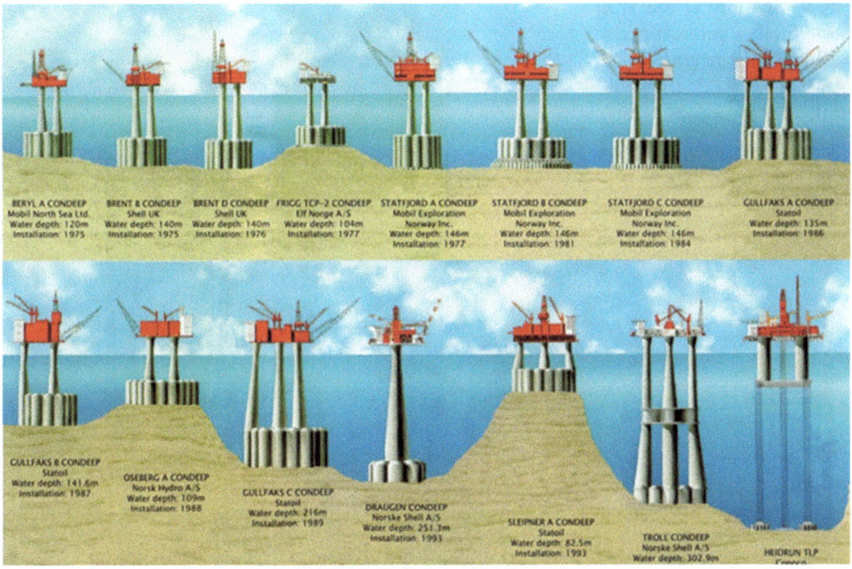

Fig. 8 Different existing Condeep platforms built by NC (Sandvik et al. 2004)

The water depth at the site was 70 m, and the original plan was to use the structure for storage. With time, the use of the structure evolved such that it served also as a field centre for nearby platforms. It was built at Hinna, south of Stavanger in Norway, were the contractors made a dry dock. The structure was built to a certain level inside the dry dock, then water was let in, and the structure was floated out were construction continued. 40 years later, oil production is still ongoing at Ekofisk—an oilfield still of large importance for the financial strength of Norway.

Those early years were important to how Norway adopted to the new era. Experience had to be gained, and the Norwegian entrepreneurship was on its toes. The Norwegian contractors Selmer and HøyerEllefsen built the Ekofisk tank on behalf of Doris. HøyerEllefsen saw the benefit of a caisson platform with vertical shafts on top of it, and the structural innovation was named Condeep. The Condeeps are shell structures, and the structural efficiency of shell structures is important for structures that float. Several platforms were to be built after the Condeep design, including the Oseberg A platform of Fig. 4. Figure 8 shows all of the Condeep platforms up to 2013. Gullfaks C and Troll A (Figs. 9 and 10) are the largest of the Condeeps.

The Condeeps were built by the Norwegian Contractors (NC), almost all of them at Hinna. NC made a significant contribution to offshore concrete structures, and also to the development of the concrete material, construction methods, construction management, contract management and many other fields of importance. They also managed all disciplines and the very important relations between all participants in the projects: client, builder, designers, universities and

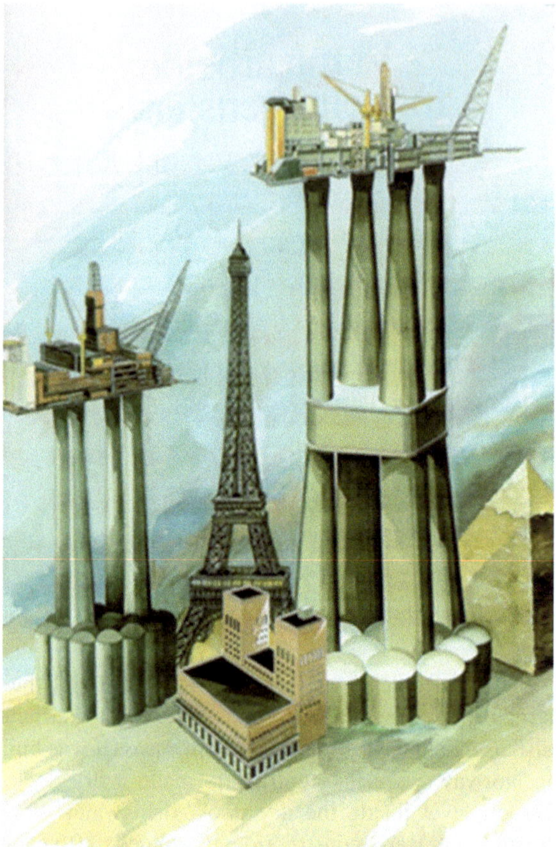

Fig. 9 Statfjord C and Troll A, with other publically well-known structures (AS Norske Shell) (Sandvik et al. 2004)

researchers, the competence of all participants were instrumental for the result (Sandvik et al. 2004; The Norwegian Concrete Association 1988, 1990). Dr. techn. Olav Olsen did most of the structural designs for NC, often in collaboration with other engineering companies. For a structural designer to work so closely with the contractor for so many years is extremely favourable for the complex projects. The work of NC is now carried out in the name of Kvaerner Concrete Solutions, for international construction projects.

Gullfaks C and Troll A, installed in 1989 and 1995 respectively, are the heaviest objects ever moved by mankind. The total mass for each of the constructions are 1.5 Mt and 1.0 Mt, respectively. Troll A is still a major supplier of gas to Europe. The project required large investment, but delivered high returns. The cost of the concrete platform is believed to be some 800 M USD (1995), a relative small fraction of the total platform cost, believed to be less than 20 %. The enormous values call for extreme engineering, and the opportunity to bring technology forward.

The Troll A platform, a gas wellhead and processing platform, is located 65 km offshore and is visible from Bergen at night and in clear weather. The base of the platform is 16,600 m^2. The lower part of the Troll A structure was subjected to a

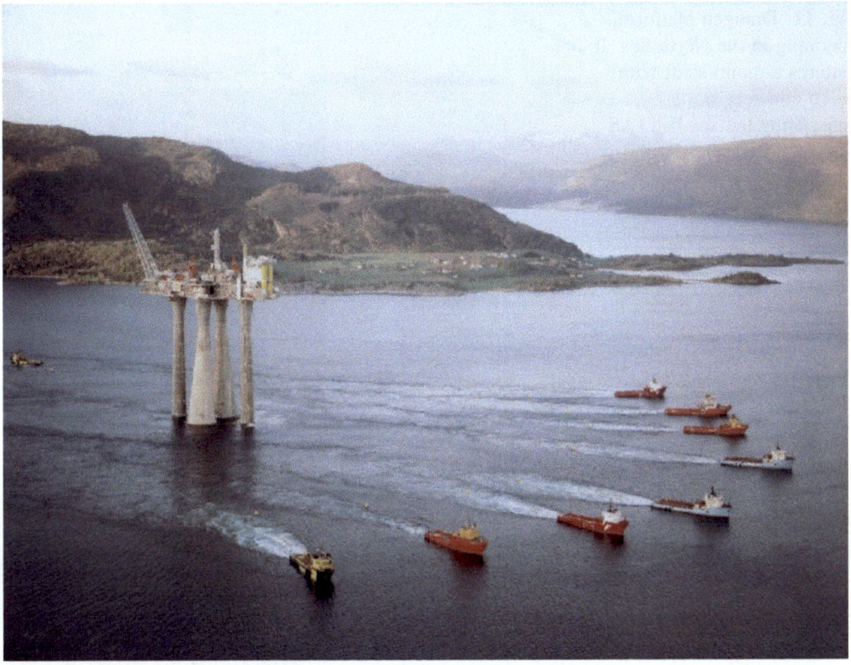

Fig. 10 The Troll A Condeep during towage, built by NC (Sandvik et al. 2004)

water pressure of 350 m during construction, and carried the topside weight of 22, 000 tonne 150 m above the sea level during towage from the construction site to the offshore installation site (Fig. 10). The need for the elevation of structure was the draft limitation of 227 m. From a hydrostatic stability point of view, it is rather challenging to carry 22,000 tonne 150 m above the sea level.

The soil condition at the site of the Troll A platform is very soft, popularly termed 'yoghurt'. For this reason, the base of the platform is equipped with 36 m long skirts to allow safe and stable foundation of the structure. The 100-year design mud-line moment is 100 GNm.

Most of the existing CGS's are resting on dense sand with short steel skirts penetrating the sand for protection against scour. For soft soils, deeper skirts like the concrete skirts of Troll A are required. The skirt pile principle was developed for Gullfaks C in the mid 1980s. The soil conditions with very soft clay, required skirts penetrating down to stiffer layers of soil. Gullfaks C has skirts penetrating 22 m into the soil. Draugen (Fig. 11) also utilises similar skirts that penetrated 9 m into the soil.

This skirt-piling principle has been used also for TLP foundations (Snorre andHeidrun), jackets (Europipe 16/11 andSleipner T) and suction anchors for floaters as an alternative to ordinary piles. Provided the soil conditions are favourable, skirt foundations have many advantages both economically and technically compared to other solutions. The advantages include:

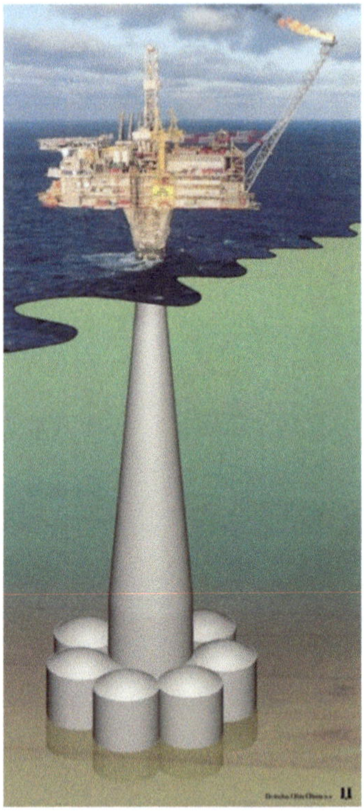

Fig. 11 Draugen platform, operating in the North Sea. It features a mono-shaft with seven concrete skirts penetrating the sea-bed (AS Norske Shell) (OO 2014)

- Lower material and fabrication cost;
- Reduction of foundation area;
- No piling or need for heavy subsea hammers;
- High position accuracy;
- Short installation time; and
- Reduced need for solid ballast.

The Norwegian experience has, since the turn of the century, been used for three Sakhalin platforms (Figs. 5 and 14); the Adriatic LNG Terminal (Fig. 17); and presently the Hebron platform being built in St. Johns, Newfoundland (The Norwegian Concrete Association et al. 2012).

The three Sakhalin platforms were built in Nahodka, south of Vladivostok, and towed to Sakhalin and installed there with a water depth of between 30 and 50 m. One challenge here is the sheet ice, constantly pushing on the concrete legs.

The Adriatic LNG Terminal was constructed in Spain (Algeciras) and towed to site, where it was placed on the sea bed.

The Norwegian experience includes also accidents. Of considerable importance was the loss of the Sleipner A platform, on the 23 August 1991, when the entire concrete substructure was lost due to rupture of a wall. Due to already agreed sale of gas, it was important for the Norwegian industry to deliver, and the second Sleipner A was designed and built on record time, meeting the installation date originally planned.

4 Materials

Material properties are an important element of design and construction of structures. For marine concrete structures, durability is important, particularly given the exposure to seawater and the risk of close contact with sea going vessels or structures. In the Arctic, robustness is essential due to ice-loading and increased need for safety in remote locations.

As the concrete structure needs to float, weight is an important parameter together with the material strength. The lower part of Troll A, for example, is subjected to hydrostatic pressure of 350 t/m^2.This is about 1,000 times the typical design load on the floor in a building or house. Realizing this, extensive research has been performed in Norway and globally on strength and weight of concrete.

The increased harshness in both the conditions and environment in which concrete is now being used, has led to a significant development on concrete compounds and reinforcement. Concrete is no longer as it used to be, neither is the reinforcement. Ultra High Performance Fiber Reinforced Concrete (UHPFRC), with strengths of possibly 200 MPa, makes new types of structures possible (UHPFRC 2013). These strengths may not be suitable for very large offshore concrete structures, but are used for smaller structures. A Japanese pedestrian bridge, the Sakata-Mirai bridge, on three intermediate supports that needed replacement exemplifies this. For the replacement bridge, no supports were required. Obviously, this same development in concrete strength is transferable to the application of the material for floating structures. However, there are some downsides to the new compounds also. One of the major ones is that both the reinforcement and aggregates comes at an increased cost and it may complicate the production process. Thus, cost has to be included in the evaluation. Nevertheless, the novel structures of tomorrow should take advantage of present advancements in materials.

Before this recent development, we may look at the Norwegian experience. The design standard calls for "any concrete platform and all appurtenances, piping and fitting to be fabricated from materials suitable for the service and life of the facilities. The design should be reviewed to ensure that the cost of the project is as low as possible over the service life. The concrete contractor shall develop a concrete mix that satisfies strength, durability requirements in their operating environment and construction methodology. Steel elements should be selected to resist the factors of design stresses, fatigue, corrosion and brittle fracture. Additional factors shall be considered with respect to hot and cold working and weld ability including resistance to lamellar tearing".

Table 1 Concrete grades

Item		ND C60	LC55
Structural material strength, ULS	f_{cn} (MPa)	36.4	33.6
Design compressive strength	f_{cd} (MPa)	29.1	26.9
Nominal structural tensile strength	f_{tn} (MPa)	2.375	2.25
Design tensile strength, ULS	f_{td} (MPa)	1.9	1.8

Table 2 Concrete modulus of elasticity

Item		ND C60	LC55
Plain concrete, ULS	E_{cn} (MPa)	29,399	
Plain concrete, SLS	E_{ck} (MPa)	30,534	23,226
Plain concrete, FLS (0.8 E_{ck})	E_{ck} (MPa)	24,427	18,581
Static, short-term loads (including reinforcement)	E_c (MPa)	35,000	25,000
Dynamic	E_{cdyn} (MPa)	40,000	

Table 3 Concrete densities

Item		ND C60	LWA C60
Plain concrete	(kN/m^3)	24.0	19.3
Reinforced concrete (150 kg/m^3)	(kN/m^3)	25.0	20.4
Reinforced concrete (300 kg/m^3)	(kN/m^3)	26.0	21.1
Reinforced concrete (500 kg/m^3)	(kN/m^3)	27.5	22.6

For concrete platforms there are two types of concrete which often are used; Normal Density (ND) and Light Weight Aggregate (LWA) concrete. The LWA-concrete may have about 20–35 % less density than the ND-concrete, which is an advantage for floating structures. However, concrete platforms are usually partly weight stable platforms and need weight in the base. Thus, it may be beneficial to use ND-concrete in the lower part and LWA in the upper part. The advantage must be compared to the possible disadvantage of dealing with two different materials at the site and during the engineering phase. In a feasibility study of an ordinary platform, the design of the concrete hull is recommended to be based on ND concrete grade C60 and LWA concrete grade LC55, both as defined by NS3473. These concrete qualities are well-proven in many countries, and are regarded as fairly easy to produce. For ordinary and pre-stressed reinforcement the grades KT500TE (NS 3570) and St 1570/1770 (EURONORM) are recommended.

The following tables give examples of material properties typically used (Tables 1, 2, 3, 4, and 5).

Additionally, Poisson's ratio $\nu = 0.20$ and the coefficient of thermal expansion $\alpha_T = 10 \times 10^{-6}\,^{\circ}C^{-1}$.

Table 4 Ordinary reinforcement

Design strength, Grade K500TE	
Yield stress (MPa)	500
Design strength, ULS (MPa)	435
Yield strain (mm/mm)	$2.5e^{-3}$
Modulus of elasticity (MPa)	$2.0e^{5}$

Nominal diameters are 8, 10, 12, 16, 20, 25 and 32 mm

Table 5 Strand properties for pre-stressed reinforcement

Design strength, Grade 260*, 15 mm (0.6″)	
Yield stress (MPa)	1670
Nominal diameter (mm)	15.2
Nominal area (mm^2)	140
Nominal mass (kg/m)	1.1
Modulus of elasticity (MPa)	$1.95e^{5}$

Tendon units: 6–7,6–12, 6–19, 6–20, 6–27 and 6–37 may be used (6–19 means 19 0.6″ strands)
* According to ASTM A 416-85

The tabulated values should be considered an example only as the application of the materials should be considered and tailored in each specific case. Research studies done on materials are generally available for everyone, although some modern concretes are protected by patents.

5 Conceptual Design

Architects shape the structure to give value and meaning for the owner. Simultaneously, the architect works to craft a statement with the structural form while maintaining a connection with the surrounding environment. The American architect Louis Sullivan stated in the industrialization period of the 19th century, that *form ever follows function.*

Marine structures are usually designed to serve a special cause or solve a specific challenge. This typically results in structures that are shaped for very physical reasons, with structural engineers principally driving the design. Bridges are one example of this. It is important to note that there has been a move towards involving architects in bridge design. This is likely because the optimal form for a structure is rarely the cheapest one to construct, and the cheapest form is rarely the most pleasing to the eye. However, given offshore structures are hardly seen by anyone due to their remote locations, for offshore and marine, structures, the important design requirements are construction phases, hydrostatic and hydrodynamic loading, and the use of the structure.

The conceptual design phase is crucial for the intended purpose. It is important to pay attention to the upsides of a marine environment and to what possibilities

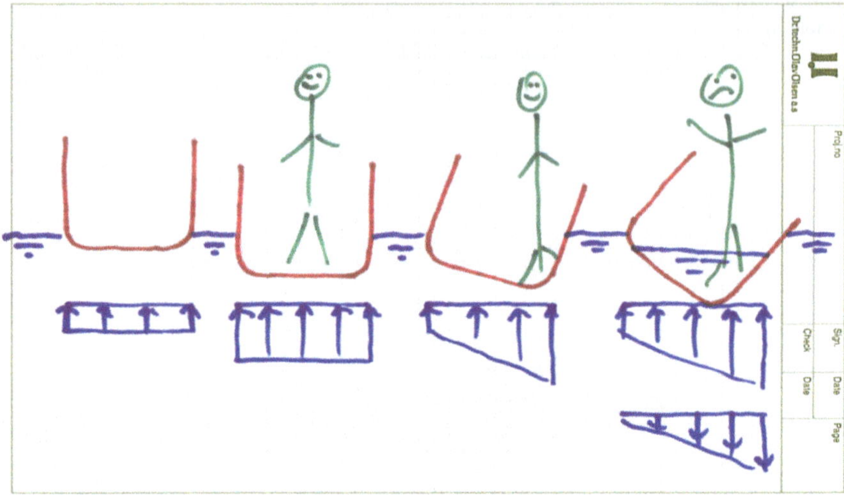

Fig. 12 Examples of different stages of floating stability (OO 2010)

the presence of water gives. Likewise, it is important to realize that the presence of marine operations may represent a threat, for both environmental issues and the safety of nearby human populations.

There are several factors influencing the design of a typical floating structure given the magnitude of scale and the high complexity. Not every aspect of the structure can be evaluated in the conceptual design phase because it simply becomes too complicated. A very important part of the conceptual design is therefore to select the most important factors that would be allowed to govern the concept design. This requires experience and skill, specifically of the structures and their functions.

To illustrate this, a few simple sketches are presented. Figure 12 illustrates the concept of hydrostatic stability and the result during unfavourable conditions. Most of us have some feeling for what hydrostatic stability is, but in the design phase, we also need to put numbers on it. The water inside the canoe gives a dramatic effect on the stability and several car ferries have capsized for similar reason. This also shows that it is not easy to cover all possible scenarios the structure may encounter during its design life.

Figure 13 shows a simplified prototype of the Condeep concept. The model was built in order to verify the installation phase, where the caisson structure had to be submerged in a controlled matter. These simple tests proved to be valuable in terms of both understanding and gaining experience with the conceptual design.

Experience gained by testing on different prototypes lead to novel installation methods. Figure 14 shows an example where the structure is submerged with an inclination. Vertical submerging implied a critical situation when the water level went from the caisson and over to the vertical legs due to the drastic reduction of vertical hydrostatic stiffness over the water plane cross sectional area. This could ultimately lead to capsizing of the construction.

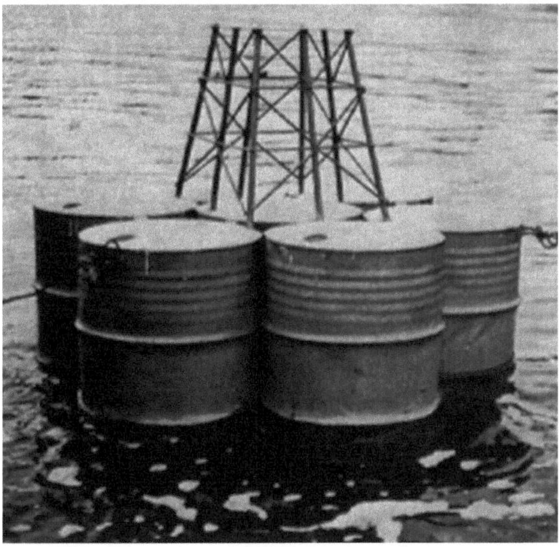

Fig. 13 Prototype to Condeep (Ellefsen 1972)

The value of simple prototypes was truly appreciated and several of them were constructed to get better *hands-on experience* with different concepts. Figure 15 shows initial tests with a different Condeep concept. Colleagues are eager to test their conceptual designs.

As a further example, Fig. 16 shows the Adriatic LNG Terminal. The structure has three main functions;

(1) Loading terminal for LNG ships that comes from mostly Qatar up through the Suez Canal;
(2) Storage facility for the offloaded LNG; and
(3) Processing of the LNG to gas that is piped into Italy.

It sits on the seabed, at a depth suitable for berthing of the large LNG tankers. A significant challenge, when designing the terminal, was to handle the relative large variations in uplift between the arrivals of the ship. The variation in uplift is caused by the various degrees of LNG content. Incidentally, one of the design requirements also was that the structure should not be visible from Venice.

Figure 17 illustrates work done by the Museum of Modern Art, MoMA, to prepare New York for climatic changes and reduce the consequences of sea level rise due to polar ice melting. The example is included here to show the variety of design consideration conceptual design may encounter. In this case, the water is the threat, and the objective is to find means of reducing the threat in the best possible way.

Different activities within the oil and gas industry has led to valuable experience and understanding of floating structures. We can now safely design and build structures for virtually any purpose for any location in the world. This also includes urban areas where the purpose of the marine structure may be different

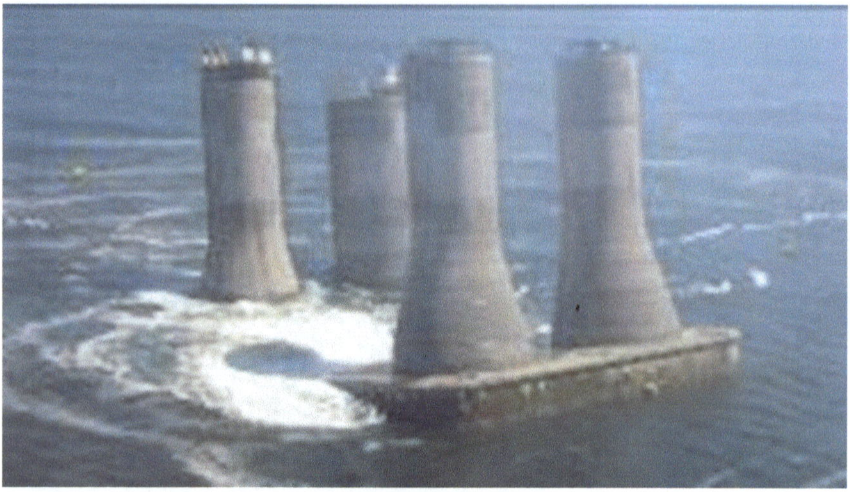

Fig. 14 Inclined installation to increase hydrostatic stiffness (Sakhalin II Platform) (SEIC 2006)

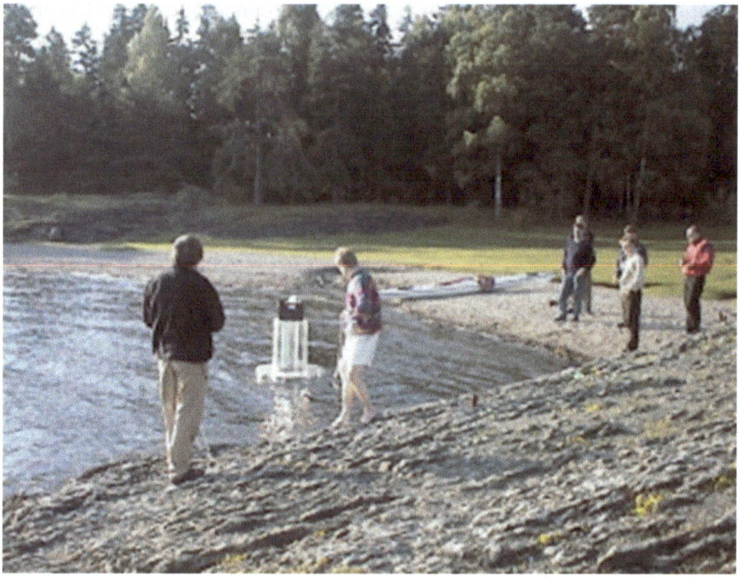

Fig. 15 How does the conceptual design work? (Olsen 2012)

from the purpose of the oil and gas industry. Even in Norway, with so few people and so much space, we build into the sea in urban areas. We build parking garages in ship dry-docks, and tow them into Oslo for installation. We build recreational floating parks and we build immersed tunnels to avoid the traffic in sight.

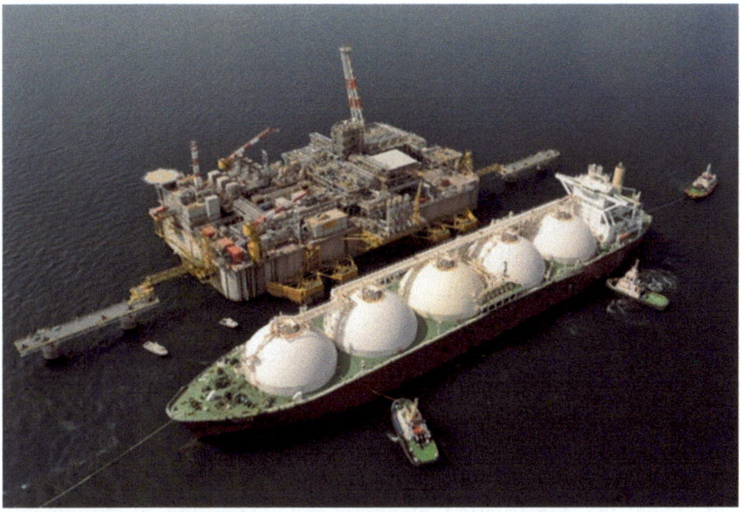

Fig. 16 The Adriatic LNG Terminal (Van Zaal 2009)

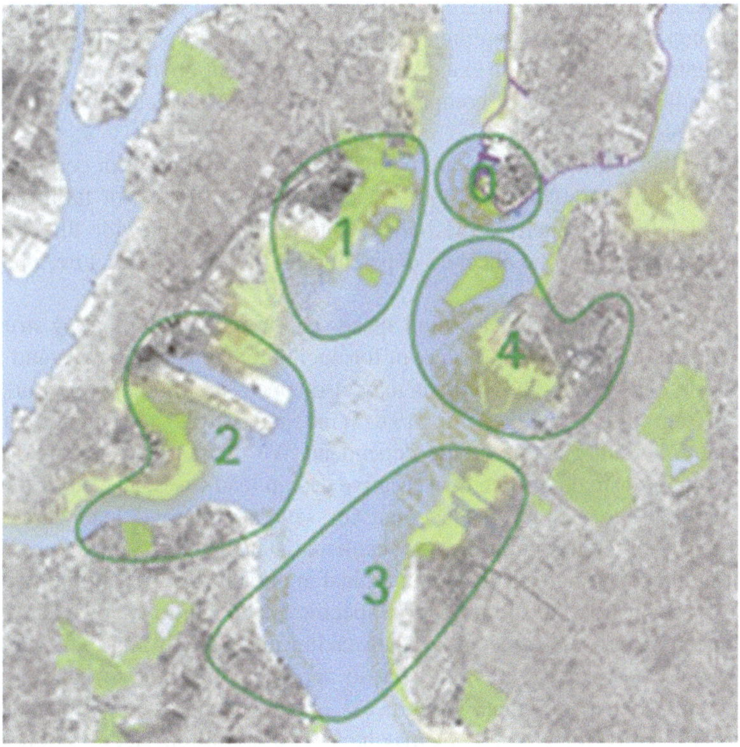

Fig. 17 On the water—Palisade Bay (Nordenson et al. 2010)

6 Detail Design

After the conceptual design, sometimes performed in many phases, the structure will be evaluated for construction, a process of many considerations. Hopefully the quality of the conceptual design is such that it does form a sound basis for detail design, which in turn will form the basis for construction. At the detail design stage, all details are described, drafted and verified. This is a considerable amount of work, and seemingly small features in the conceptual design may result in major complications in the detail design.

Designing marine concrete structures is in many respects very similar to designing ordinary land-based concrete structures. Reinforcement need to be calculated and drafted, so that the construction can take place. It should always be kept in mind that the reinforcement content typically is very large in marine structures, as the structure is made slender to float and to minimize hydrodynamic loads. Typical values may be 300–400 kg/m^3, locally as high as 1,000 kg/m^3 has happened. Thus, the construction planning and integration of design and construction is important. Another significant difference between land-based and offshore structures is also the environmental design criterias as well as the increased number of both operational and accidental states.

The operation states are different distribution in ballast or other changes in the platform configuration as loading of equipment, lifting operations or transport phases. Figure 18 illustrates an example of ballast distribution in a semi-submersible platform hull. The number of compartments also give an indication of all the different possible configurations.

Some of the accidental states are typically induced by failure in compartments due to collision; equipment failure that may lead to explosions; or falling objects that may potentially penetrate the hull. Offshore standards provide directions on what states will need to be checked depending on the type of structure, operation and location.

On land we account for mean wind speeds and gusts, but the floating structures have additional requirements to account for, including the need to withstand forces from current and waves. These forces may also appear in different magnitudes and from different directions (see Fig. 19). Wind and waves are often treated as somewhat coherent, but this is site dependant. We rely on site information and offshore standards to determine conservative design load cases. The environmental design aspects cases alone can account for several thousand load cases when applied to the different operation and accidental states.

Finite Element Analysis (FEA) is a typical approach to verifying the structural integrity of a construction submitted to a specific load condition. However, this is a time-consuming process, especially when taking into account that both wind and waves should be treated with a stochastic approach. In order to handle all the numbers we need a tool that will process all the results of the finite element analyses. Typically, the global FEA may have several millions degrees of freedom. To obtain an efficient design process, one needs to use post-processors that find the

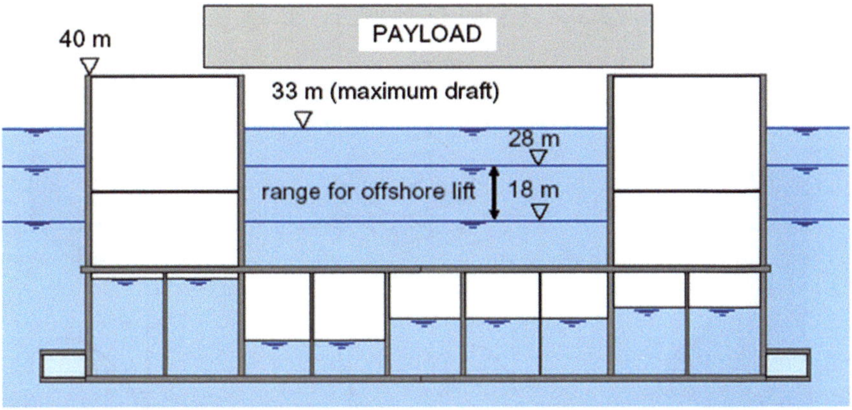

Fig. 18 An illustration of possible trim states in the ballast compartments of a typical semisubmersible hull (OO 2007)

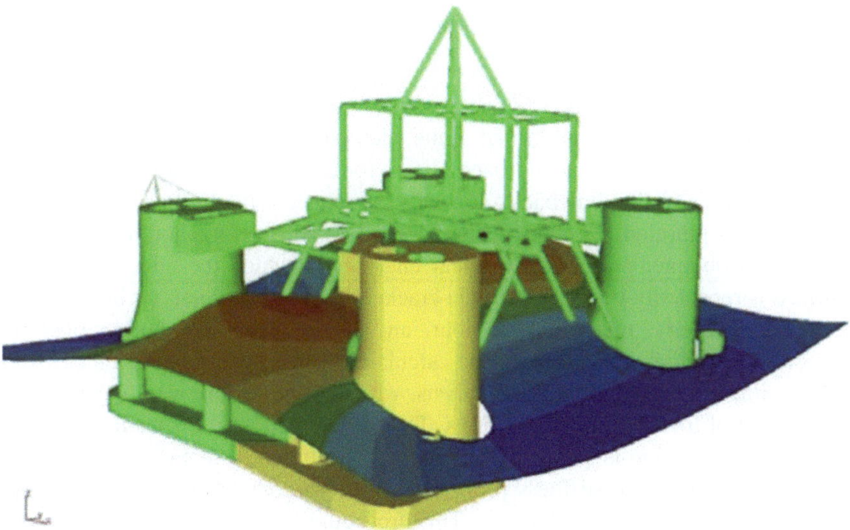

Fig. 19 Irregular wave field on a typical semisubmersible platform where hydrodynamic forces are mapped onto panel model in order to investigate structural forces in time domain (OO 2007)

least favourable load combination at every point of the structure, for stresses in the concrete and reinforcement layers, of which there are several, and then perform more detailed studies.

Traditionally, structures have been analyzed assuming linearly elastic response. Reinforced concrete, however, does not respond linearly elastic, and this is recognized in the sectional design when checking against codes. Consequently, there is a contradiction in the traditional design.

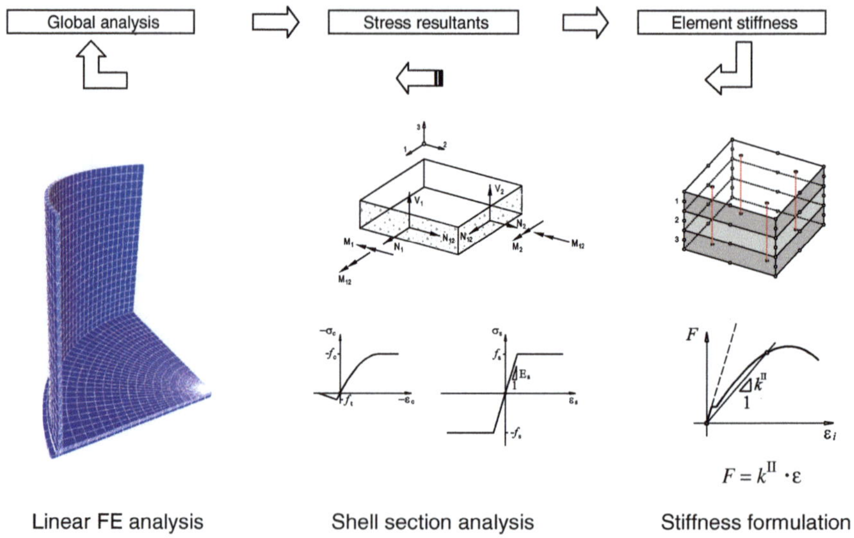

Fig. 20 Iterative approach for predicting nonlinear behavior of shell structures (OO 2014)

To remedy this contradiction, we now have, after extensive development, the tool for running non-linear global finite element analyses (the tool is named ShellDesign). The FE-based approach allows for a global non-linear response that eliminates some of the drawbacks normally associated with nonlinear simulations like time consumption. By the so-called consistent stiffness method, the nonlinear response is obtained by iterative, linear analyses in which the effective element stiffness is repeatedly updated from the cracked shell section analysis based on the sectional geometry, material behaviour, and linear strain distribution over the section and equilibrium of forces. The calculated stiffness parameters are fed back into the linear—elastic structural analysis, which is the re-run, resulting in a new distribution of forces and deformations. This is repeated until a specified stiffness convergence criterion is satisfied. The process is illustrated in Fig. 20.

At convergence, the stiffness properties in all sections are consistent with the actual load level and material layout. This procedure is possible due to the use of a special linear elastic element type. Despite the progress in development of complete constitutive models for reinforced concrete shells, current commercial nonlinear FEM programs still have limited applicability in practical design as both speed and integrated operations are such important parameters. For a more elaborate description, refer to (Nyhus 2014).

The Modified Compression Field method of Prof. Michael P. Collins and his colleagues at the University of Toronto is also included. This method gives better predictions of structural response (Collins and Mitchell 1991), and is particularly important for structures that are unlike structures of common experience. The tool allows for both optimizing the design of the structure and the possibility to document strength under severe conditions.

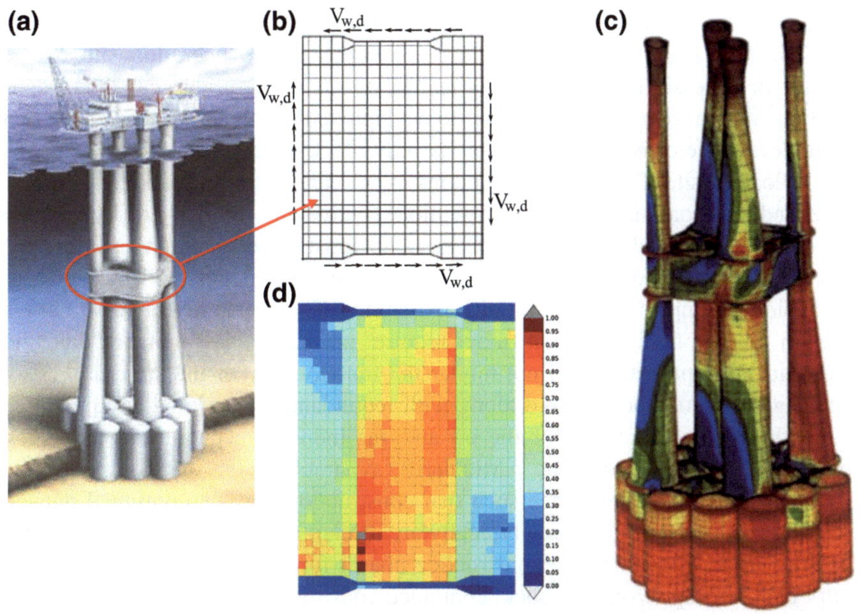

Fig. 21 Results from nonlinear analysis of **a** The Troll ARiegel, **b** Simplified geometry and applied load. **c** Mapped stresses on structure. **d** Contour plot of the concrete utilization (OO 2014)

ShellDesign is therefore a state-of-the-art design tool for analysis of concrete structures and includes innovative advanced calculation procedures. The procedures have been shown to be very useful in real problems. Figure 21 shows one of these, where it was possible to document sufficient strength of the Troll A platform subjected to new loading conditions following a refit and an extended life with the consistent stiffness method. Comparing the results of the traditional linear design methodology with the results of the ShellDesign methodology is educational, and it makes good sense to achieve consistency in design. This is elaborated on in (Nyhus 2014). The owner of the Troll A platform was very happy it was possible to document sufficient strength, and said that if the results had not supported the continued use of the structure, the alternative would have been to build a new platform.

The detail design of large structures subjected to a great number of loads is a considerable task. Documentation is required, both for construction, operation and for everyone involved, including authorities.

7 Construction

Constructing concrete structures is not a novel exercise with most counties having the resources and expertise to build them. However, marine concrete structures have some peculiarities that should be noted.

First, the structure is likely to be built at a different location to where it will be installed. This may be due to weather conditions at the operation site or the distance or accessibility of materials and labour. This may also call for a construction site that permits the structure to be deployed or towed out when built to a certain level. The construction site may be a dry dock, a barge, a skidding ramp, or even a floating site. For smaller structures even crane lifting may be an option.

Second, the floating phase needs attention. Hydrostatic stability and hydrostatic loads are important elements that the designer must understand. In addition, the structure may need to be operated, and mechanical equipment will be required. The typical examples here are pumps and cranes. The structure may also have a function in its final use that may require mechanical equipment, but the mechanical equipment should not hinder the functionality of the structure. Also, water tightness may be required along with redundancy for critical equipment.

For the larger structures for the oil and gas industry, typically the dry dock option is selected. It would be particularly favourable if the dry dock could be located within a sheltered place such that construction may be continued outside the dry dock, as these structures typically are too large to be completed in the dry dock. Figure 22 illustrates the construction phases for such a production scenario: in dry dock; towage from dry dock; continued construction outside the dry dock in a floating site; coupling of the substructure with the topside (termed 'deckmating'); tow out to field and installation and operation.

There are some important lessons for successful design and construction of these floating structures. Since weight is so important for a structure that floats, the design should be made slender through structural design, such as using shells as structural elements. In addition, the walls are usually made as thin as possible, with the reinforcement content thus set to a higher quantity. This calls for extensive construction planning. It is very different to build a concrete structure with steel reinforcement of 100 kg/m^3, as compared to 400 kg/m^3.

Integrating design and construction is important because the construction influences the design so strongly, and vice versa. To add to the challenge, detail design is often not completed for the entire structure before construction starts.

Prior to the start of construction, a design of the entire structure is performed to define all wall thicknesses and pre-stressing layouts as well as overall quantities and layout of reinforcement. Typically, this design is based on hand calculations and previous experience, though it may include finite element analyses. Overviewing and estimating the design loads is essential. This phase also includes an overall design of all the mechanical systems that should go inside the concrete structure.

Further local design and detailing of rebar arrangements and the mechanical systems for the lower parts (the first casting sequences) have to be completed before any construction work begins. If the project includes development of a construction site, this design work is normally conducted concurrent to the completion of the construction site. Dependent on the duration of this work, this phase may be on critical path of the execution schedule.

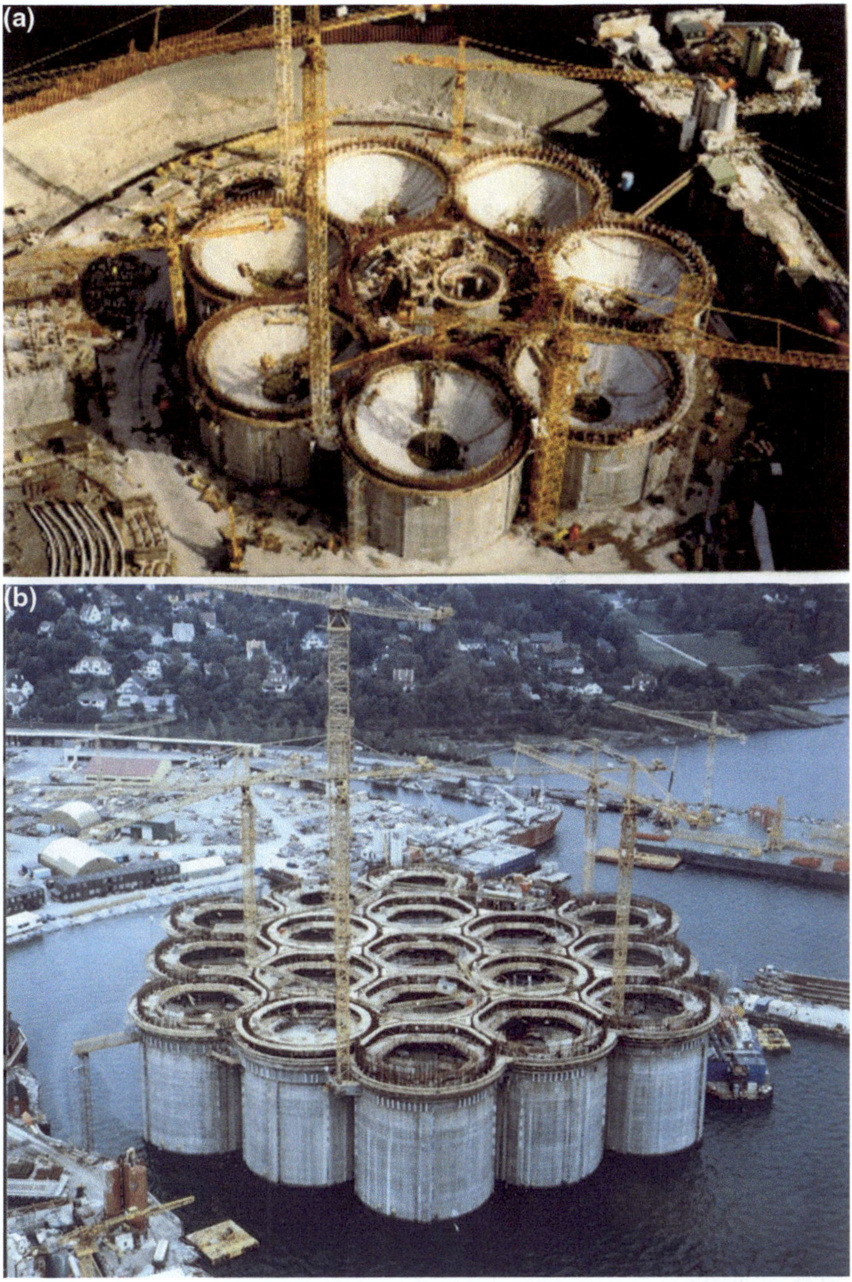

Fig. 22 Typical construction of large concrete offshore structures (Sandvik et al. 2004). **a** Construction in dry dock (Draugen)(NC). **b** Float out from dry dock (Troll)(NC). **c** Construction at wetsite (Draugen)(NC). **d** Installation of topside (Draugen, deck mating)(NC). **e** Tow to offshore installation site (Draugen)(NC). **f** Installation at offshore location and operation (Troll)(NC)

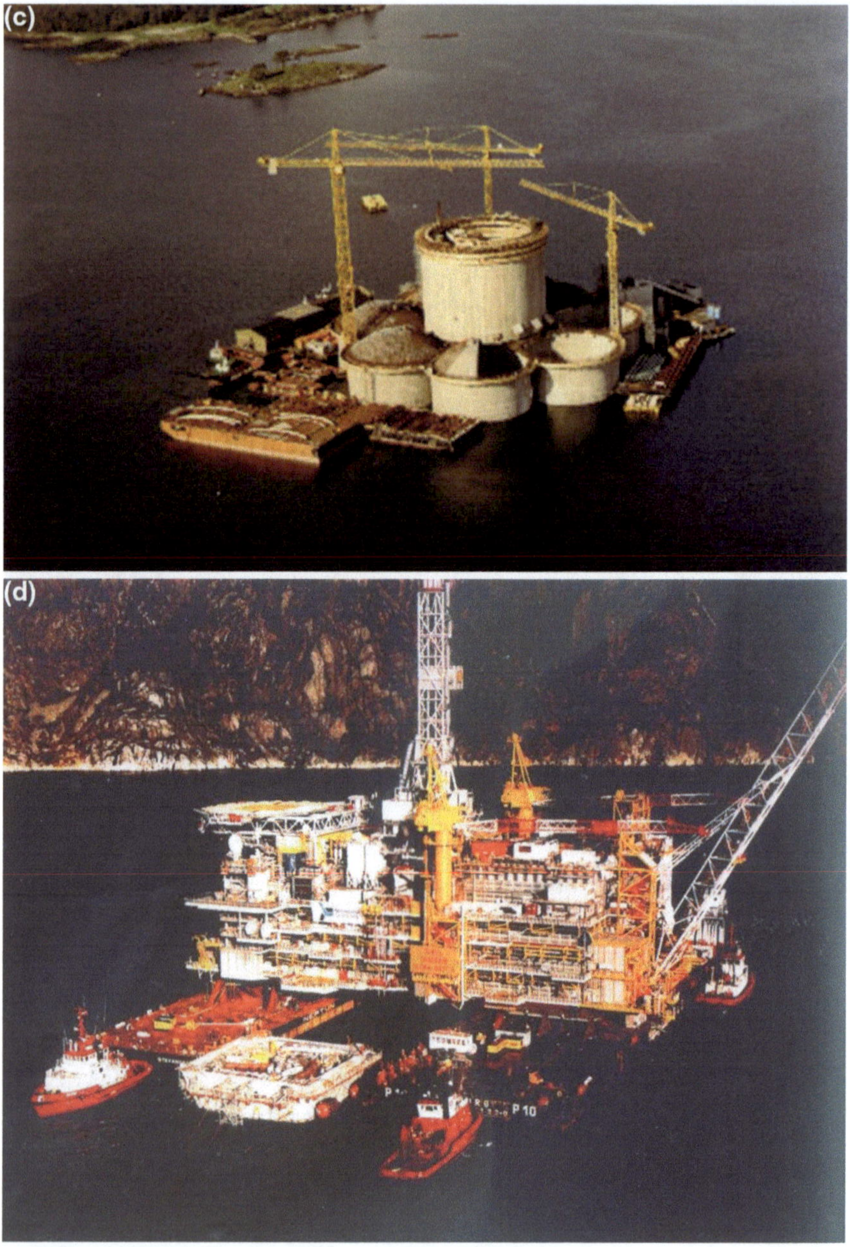

Fig. 22 continued

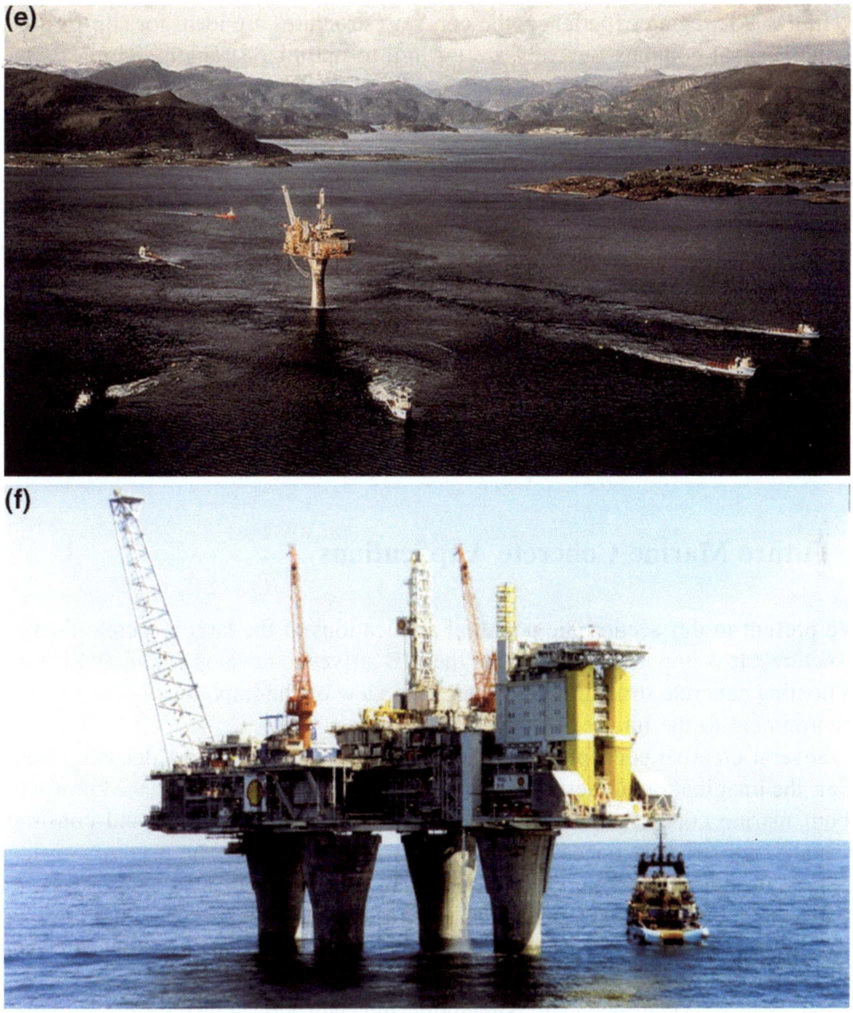

Fig. 22 continued

The remaining part of the concrete and mechanical systems detail design will often be performed with an overlap with construction work. This part of detail design is not normally on the critical path of the execution schedule. A typical medium size concrete platform built in concrete will require a fabrication dock with an area of some 140 × 140 m, and a water depth of 10–12 m. At Hinna the dock sizes were 200 × 200 m and depths 9, 6 and 14 m. The drydocks of Algeciras (Spain), Vostochny (Russia) and Bull Arm (Canada) are 20, 14 and 16 m deep, respectively.

In the Norwegian experience, the concrete structures are ideal for slip-forming as the vertical columns typically are tapered to optimize strength and mass. Slip forming is a continuous construction of walls, and the technique is developed so far that the geometry may be altered to tailor the structure to the need. Even the lower hull with an assumed height of some 10–12 m will most likely be slip-formed. The mechanical outfitting will be placed during construction, in sequence with other construction activities. All steel will be attached to the concrete at preinstalled embedment plates. Such plates are placed during the slip forming, and after slip forming steel structures will be welded onto embedded steel plates. Penetrations through walls or slabs will also be placed during casting.

The Norwegian experience with the marine concrete structures is generated through 40 years of construction and design, and through development and research. This total experience positions the local industry well to take on the future applications of marine concrete construction.

8 Future Marine Concrete Applications

We present in this section the potential applications of the large concrete floating structures. It is important to note that the FIB strives to develop useful application of floating concrete structures, and shares the view on the importance of the marine environment to the future.

Several creative concepts have been presented during the last decades. However, the imagination of these proposals appear limited. Fortunately, we know a lot about marine concrete structures, and are actually able to design and construct virtually any structure conceived. Despite this, there remains the need to consider cost, environmental issues against the actual need and demand for the structure. Given that there is a greater number of countries with strong financial abilities and a desire to colonise adjacent ocean space, there is a large potential for larger and novel concepts, bring us examples of construction unlike what has been the tradition. Nations such as Japan, Singapore and Monaco, are now contemplating urban spaces within the "free" ocean space available in the vicinity. However, urban areas have a different design specification as compared to the oil and gas offshore structure. Nevertheless, the experience garnered from remote oil and gas projects may be transferred in order to uncover potential and novel solutions.

As an example, Fig. 23 illustrates a facility bridging land and the sea. In this case the concept is proposed to be sited outside Monaco, a place where land is scarce and, consequently, the cost of land is very high. The base of the structure including a portion of the legs, may be constructed as the offshore concrete platforms, at a suitable place and then towed to location for installation. That way one can build without disrupting the city, and construction can occur at a remote location where the necessary construction experience is located. One may be able to avoid one of the major problems in urban development: noise during the construction phase.

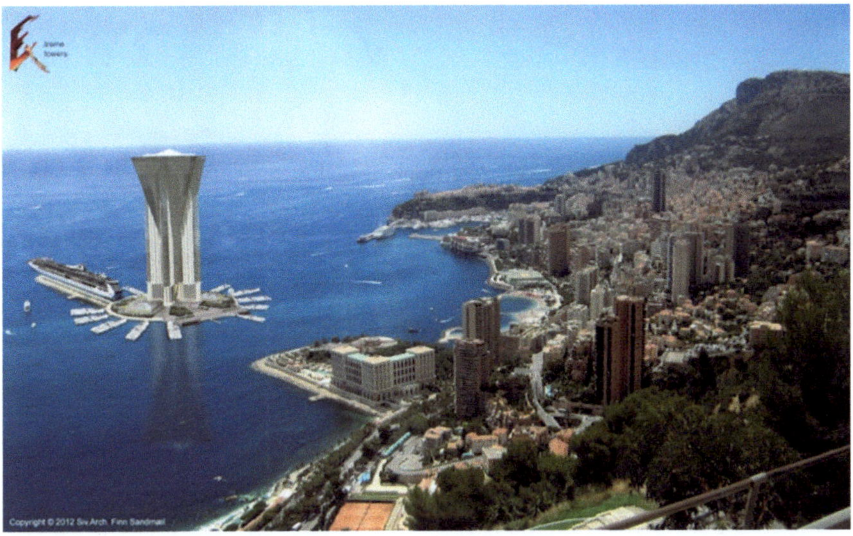

Fig. 23 Yogamid, by Architekt Finn Sandmæl—urban development where space is scarce (Sandmæl 2013)

In terms of size, the Troll A platform, for reference, is close to 500 m tall, but the shallow Mediterranean Ocean will not allow that kind of towage, so parts of this structure would have to be built on site. The base however, may even come in parts allowing the use of smaller construction yards and towage through canals.

Also in densely populated areas, the shoreline is often used for railroads and highways. Historically, one can see the reason for this as it relates to where people were and where construction was convenient. However, given that coastal land use is increasingly captured for leisure activities and for residential purposes, roads and railroads are being moved to tunnels, either on land or in the sea. Moving the new facilities to another land-based location may easily be a more favourable option than trying to move the existing infrastructure.

However, what if more infrastructure developments are needed in the already densely populated areas, or if it is simply not possible to follow the shoreline? A typical example may be straight crossings. They also concern infrastructure in water, and the principles may be used for urban infrastructure as well. There are five ways of typical strait crossings; bridge, floating bridge, submerged floating tunnel, immersed tunnel, and subsea tunnel. These strait crossings are shown in Fig. 24.

The bridge, immersed tunnel and the subsea bridge are commonly used. The floating bridge however, is utilized more rarely. We have two of these in Norway, the Bergsøysund Bridge and the Nordhordaland Bridge. The Submerged Floating Tunnel (SFT) has never been built for traffic. There is one built to protect gas pipes on the west coast of Norway (Fig. 3). The purpose is to bridge the subsea topography and to protect the gas pipes from the harsh environment. Submerged

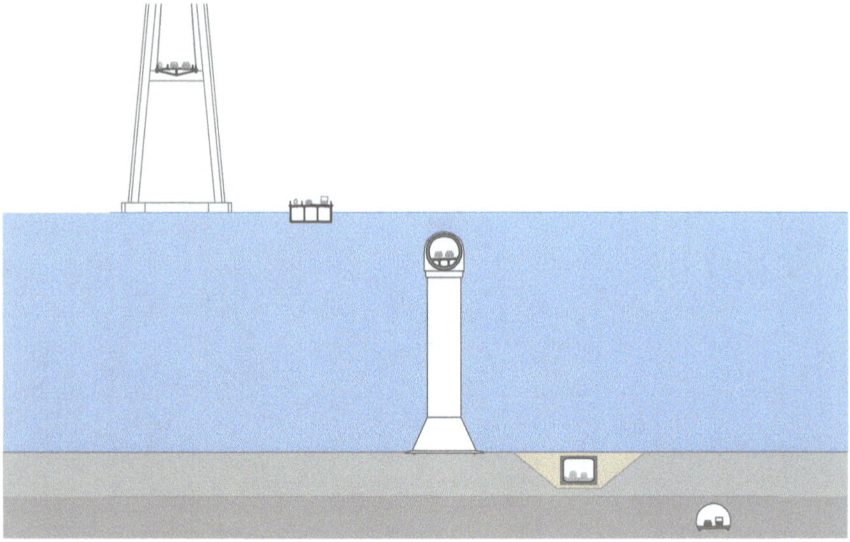

Fig. 24 Strait crossings (OO 2014)

floating tunnels may be used for strait crossings, in an environmentally friendly way. Figure 25 illustrates a submerged floating tunnel, here shown approaching the *basement* of a town for a car-free environment, (Muttoni et al. 2001; Haugerud et al. 2001). Ships can pass above, and the bridge/traffic is invisible. The bridge may be built at a different location and towed to site in order to avoid conflict with ongoing activities.

Both local and national authorities may also be a source for inspiration on novel concepts and development. For instance, there is a political decision in Norway, somewhat neglecting the normal economical constraint, to build a ferry-free E39 between Kristiansand and Trondheim, some 1,100 km. There are eight major strait crossings on the route. One of these, across Sognefjorden, represents major challenges: The width of the fjord is 3.7 km wide and 1.3 km deep. For a bit of perspective, the width is about twice the length of the longest suspension bridge in the world and the depth is more than four times deeper than the depth that the Troll A platform is situated at. In addition, there are large ships passing frequently. Further details on the SFT study for this crossing may be found in (Brandsegg et al. 2013).

A submerged body of an SFT needs to be supported vertically, due to variable weight. This support may be by hanging in pontoons or by tension legs anchored to the seabed. For the 1.3 km deep Sognefjord, the pontoon solution is relatively obvious.

For longer crossings, it may be desirable to use several pipes in the tunnel. Having one or two second pipes allow for better safety and evacuation options, but they also increase global strength. Additionally, more than one pipe also

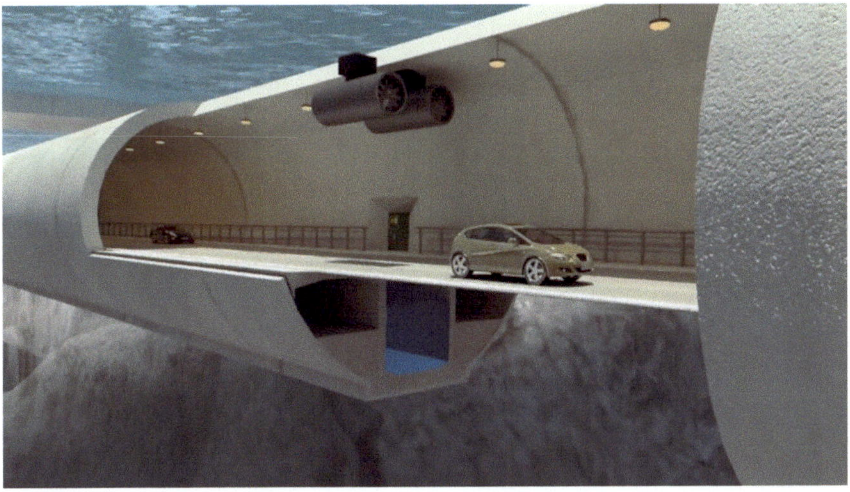

Fig. 25 Cross-sectional cut in the SFT proposed for E39 through Sognefjorden (SVV 2013)

significantly increases the possible to alter the global response of the system. Figures 26 and 27 show an option where two main pipes, cross-braced by diagonals, is utilized. This allows for efficient and safe operations with construction in dry docks.

A hybrid of tunnel and structure may be considered for ultra long subsea tunnels, as invented by Allan Sharp (Olsen et al. 2013). There is a limitation on

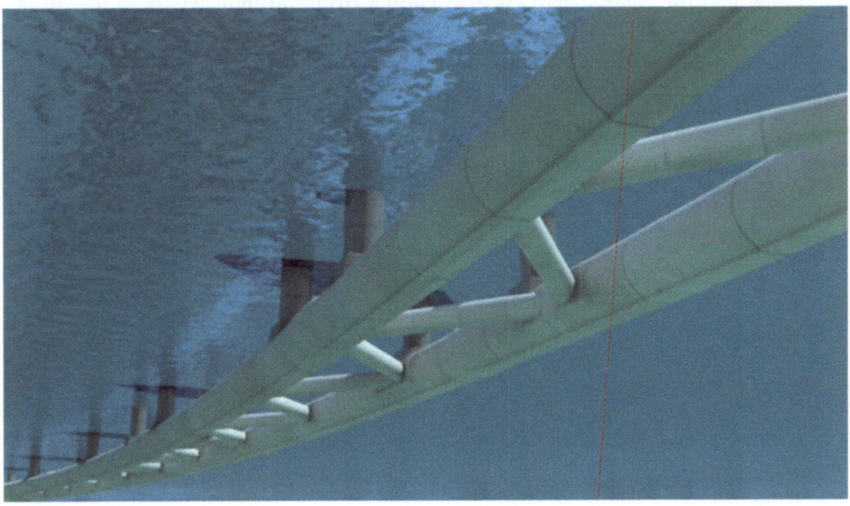

Fig. 26 SFT at Sognefjord, two separate tunnels with escape routes between. Pontoons visible as shadows (SVV 2013)

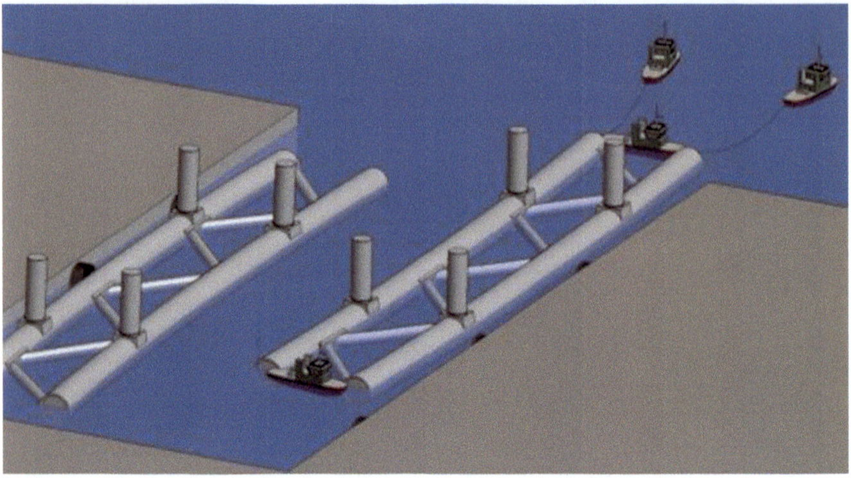

Fig. 27 Proposed construction method in dry dock (OO and Reinertsen 2012)

how long subsea tunnels may be built. The duration of construction, financing, cooling, and ventilation are the most important. By suggesting intermediate access shafts, it can be demonstrated that the length limitation can be avoided. This is of great significance for the transport of the world, possible locations are from Japan to the mainland, South Korea to China, Mainland China to Taiwan, and cross the Bohai Sea in China.

The concept, as shown in Fig. 28, is simple. It combines the well-known technologies of tunnelling and offshore structures. The interface between the two (the tunnel and the access shaft) is also known from the industry, both the oil and gas industry and the hydroelectric energy industry.

Nevertheless, straight crossings are not the only challenge that the future awaits. For instance, there are requirements to remove offshore steel platforms (OSPAR Convention) after the end of their service life. Until recently, most of the platforms have been in operation, but in the next years, most of them are closing

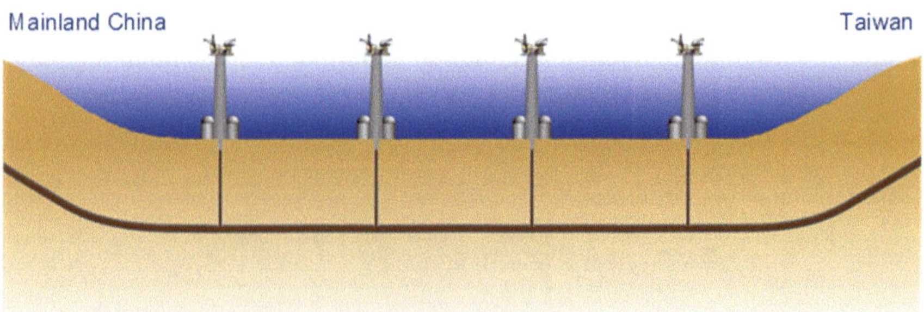

Fig. 28 Ultra-long subsea tunnel layout (Olsen et al. 2013)

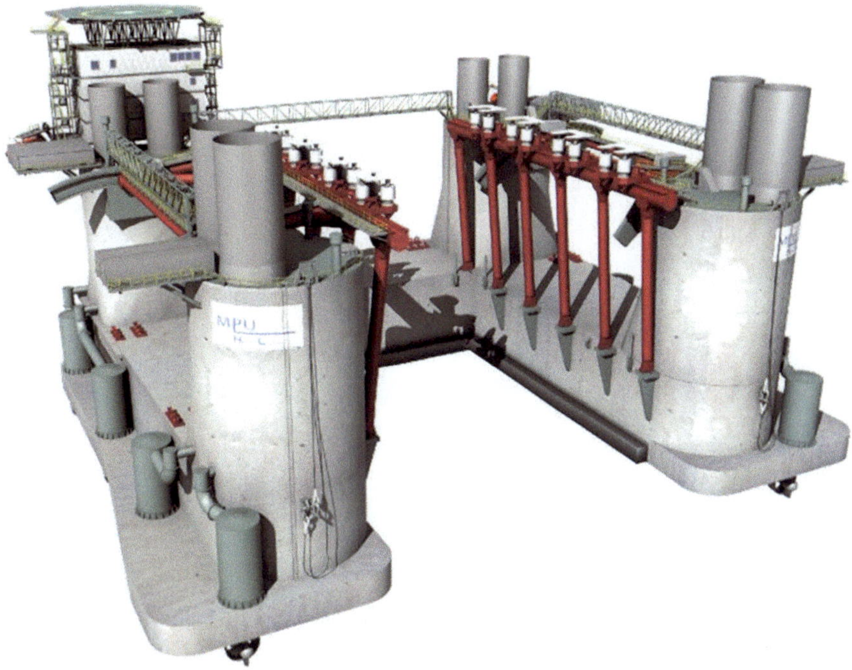

Fig. 29 Heavy lifter for removing and installing offshore structures (MPU 2007)

up to the end of their service life and have to be decommissioned. This operation is no easy task and there have been a few untraditional approaches. As an example, the Norwegian energy company Statoil tried to auction the Huldra platform at finn.no, the Norwegian equivalent of eBay, for less than a dollar. The catch, of course, was that the buyer had to remove the platform safely from its current site. Due to the complexity and cost of such an operation, they had no luck in trying to sell the platform.

For the North Sea alone, the decommission market represents a potential market value of some 10 billion US$. This market potential initiated the development of a robust Heavy Lifter that utilizes the simple principle of Archimedes in order to lift straight up. It is very much easier to lift straight up as compared to lifting up from a distance. Managing this you need to get underneath the object to be lifted, not easy on a platform standing on its legs. The solution is a robust concrete hull that is strong enough to manage with three pontoons, U-shaped, and which at the same time can resist hydrostatic loads to carry the load. The capacity is 12,000 tons, and the U-shaped semi-submersible is planned to be self-propelled. The design is shown in Figs. 29 and 30, and illustrates how the water and proper design may be combined to create a useful product.

SEMO is the name of another concept were the buoyancy is used to create an optimal design. The SEMO is intended for oil and gas production, and is a robust

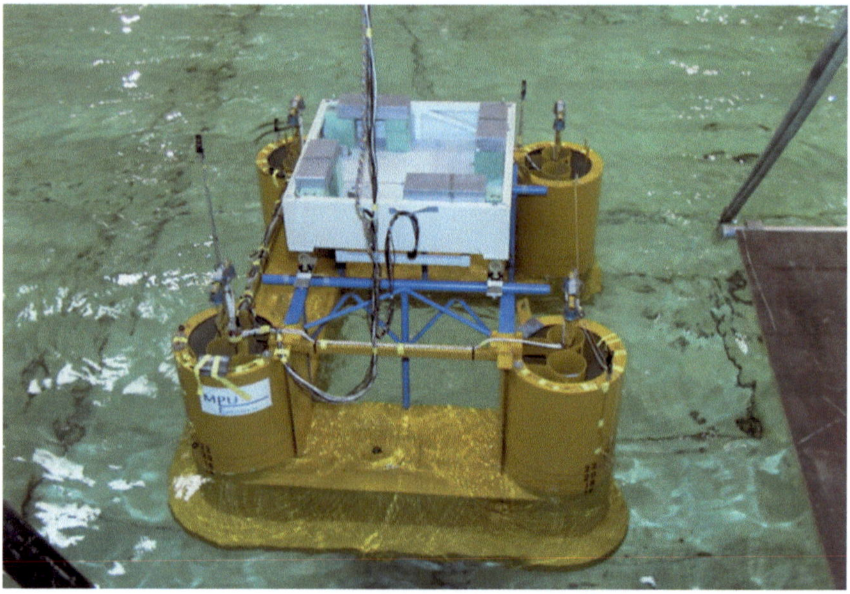

Fig. 30 The Heavy lifter in the tank-test lifting 12,000 tons (MPU 1999)

floater that can survive in arctic conditions. The experience from the hydrodynamic tests of the Heavy Lifter in Figs. 29 and 30 provided robust confidence in the SEMO, shown in Fig. 31. The SEMO is a floating mono-hull structure.

The incentive of the design was the complexity of the moored ship-shaped solution, currently a popular option for offshore field development. Such ships are swivelling around a turret that is moored to the seabed. The design philosophy of the SEMO is that it may be round, and not ship-shaped, as it is not going anywhere. It may also be considered an enlarged turret, without the complicating body of a ship attached to it. Advanced steel structures for the marine environment require competence and capacity not easily available. Building the SEMO in concrete may employ the local construction industry and provide interesting opportunities with regard to local fabrication and assembly. For many countries, it is important to build new industry and further develop the economy, as was the case for Norway when oil and gas was found on the Norwegian Continental Shelf. The fabrication of a SEMO may give significant amount of work locally, which could give such a concept a political advantage compared to solutions built in other countries.

Marine concrete structures are not necessarily huge and complex. One recent concept for the oil and gas industry is a simple floating storage structure, depicted in Fig. 32. Such structures may be utilized in order to save space on land or to enable more efficient transport logistics. They may be made to any size desired, and easily moored. The purpose may be as simple as simply storage of oil. Other usages may include various processing units.

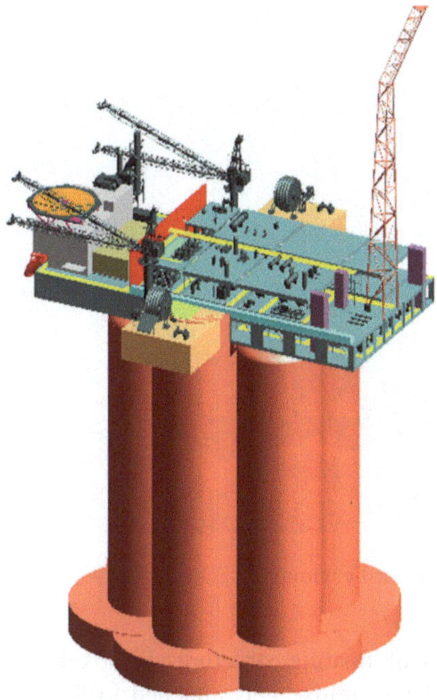

Fig. 31 SEMO, a robust concept for oil and gas production, possibly also in the Arctic (MPU 2007)

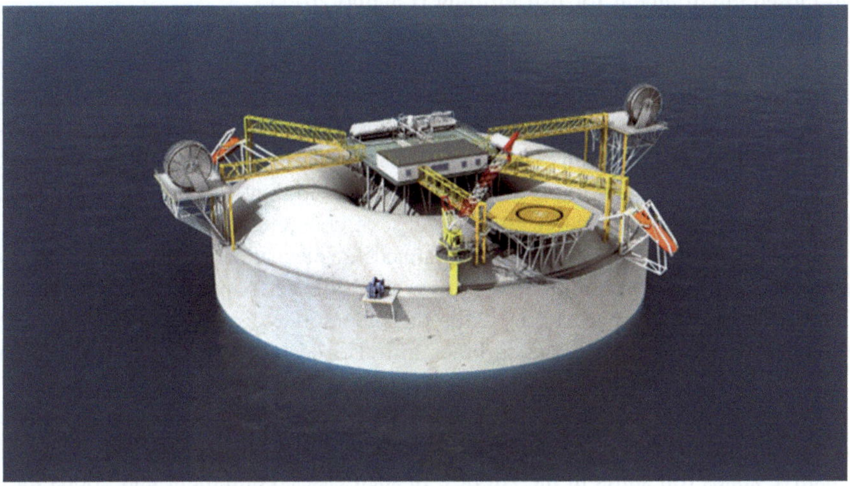

Fig. 32 Simple floating storage structure (OO 2010)

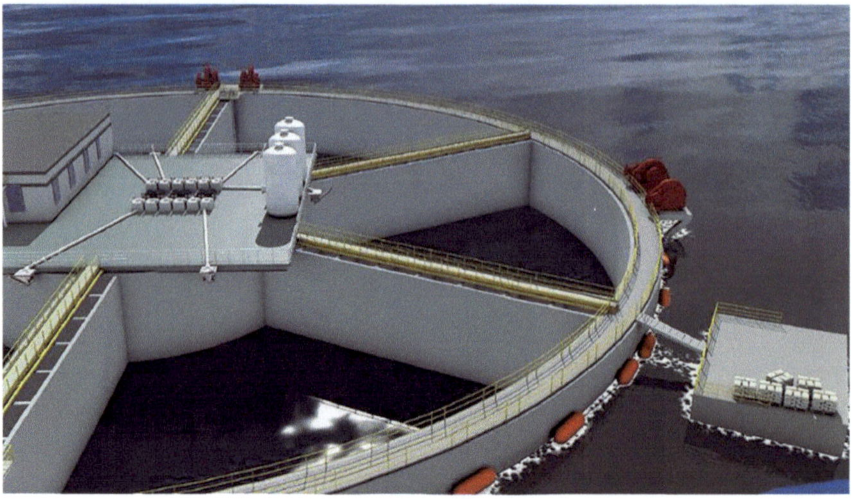

Fig. 33 Fish farming in a closed environment (OO 2010)

Another application of floating structures is novel solutions for fish farms. Present fish farms typically utilise nets attached to floating rings, a rather inexpensive structure. However, there are environmental challenges associated with the design of a fish farm, such as whether the fish should be allowed to escape and mingle with wild fish, or whether contaminated water may enter the open net. Figure 33 illustrates how a marine concrete structure may be utilized as a floating, but closed environment for fish farming (Landbø 2011).

We may also imagine mobile power plants, even nuclear perhaps, and renewable energy production. Offshore structures for renewable energy have been under rapid development the last decade. There is a few operating floating solutions, but several are under development. Figure 34 illustrates the OO-star concept developed at Dr.techn.Olav Olsen. The concept may be built in either steel or concrete. A hybrid foundation solution is also being considered, but the concrete solution is the one being prioritized due to the low cost of concrete.

Figure 35 illustrates wind turbines planned for electrification of a platform, to avoid burning gas. The solution features a gravity base foundation, similar to the Condeep concept that was utilized for the older oil & gas platforms. Like the previous platforms, these wind turbines are also designed to be slip formed in sheltered waters and towed to site (Vici Ventus 2009).

Are there other potential applications? We may imagine terminals placed in catastrophe-prone areas (earthquake, tsunamis), war zones, and environmental sensitive zones (oil spills in the Arctic). In such cases, equipment and goods may be stored on board the terminals, and work force hastily shipped or flown in when activated.

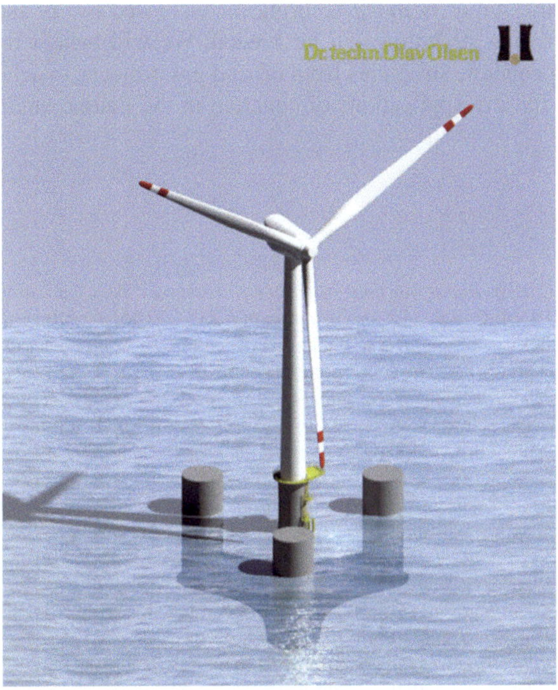

Fig. 34 Floating wind turbine (OO 2014; Landbø 2013)

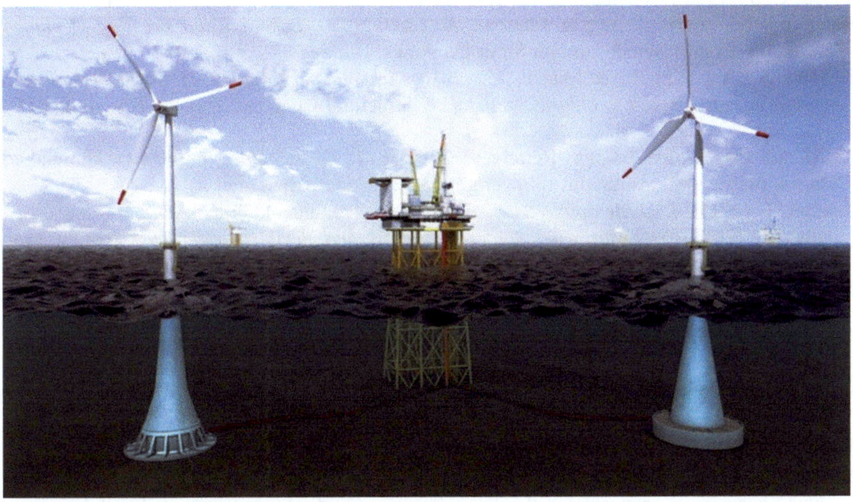

Fig. 35 Concrete structures for offshore wind turbines (Vici Ventus 2009)

In general, the future is likely to bring an increased use of the marine environment and floating structures will be needed. We will benefit from the experience of floating concrete structures from oil and gas projects, to achieve robust and safe structures for any applications of concrete in the marine environment.

References

ACI Committee. (2010). *Report on float-in concrete structures,* ACI Committee 357.

Anderson, A. R. (1976). *Design and concstruction of a 375 000 bbl pre-stressed concrete floating LPG storage facility for the java sea.* Offshore Technology Conference, OTC 2487.

Brandsegg, A. S. et al. (2013). The 6th Conference on*Strait Crossings, extreme crossings and new technologies: Development of a submerged floating tunnel concept for crossing the Sognefjord.*, Bergen, Norway. Arranged by the Norwegian Road Administration and Tekna.

Collins, M. P., & Mitchell, D., 1991. *Pre-stressed concrete structures.* New York: Prentice-Hall Inc.

Derrington, J. (1977). *Pre-stressed concrete platforms for process plants.* London: The Concrete Society in Association with the Royal Institute of Naval Architects.

Ellefsen, H. (1972). *Photo of the Condeep concept.* The contractor Høyer Ellefsen was the inventor of the Condeep.

ESA. (2012). *ESA—ship arctic.* Retreived from December 1, 2014, from http://www.esa.int/spaceinimages/Images/2012/07/Ship_Arctic

FIB. (2009). *Concrete structures for oil and gas fields in hostile marine environments.* Lausanne, Switzerland: Federation International du béton.

Fjeld, et al. (1994). *The north sea concrete platforms—20 years of experience.* Houston: OTC.

Gazprom. (2013). *Gazprom—Sakhalin II.* Retreived December 1, 2014, from http://www.gazprom.com/f/posts/91/279421/dsc_4851_1.jpg

Gloyd, C. S. (1988). Concrete floating bridges. *Concrete International10*(5), 17–24.

Haugerud, S., Olsen, T. O., & Muttoni, A. (2001). *The Lake Lugano crossing—technical solutions.* Fourth Symposium on Strait Crossings, Norway.

Helland, S., Aarstein, R., & Maage, M. (2010). *In-field performance of North Sea offshore platforms with regard to chloride resistance.* Structural Concrete, Thomas Telford and FIB, 11, No1.

KHM. (2013). *Gokstadskipet.* Retreived October 1, 2014, from http://www.khm.uio.no/besok-oss/vikingskipshuset/utstillinger/gokstad/2-gokstadskipet.html

Landbø, T. (2011). *Lukkede oppdrettstanker i betong.* Trondheim: TEKMAR.

Landbø, T. (2013). *OO star wind floater—a robust and flexible concept for floating wind.* Stavanger: s.n.

Morgan, R. G. (1975). *History of and experience with concrete ships.* Berkeley, California: Ben C. Gerwick jr.

Morgan, R. G. (1977). *Development of the concrete hull. Concrete afloat.* London: The Concrete Society in association with the Toyal Institution of Naval Architects.

MPU. (1999). *Photo taken during wave tank testing of the MPU Heavy Lifter.* OSLO: MPU Enterprise.

MPU. (2007). *Illustration created by resources at MPU.* Norway: MPU Enterprise.

Muttoni, A., Haugerud, S. A., & Olsen, T. O. (2001). *The New AlpTransit railway across the alps—a crossing proposal for the Lake Lugano.* Fourth Symposium on Strait Crossings, Norway.

Nilssen, S. (2008). *Photo of Oseberg A.* Norway: Hydro.

Nordenson, G., Seavitt, C., & Yarinsky, A. (2010). *On the water—Palisade Bay.* New York: MoMA.

NS. (1998). *NS 3473 Norwegian standard: Concrete structures—design rules* (5th edition). Norwegian: Standard Norge.

Nyhus, B. S. (2014). Consistent practical design of concrete structures. *fib Structural Concrete in the fall of 2014* (to be published).

Olsen, T. O. (2012). *Photos from personal gallery*. Oslo: Dr. Techn. Olavolsen AS.

Olsen, T. O., Sharp, A., & Mainwaring, G. (2013). *Ultra-long undersea tunnels*. Bergen: Strait Crossings, Extreme Crossings and New Technologie.

OO. (2007). *Illustrations created by resources at Dr.techn. Olav Olsen*. Oslo: Dr. techn. Olav Olsen.

OO. (2010). *Illustration produced by resources at Dr.techn.Olav olsen AS*. Oslo: Dr.techn.Olav olsen AS.

OO. (2012). *Photoes taken by representatives of Dr.techn. Olav Olsen AS*. Oslo: Dr.techn.Olav olsen AS.

OO. (2014). *Illustration produced by resources at Dr.techn.Olavolsen AS*. Oslo: Dr.techn. Olav Olsen AS.

OO., & Reinertsen. (2012). *Illustration created during collaborative work between Dr.techn. Olav Olsen and Reinertsen*. Oslo: Dr.techn. Olav Olsen and Reinertsen.

Sandmæl, F. (2013). *Yogamid, located outside Monaco, private communication*.

Sandvik, K. et al. (2004). *Offshore structures—a new challenge*. XIV National Conference on Structural Engineering, Acapulco.

SEIC. (2006). *Photo taken by resources at Sakhalin Energy Investment Company Ltd*. Russia: SEIC.

Selmer, F. (1988). *Statpipe—shore approach immersed concrete tunnel*. Skanska Norway AS: Ingeniør F. Selmer A/S.

SVV. (2013). *3D visualization of the proposed floating tunnel across Sognefjorden*. Norwegian: Statens Vegvesen.

The Norwegian Concrete Association. (1988). *Norwegian concrete engineering. Concrete for the world*. Oslo: Norwegian Concrete Association.

The Norwegian Concrete Association. (1990). *Norwegian Concrete engineering. Concrete for the world*. Oslo: Norwegian Concrete Association.

The Norwegian Concrete Association, Byggeindustrien and the Norwegian fib delegation. (2012). *CONCRETE under the Northern lights*. www.bygg.no

UHPFRC 2013 (2013). *Proceedings of the RILEM-fib-AFGC International Symposium on Ultra-Hibh Performance Fibre-Reinforced Concrete UHPFRC*, Marsielle: Edited by Toutlemonde and Resplendino. RILEM Publications S.A.R.L.

Valenchon, C., & Nagel, R. (1995). *The N'kossa concrete oil production barge*. OMAE, 14th International Conference, Copenhagen.

Van Zaal, R. (2009). *Airial photo of the Adriatic LNG terminal*. Italy: LNG Terminal in the Adriatic.

Vici Ventus. (2009). *Offshore wind turbines: Concrete foundations*. Vici Ventas Retreived January 10, 2014, from http://www.viciventus.no/getfile.php/Dokumenter/GBF_product_sheet_231111.pdf

Mega-Float

M. Fujikubo and H. Suzuki

Abstract The *Mega-Float* project is a pontoon-type Very Large Floating Structure that was designed for deployment in protected waters such as in a large bay. It consists of a floating structure, a mooring system, and access infrastructure. While construction of a breakwater was contemplated, the existing breakwater around the site was found to be sufficient. Unlike conventional ships where the response is dominated by rigid-body motions, the response of the *Mega-Float* is dominated by hydroelastic responses because of its thin mat-like configuration. A 6-year research project was conducted in Japan with the aim of developing the fundamental design, construction, and operational technologies in Phase I (1995–1997) and a corroborative study on the use of the Mega-Float as an airport in Phase II (1998–2001). This chapter gives an overview of the Mega-float research project including its purpose and achievements.

1 Introduction

Japan is an island nation with a scarcity of land for development. Owing to Japan's needs for ocean space utilization, the Technological Research Association of Mega-Float (TRAM) was founded in 1995 with the support of the Ministry of Land, Infrastructure, Transport and Tourism, Japan. TRAM conducted research and development on the Mega-Float until 2001. Mega-Float is a pontoon-type VLFS which is cost-effective, competitive and suitable for development in

M. Fujikubo (✉)
Department of Naval Architecture and Ocean Engineering, Graduate School of Engineering, Osaka University, 2-1 Yamadaoka, Suita, Osaka 565-0871, Japan
e-mail: fujikubo@naoe.eng.osaka-u.ac.jp

H. Suzuki
Department of Ocean Technology, Policy, and Environment, Graduate School of Frontier Sciences, University of Tokyo, 5-1-5 Kashiwanoha, Kashiwa, Chiba 277-8563, Japan
e-mail: suzukih@k.u-tokyo.ac.jp

C.M. Wang and B.T. Wang (eds.), *Large Floating Structures*, Ocean Engineering & Oceanography 3,
DOI 10.1007/978-981-287-137-4_8

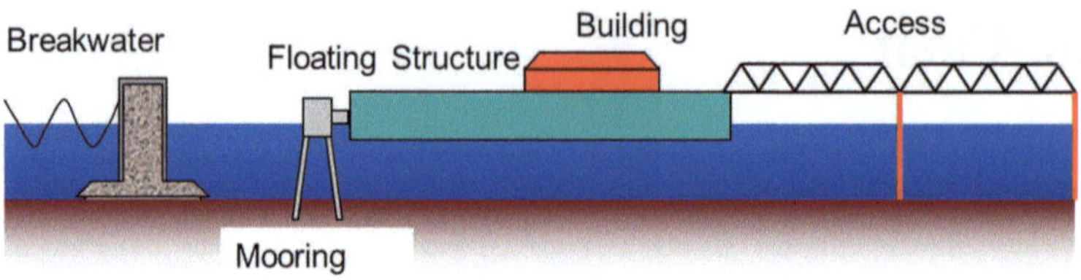

Fig. 1 Concept of Mega-Float

protected waters such as in a large bay. The concept of the Mega-Float is shown in Fig. 1. It consists of a floating structure, a mooring system and access infrastructure. If necessary, breakwater construction is considered.

The Mega-Float research project was performed in two phases (Isobe 1999). In Phase I (1995–1997), the fundamental design, construction and operational technologies were developed, while in Phase II (1998–2001), a corroborative study on the use of the Mega-Float as an airport was performed. VLFSs are so large in the horizontal plane that dynamical elastic deformations are dominant as compared to rigid motions under wave action. Interactions between hydrodynamic pressure and elastic deformation are therefore essential for their dynamic responses of VLFSs. Many kinds of numerical methods have been developed to perform the hydroelastic analysis of pontoon-type VLFSs. Onsite experiments were performed in the Tokyo bay to demonstrate the soundness of the technology with a Phase-I demonstration platform measuring 300 × 60 × 2 m, and a Phase-II demonstration platform measuring 1000 × 60 m (max. 121 m) × 3 m, as shown in Fig. 2. Takeoffs and landings of small aircrafts were successfully demonstrated in the Phase II experiment.

This chapter gives an overview of the Mega-float project including its purpose and general achievements. Also, more details on several key technologies developed during the Mega-Float project are presented.

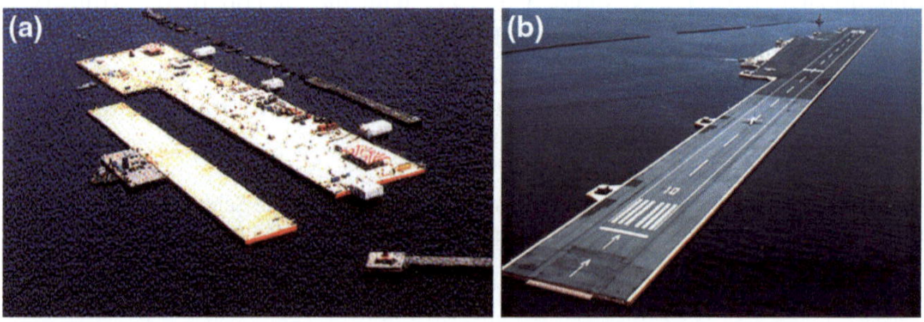

Fig. 2 Mega-float experimental models. **a** Phase I model. **b** Phase II model

2 Overview of Mega-Float Project

The concept driving the design of the Mega-Float was that of a very large mat-like structure used as a floating platform for various human activities. Owing to its unprecedented length envisaged for the structure, the geometrical configurations and an expected long-term service period, there were many technological challenges associated with the design, construction and operation of the Mega-Float.

The largest single-body floating structure currently existing in Japan is the Kamigoto Floating Oil Storage Base, which is 390 m in length and 97 m in width, as described in Chap. 4. The entire oil storage structure could be constructed in a ship building dockyard by using conventional large-scale ship-building technologies. On the other hand, the entirety of the Mega-Float far exceeds the size of the building dockyards and therefore modular units of the Mega-Float were constructed and towed to the site and assembled at sea. This required the technological feasibility of the large-scale assembly work at sea to be investigated against new standards of construction quality, safety, efficiency and cost.

This assessment needed to take into account the structural characteristics of the proposed project. In the case of conventional ships and floating offshore structures, the dynamic response is dominated by rigid-body motions, whereas the dynamic response of VLFSs is dominated by the hydroelastic response. The fluid-structure interaction effects needed to be considered at every design stage of the floating structure. This meant that hydroelastic response analysis programs need to be developed with various complexities and levels of modeling.

Long term durability with minimal maintenance cost was another challenging design aspect that had to be solved in order for the Mega-Float to be viable as a floating piece of infrastructure such as an airport, bridge, and storage base. The assessment of the environmental impact due to the presence of the floating structures was also an essential process to ensure that this novel type of structure gained acceptance in society.

Figure 3 summaries the key research areas for realizing the Mega-Float (Sato 2003). In addition to the technological subjects, the development of the safety and design guidelines and the establishment of approval process by government were included in the major subjects. A 100-year service life was taken as the target life for long-term operational technologies.

The budgets and activities of the 6-year Mega-Float project are shown in Table 1. The research and development programme comprised two phases. In Phase I, the fundamental technologies of design, construction and operation were developed. The research subjects are divided into five groups:

(1) Floating structure design technology
(2) Floating construction work technology
(3) Long service duration technology
(4) Operational function ensuring technology
(5) Environmental impact assessment technology

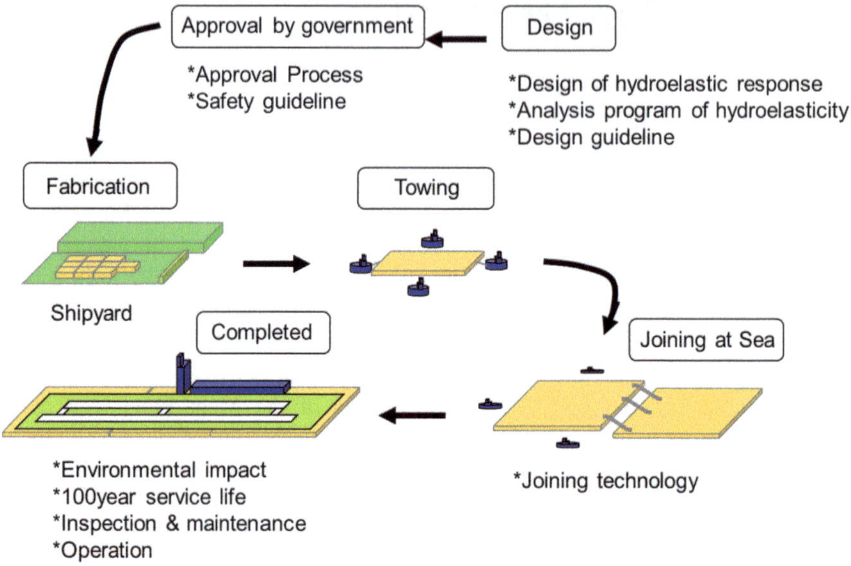

Fig. 3 Research areas for realizing Mega-Float (Sato 2003)

Table 1 Budgets and subjects of phase-I and phase-II research

	Phase I (1995–1997)	Phase I (1998–2000)
Objective	Establish basic technology	Establish airport construction technology
Experiment	300 m long model	1,000 m long model
	Joining of units at sea	Joining of units at sea
Research	• Design	• ILS test
	• Fabrication and joining at sea	• Landing and take off of airplane
	• Operational requirement	• Concept study
	• Environmental impact	• Legal aspect
Budget	\$68.2 million	\$103.6 million

Onsite experiment using a 300 m long model, as shown in Fig. 2a, was conducted and the soundness of the developed technologies is demonstrated.

Following the successful completion of the Phase-I research programme, the Phase II was conducted as a corroborative study on the use of the Mega-Float as an airport (Sasajima 2002). The 1000 m long × 60 m wide (maximum 121 m wide) × 3 m deep floating structure with a 900 m long × 25 m wide runway was built off Yokosuka city for the planned experiments. The airport model consisted of a floating structure, a dolphin mooring system, and existing breakwater, and was designed by using the results of Phase-I, i.e. hydroelastic response analysis, local structural analysis, and studies on the use of VLFS for the airport. The 300 m long

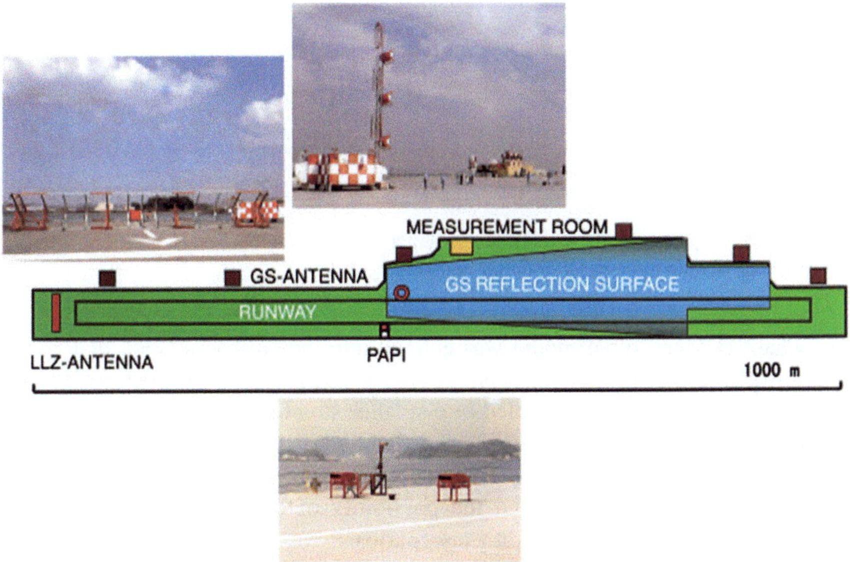

Fig. 4 Airport equipment of Phase-II experimental model (Sasajima 2002)

floating model of Phase-I was utilized for the 121 m wide part. Figure 2b shows a photograph of the Phase-II experimental site.

Amongst many possible applications of the Mega-Float, the floating airport was selected as a target application. This was because it required a very severe operational limit when compared to other applications. Slight dynamic responses of the elastic floating structure can sensitively cause directional deviations of electronic or light beams emitted by the sensitive airport electronic and optical devices. Figure 4 shows the arrangement of the equipment to check the functions of the airport. The function of the Instruments of Landing System (ILS) was verified, which consists of a Localizer (LLZ) for sending out electronic waves to indicate horizontal approach directions, a Glide Slope (GS) including vertical approach angles, and the Precision Approach Pass Indicator (PAPI) which is the optical device to give an airplane a set of 2 or 4 different-colored light beams to indicate the approach angle (Sasajima 2002). An airport function simulation program was developed to check the influence of the deformation of the experimental model of Phase-II. It was confirmed that the actual measurements coincided with the results of the analysis of the experimental model. Figure 5a shows a snapshot of the actual measurement by an approaching airplane, and Fig. 5b the take-off and landing experiments.

The research into the technologies (1)–(5) as mentioned above were continued in Phase II as well to obtain further advances in each categories. The major achievements of the research are presented in the following sections. These findings are expected to form the basis for the design, construction and operation of the prototype floating airport and other structures in future. In 1999, the 1,000 m

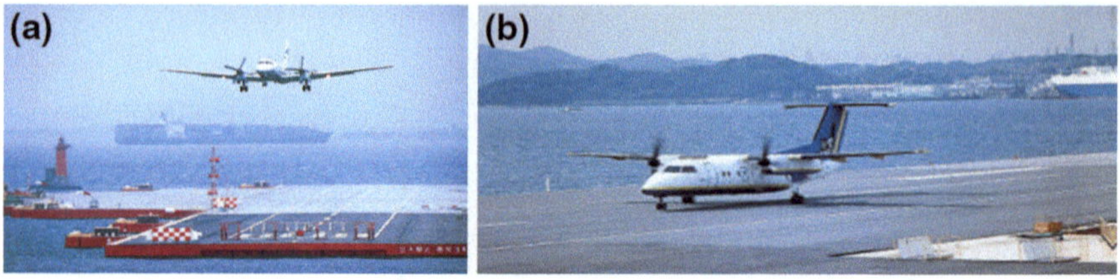

Fig. 5 Airport equipment of Phase-II experimental model (Sasajima 2002). **a** Actual measurement by approaching airplane. **b** Take-off and landing test

long Phase-II Mega-Float model was entered into the Guinness Book of Records as the world largest man-made floating island. After the Mega-Float project, the floating structure was disassembled into smaller structures and reused for various applications, such as a floating fishing park and a floating pier. This demonstrated one of the advantages of floating structures—a versatility that "future-proofs" the investment in the structure given the possibility of dismantling and separating individual modules for reuse in other scenarios.

3 Hydroelastic Behavior

3.1 General Characteristics of Hydroelastic Behavior

Unlike conventional ships and floating offshore structures, an elastic response is dominant for VLFS. A comparison of static response under a concentrated load is illustrated schematically in Fig. 6. As a rational measure to distinguish VLFSs from conventional ships and floating offshore structures in terms of global response, Suzuki et al. (1996) derived analytically a characteristic length λ_c from a uniform beam model on an elastic foundation. This characteristic length λ_c is given by

$$\lambda_c = 2\pi \left(\frac{EI}{k_c} \right)^{1/4} \tag{1}$$

where *EI* is the bending stiffness of beam and k_c the spring constant of hydrostatic restoring force. λ_c corresponds to the length of locally deflected region by a static concentrated load, as shown in Fig. 6. This indicates that the influence of an applied load on the elastic deformation is limited within the region of the length λ_c. If the length of structure is smaller than the characteristic length, the response is dominated by rigid-body motions, whereas if it is larger than the characteristic length, as typically in VLFS, the response is dominated by elastic deformations. The relationship between the wavelength and the characteristic length is another important factor on the global response of floating structures. If the wavelength is smaller than the characteristic length, the wave exciting forces alternates in the

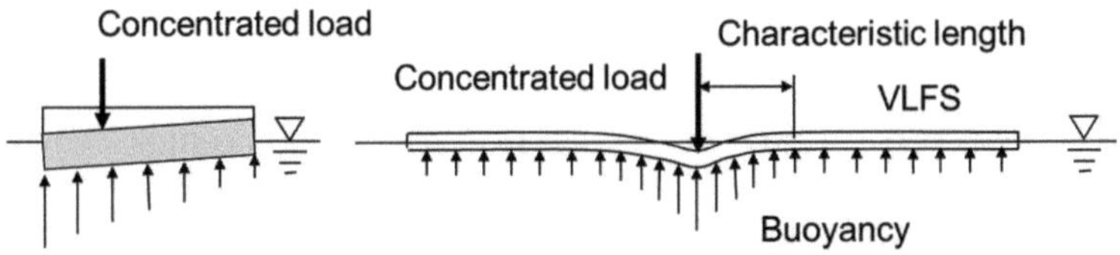

Fig. 6 Global response under a static load

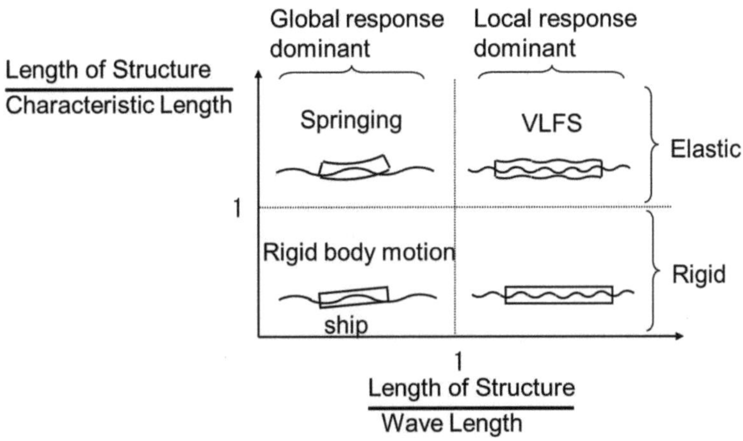

Fig. 7 Mapping of global response of floating structures

range of the length λ_c and the load effect cancels each other; thereby resulting in a smaller global response. On the other hand, if it is larger than the characteristic length, the global response becomes significant. These characteristics of the global response with respect to the characteristic length are summarized in the map given in Fig. 7. In summary, VLFS can be characterized by its huge structural size compared not only to the wave length but also to the characteristic length λ_c. Equation (1) may be used to define VLFS in the sense of mechanical behavior.

3.2 Analysis and Design Tool for Hydroelastic Behavior

The dynamic behavior of the Mega-Float is characterized by its hydroelastic behavior. Several computer programs for the hydroelastic global response analysis of a pontoon-type VLFS were developed with various complexities and levels of the fluid and structure modeling, both inside and outside TRAM (TRAM 1999a, 2000, 2001). Some of the analysis programs developed for global-response analysis are presented in Table 2. These ranges from the simplest 1-D beam and 2-D fluid models (Program A) to the most complex 3-D FE and 3-D fluid models capable of considering the variable draft and water depth and the arbitrary planar

Table 2 Analysis programs developed for global response analysis

Program	A	B	C	D	E
Fluid domain	2-D DD	3-D DD	3-D DD	3-D BEM and FEM	3-D DD and FEM
Water depth	Uniform	Uniform	Uniform	Uniform	Variable
Draft	Uniform	Uniform	Uniform	Uniform	Variable
Structure	Beam	Plate	Plate	FEM	FEM
Shape		Rectangular	Combination of rectangular	Arbitrary	Arbitrary
Stiffness	Uniform	Uniform	Uniform	Variable	Variable
Mass	Uniform	Uniform	Uniform	Variable	Variable
Breakwater			Considered		Considered
References	Yamashita and Harada (1997)	Ohmatsu (1997)	Ohmatu (2000), Utsunomiya and Watanabe (1998)	Yasuzawa et al. (1997)	Seto et al. (2003)
				Yago and Endo (1996)	

DD domain decomposition method

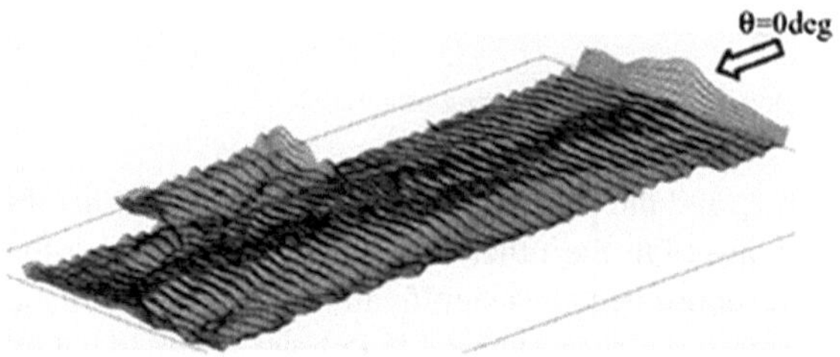

Fig. 8 Deflection amplitude distribution of 4,000 m long floating airport model (wave period = 6 s)

shape of floating structure and fluid boundaries including a breakwater scenario (Program E). An example of the deflection amplitude distribution of a 4,000 m long floating airport model in longitudinal waves, calculated by the program E developed by Seto et al. (2003), is shown in Fig. 8.

A uniform beam or plate model is effective for determining the global strength and stiffness of structures required to meet strength and operational criteria at the initial design stage. For the subsequent design stage, more detailed structural models that can deal with structural member behaviors are needed. Figure 9 shows a typical structural arrangement of a pontoon-type VLFS. It consists of a deck, a bottom and longitudinal and transverse bulkheads of a grillage form. A large opening may be located at the bulkheads for the usage of internal spaces for various functions such as a transportation system for airport passengers. A

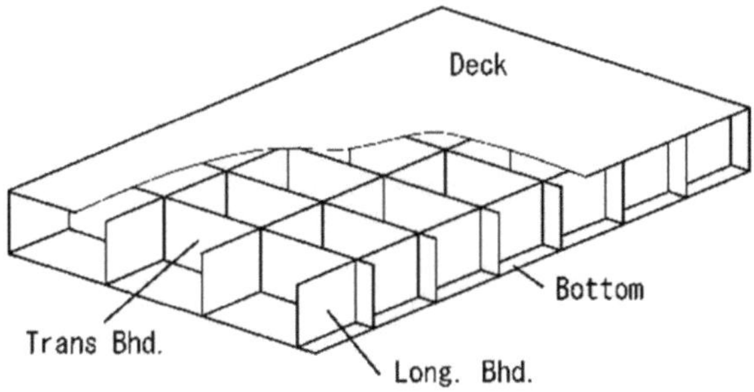

Fig. 9 Typical structural arrangement of a Pontoon-type VLFS

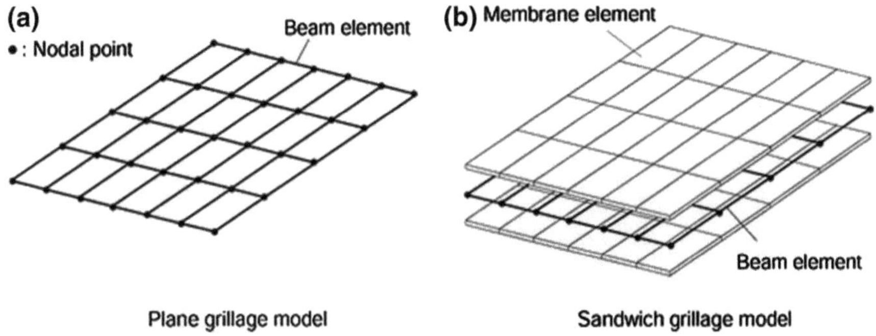

Fig. 10 Typical structural arrangement of a Pontoon-type VLFS

3D-shell FE model is the most accurate model for structural analyses, but owing to a huge structural size of VLFSs, the application of the 3D-shell FE model to the entire structure is limited. The grillage models as shown in Fig. 10 are simpler and a more rational approach to modeling VLFSs. They can account for both bending and vertical shear stiffnesses in a rational manner including the effect of openings. The details about the grillage models are found in Fujikubo and Yao (2001). For fatigue strength assessment, a more detailed structural model is required. Inoue (2003) proposed a method for predicting the stress response of local structures of a VLFS in multi-directional waves by using zooming FE analyses of local structures (see Fig. 11) under the representative loads effects calculated by the hydroelastic global response analysis.

The hierarchical system of structural response analysis from the global hydroelastic response analysis to the local stress analysis, developed in the Mega-Float project, can be applied to any type of VLFSs.

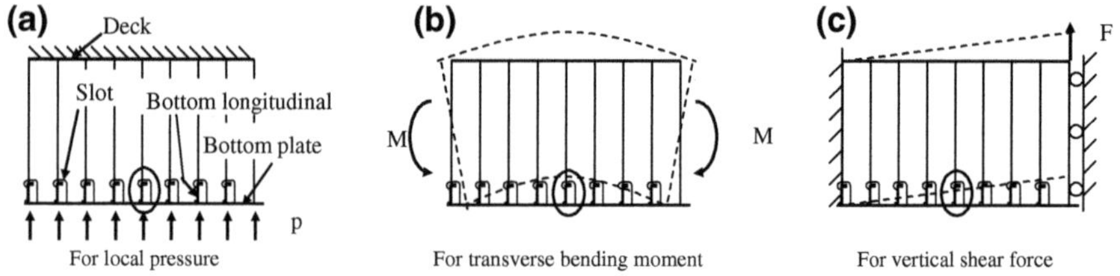

Fig. 11 Local model to calculate stress factors of slots of bottom longitudinals (Inoue 2003)

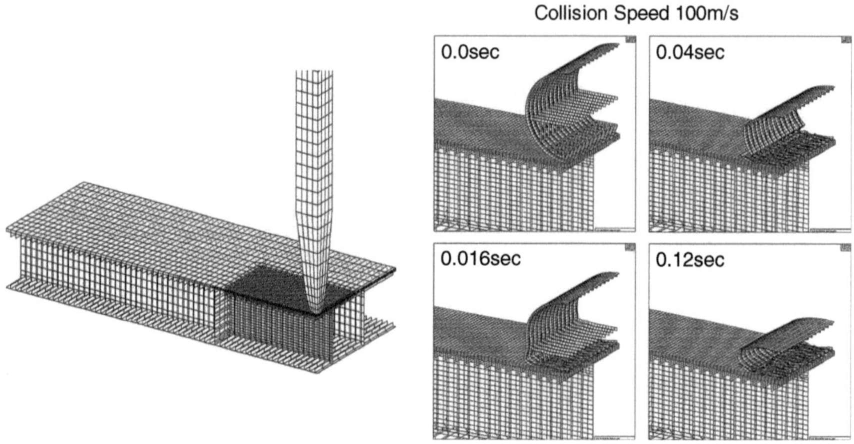

Fig. 12 Simulation model and results crash of fuselage (TRAM 1997)

4 Design Methodology

4.1 Limit States

In order to ensure the structural integrity of VLFS, failure modes and associated design limit states are to be adequately identified from both component and system levels. Table 3 summarizes the limit states defined in the safety guideline of Mega-Float (TRAM 1999b).

The allowable strength limit state is intended to confirm that a structure has strength for safe use throughout the service period with an appropriate reserve. Buckling and yielding strengths of principal structural members are evaluated based on their hydroelastic behaviors. Mega-Float type VLFS is characterized by its thin mat-like configuration, and the deck and bottom panels are subjected to combined biaxial and shear loads of similar order of magnitude. The buckling and yielding under combined load actions should be more carefully checked when compared to conventional ships. Because of the relatively very small depth of the

Table 3 Design limit states for structural safety

Limit states	Definition	Characteristic value of environmental loads	
		Case A	Case B
Allowable strength limit state	A state in which a structure or principal structural member reaches strength limit state, where a slight damage will not hinder the use of the structure	Class 1	
			Class 1
Fatigue limit state	A state in which cracks initiate and propagate due to fatigue caused by a repeated load and the load bearing capacity of the member starts to decrease	Expected load frequency	
Ultimate strength limit state	A state in which a structure or a principal structural member reaches a full collapse state, where a structural damage dangerous to human lives, a sinking or a drifting occurs	Class 2	
			Class 2
State after partial damage	A state in which load redistribution due to some damage makes the other intact structural members or structure damaged	Class 0	

Note
Case A: wave, wind, current and others excluding Case B
Case B: earthquake, seaquake, tsunami and storm surge
Class 0: return period of 2 years
Class 1: return period of at least 2 times service year
Class 2: return period of at least 100 times service years

structure, the shear strength of bulkheads should be also carefully examined, particularly when a large opening is present in the bulkheads.

Fatigue strength limit state requires the rational design and maintenance in order to prevent the failure due to repeated loads. Operational loads may have more significant impact on the fatigue strength than environmental loads in some parts of VLFS depending on the operational and environmental conditions. For instance, the repeated load due to landing and takeoff of airplanes is a major load effect on the runway part. These operational loads are more predictable and thus less uncertain than environmental loads in general. The safety measure for the fatigue strength limit state should be adequately made by considering these characteristics in operation and structural location.

Two types of system-level limit state checks are specified for the Mega-Float. One is the ultimate strength limit state which intends to confirm that a progressive collapse of an intact structural system under abnormal load effects due to waves, earthquakes, and tsunamis does not achieve a full collapse state where structural damage may lead to a loss of human life, sinking, or drifting. The other is the residual strength check of a structural system in a damaged condition, such as due to fuselage crashes and ship collisions. Figure 12 shows the simulation results from a vertical fall of the fuselage.

From the novel nature of VLFS, these system-level checks require an identification of failure scenarios and an evaluation of associated risks, which can give a quantitative measure of the safety. From this viewpoint, the recommendation of

Table 4 Typical serviceability criteria of Mega-Float (TRAM 1999a, 2000, 2001)

Facility	Criteria	Rule
Runway	Slope longitudinal <1.0°, transverse <1.5°, radius of curvature >30,000 m	Airport facility design standard
Taxiway	Slope longitudinal <1.5°, transverse <1.5°, radius of curvature >3,000 m	Airport facility design standard
ILS/GS	Misalignment <0.144°	Civil aeronautics law
PAPI	Misalignment <0.1°	Civil aeronautics law

risk-based evaluation for overall safety was included in the safety guidelines of Mega-Float. Some examples of the overall safety evaluation are presented in Sect. 4.3. In addition to the strength limit states, serviceability limit states are considered depending of the application of the Mega-Float.

The characteristic values of environmental loads corresponding to each limit states are indicated in Table 3. The simultaneous occurrence of the extreme wave, wind and other environmental loads and the abnormal loads such as earthquake and tsunami are not considered because of small probability of occurrence. The Class 1 environmental loads are considered for serviceability limit states.

4.2 Design Criteria

Serviceability and safety criteria were key issues in the design of the Mega-Float. Design-criteria development for the Mega-Float was needed for its intended use as an airport. The basic philosophy related to the serviceability criteria was that Mega-Float's runway had to be equivalent with and conformed to codes for land-based runways. Existing standards and rules, such as the Airport Facility Design Standard, Civil Aeronautics Law, Building Standard Law, and the Standard for Cranes, were investigated. The influence of elastic responses on an Instrument Landing System (ILS), Precision Approach Pass Indicator (PAPI), and Future Air Navigation System (FANS) was investigated using an airline flight simulator. Fluctuations in angles of Glide Slope (GS) signals due to hydroelastic responses of runways were simulated and the influence on controllability of airplanes was investigated (TRAM 1999a, 2000, 2001). Table 4 shows the typical serviceability criteria of Mega-Float that were derived from the research. One of the criteria that was the most difficult to satisfy during the study was that the minimum radius of curvature for the runway should be 30,000 m as reported by Sato (2003).

The study produced the Technical Guidelines of Mega-Float (1999b) that summarized the research findings. Table 5 shows the contents of the guideline. It covers the design guideline for a floating structure, a mooring system, wave control facilities and their quality control and maintenance. In relation to the structural safety criteria of the floating structure, the design load conditions and the associated safety factors are specified. Considering the novel nature of the structure, it is recommended for the allowable strength limit state, which employs a

Table 5 Contents of technical guidelines for Mega-Float (TRAM 1999b)

Volume 1	General rules
Volume 2	Environmental impact assessment
Volume 3	Materials
Volume 4	Design load
Volume 5	Hull structures
Volume 6	Station keeping facility
Volume 7	Wave control facility
Volume 8	Disaster prevention measures
Volume 9	Quality control for construction works
Volume 10	Maintenance and inspection
Volume 11	Overall safety evaluation

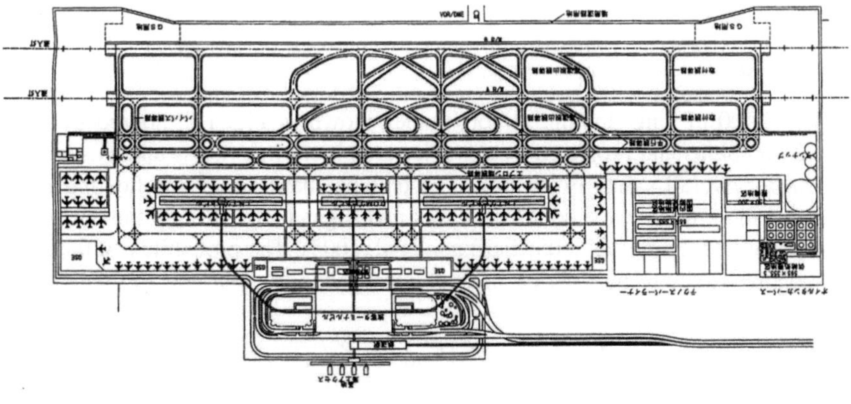

Fig. 13 4,000 m class floating airport model for design and simulation

partial safety factor format, to verify the load and strength factors using the structural reliability approach when considered to be necessary.

Based on the guidelines, a 4,000 m generic airport, shown in Fig. 13 was designed (Sato 2003; TRAM 2000). One notable item in the guideline was the recommendation of risk-based evaluation for overall safety.

4.3 Overall Safety Evaluation

The aim of overall safety evaluation was to identify the catastrophic failure scenarios on a system level, to quantify the associate risks, and to mitigate the risks when they exceed the acceptable level. This evaluation had to encompass the various sub-systems of the Mega-Float, such as a floating structure, a mooring system, a breakwater and an access system. The worst system failure scenarios were identified from possible scenarios by expert investigation (Fig. 14).

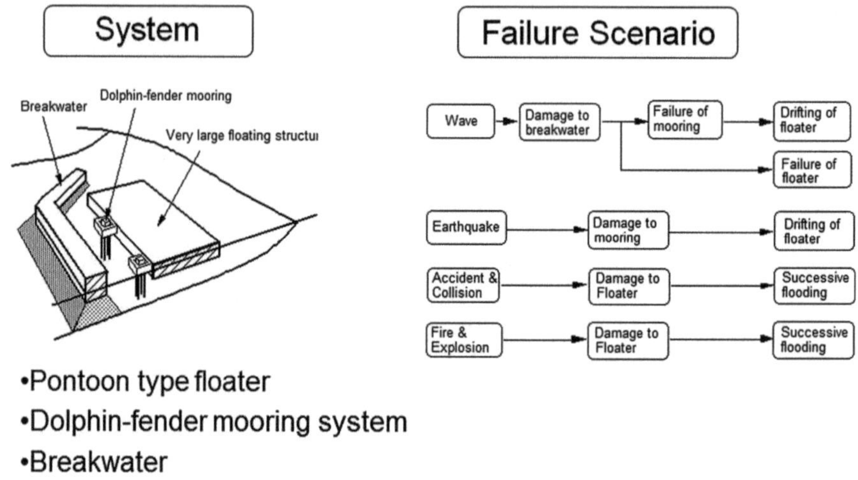

Fig. 14 System failure scenario of Mega-Float

The 4,000 m long floating airport model shown in Fig. 13, moored by more than 30 dolphins and protected by a gravity-type breakwater, was taken as a target system for overall safety evaluation. The collapse behavior of the floating structure, due to extreme wave conditions, was investigated considering the damage to the breakwater (Fujikubo 2005; Fujikubo et al. 2005). The probability of failure of the floating structure in a period of 200 year service life, P_f^s, was evaluated as

$$P_f^s = \left(1 - P_f^b\right) P_{intact}^s + P_f^b P_{damaged}^s \tag{2}$$

where P_f^b is the probability of the overturning of breakwater, and P_{intact}^s and $P_{damaged}^s$ are the conditional failure probability of the floating structure in the intact and damaged conditions of breakwater, respectively. For progressive failures of the mooring system, which may lead to a drifting of Mega-Float, the collapse behavior of a single dolphin was investigated by pushover analysis. Based on these results, the system failure was investigated (Kato et al. 2001)

On the other hand, the target safety level for a VLFS when it is used as an international airport was discussed by Suzuki (2001). First, the safety level of several activities in Japan, such as transportation, accidents, and natural disasters, was examined and expressed in terms of fatal accidental rate (FAR, the number of victims per 100 million man-hours of activity). By considering the particular features of a VLFS, such as a public use, the involuntary nature of risk, and gross errors, the value of 1×10^{-6} was derived as an allowable annual failure probability of the VLFS, as shown in the risk matrix of Fig. 15. By comparing with the present levels of the risk of loss-of-life investigated by the ISSC2000 (2000), the proposed value almost coincides with the safety level for a hospital, a school, and other public facilities. It is interesting to note that the same value for annual target reliability was arrived at independently for Mobile Offshore Base (Bhattacharya et al. 2001).

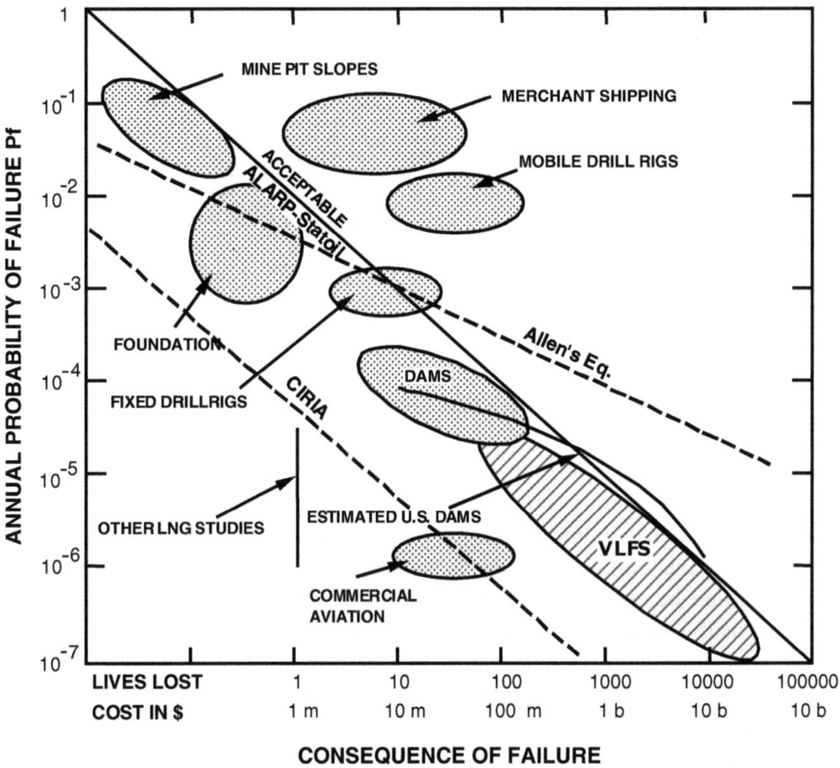

Fig. 15 Target safety level tentatively evaluated for Mega-Float airport (Suzuki 2001)

By comparing with the target safety level, the failure probability of the floating structure and that of the mooring system were found to be within the acceptable level.

5 Floating Construction Work Technology

The construction of the Mega-Float needs the towing of its component units from the building dock to the construction site and assemblage at sea, as illustrated in Fig. 16. Experiments to test the towing of Mega-Float units were carried out during the construction of onsite experimental models (Hara et al. 2003). The effects of unit size and wave conditions on the strain induced in the towed unit were investigated. In Phase II experiments, a 300 m long, 60 m wide and 2 m deep unit was towed 195 miles in the Pacific Ocean from Tsu to the experimental site off Yokosuka. The numerical method for predicting the response of units in wave conditions was verified and a critical wave condition was identified during the operation. Towing resistance and drift force were also investigated (TRAM 2000).

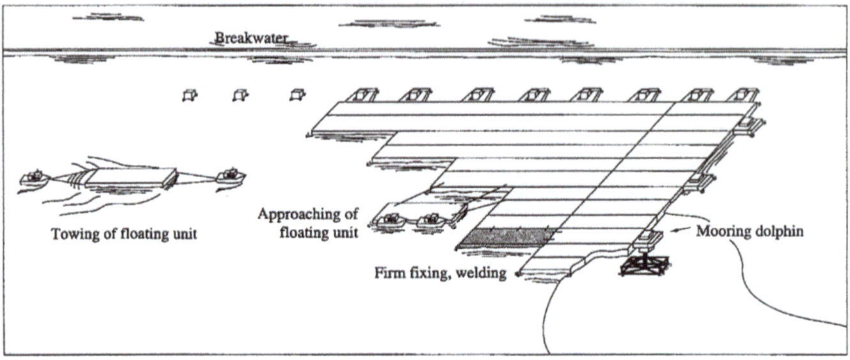

Fig. 16 Conceptual illustration of Mega-Float construction (Torii et al. 2000)

The floating units can be connected either by welding or mechanical joints. In the former method, there is no structural discontinuity and the stress is well distributed, while high reliability of the welding work is required. In the latter, the mechanical connection work takes a comparatively short time and detachable mechanism can be employed. On the other hand, although the units are unified macroscopically, partial discontinuities remain at the connections, which may lead to local concentration of displacements and loads at joint pieces. The mechanical connection is therefore effective only when these drawbacks are acceptable and detachability is required. The welding connection is more suitable for constructing integrated and continuous structures intended for long service life like floating airports. For these reasons, the welding connection was accepted in the Mega-Float.

Special devices for pulling and connecting floating units were developed as shown in Fig. 17. One essential problem is how the welding work could be performed beneath the water level. There are basically two kinds of underwater welding; i.e. dry welding and wet welding. Since the latter requires divers, special welding rods, highly skilled welders and resulting high cost, a bottom-attached air chamber and draining with compressed air were developed (Torii et al. 2000).

The influence of both wave conditions and unit-joining sequences on the responses of structure and performance of construction were investigated (TRAM

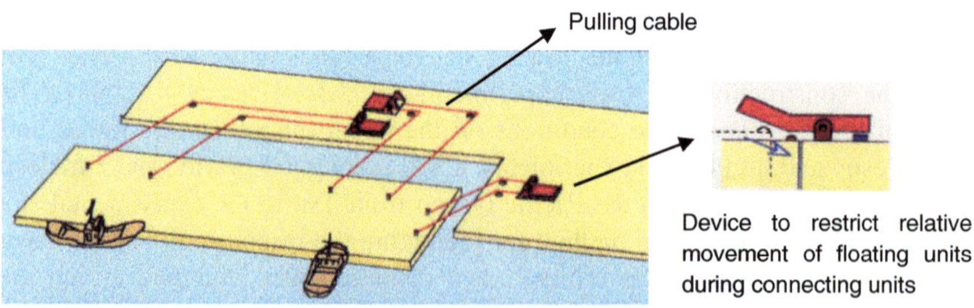

Fig. 17 Method of joining Mega-Float units afloat

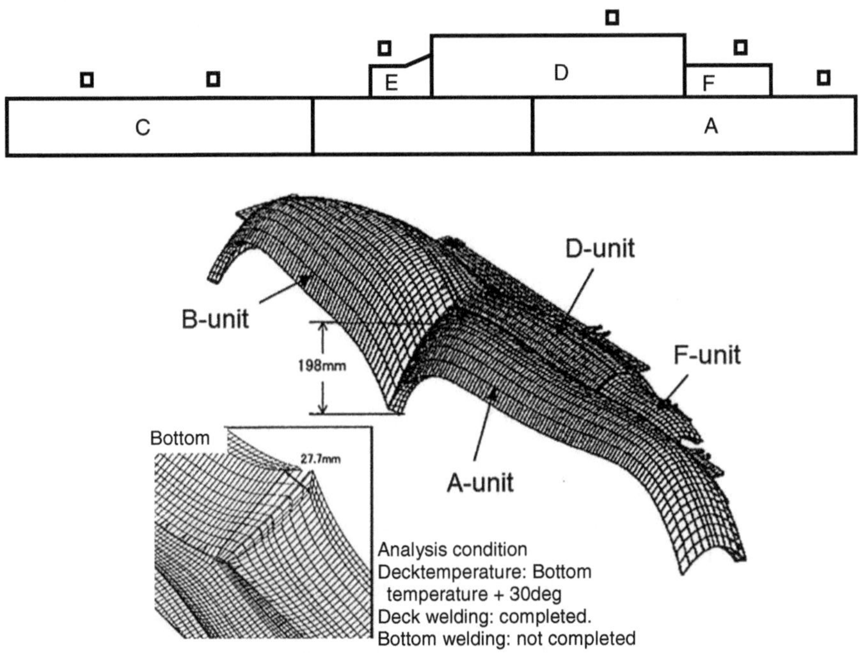

Fig. 18 Calculated deformation of Phase 2 model due to sunlight (Yamashita et al. 2003)

2000). Figure 18 shows the thermal deformation of the Phase 2 model due to sunlight (Yamashita et al. 2003). The estimated maximum deformation of the model was 190 mm. The developed joining technology was tested in the construction of Phase I and Phase II experimental models, with successful achievement.

6 Long Service Duration Technology

A service life of 100 years is expected for VLFS as bases for infrastructure facilities intended for permanent use. Hence, a great deal of importance is placed on technologies to give durability to the structure. These include a corrosion protection system, application of new material, long-term monitoring technology and underwater repair technology (Torii et al. 2000).

Once completed, VLFS cannot be returned to a shipbuilding dock for maintenance. For this reason, a service life basically of 100 years was envisaged in the study of its corrosion protection specifications, which are shown in Fig. 19. The characteristic features of this corrosion protection strategy include: the use of titanium cladding steel plate at the splash zone; combinations of paint coating and sacrificial anodes applied to the bottom and internal ballast tank; and dehumidifying for the void sections. Titanium was chosen from the proven performance

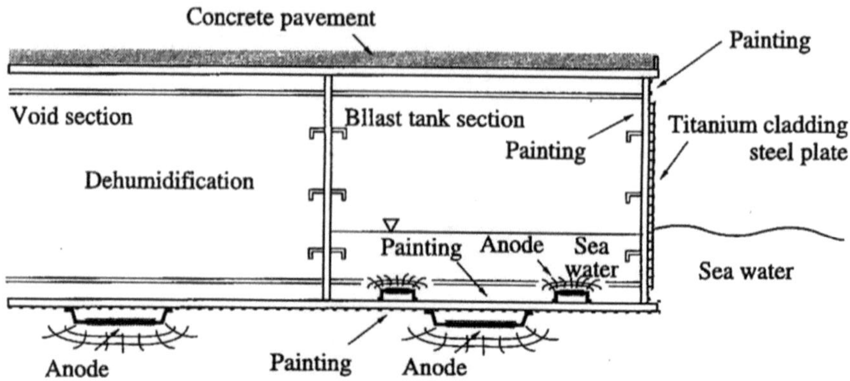

Fig. 19 Corrosion protection system of Mega-Float (Torii et al. 2000)

through its application to the piers of the Trans-Tokyo Bay Roadway Steel Bridge. However, the Mega-Float was the first project to apply the same approach to the welded joints between floating units constructed at sea.

Problems involved in the application of the titanium to the splash zone of the floating structure were investigated from the following aspects: (1) lining work test for verifying workability in the offshore environment with wave motions; (2) development of an automatic vertical welder for the offshore use; and (3) tests of offshore repair work methods. Figure 20 shows an offshore lining work test. For stable and quick welding work of titanium under the offshore working conditions, the double-torch plasma welding method was introduced for the automatic welders. In the tests of offshore repair work, an artificial defect was created on a titanium-cladding plate welded on to the floating structure and a titanium plate patch was welded to cover the defect. It was confirmed that the corrosion resistant materials could be welded with satisfactory quality when a dry chamber is provided to create an atmospheric condition.

7 Environmental Impact Study

A concern of the public in general about large-scale development projects like Mega-Float is its consequential environmental impact. Research into the environmental impact of Mega-Float was carried out to answer this question (TRAM 1999a, b, c, 2001, 2002). The study involved the physical environment, such as flow, salinity, and temperature around the Mega-Float. Water circulation was investigated by numerical simulation and field measurements during the Phase II experiments. The ecosystem around the Mega-Float and the water quality around and below the Mega-Float were investigated both experimentally and numerically. Few changes were observed in various indices of either the physical or ecological environments, except for a small reduction in dissolved oxygen (DO) adjacent to the floating structure as a result of the

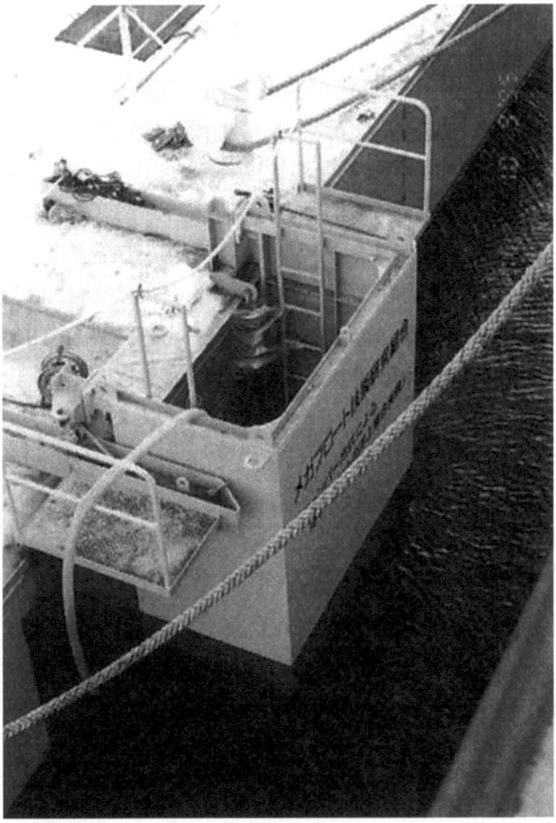

Fig. 20 Offshore lining work test for demonstration (Torii et al. 2000)

activity of attached organisms. Interestingly, organisms adhering to floating body have a fish-attracting effect. However, excessive organisms should be prevented from fouling the Mega-Float. Since it was verified in Phase I that forming an air layer on the bottom surface prevented organisms from fouling the bottom surface, it was applied to Phase II research. It was confirmed that the idea was effective for the environment around the floating structure (Sasajima 2002; Sato 2003).

8 Legal Jurisdiction and Government Approval

It was envisaged that a proposal for Mega-Float infrastructure project would be evaluated and approved by an authority solely responsible for overseeing the entirety or part of the Mega-Float. As the approvals process through the government channels and the legal challenges the project might face would differ depending on the relevant piece of legislation that applied, a survey of the various regulatory regimes were investigated.

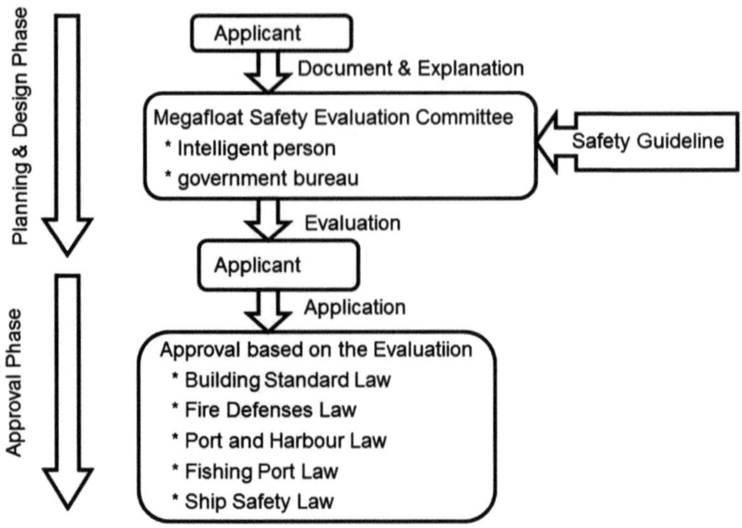

Fig. 21 Approval process in government

The general layout and planning approval of Mega-Float had to be compliant with regulatory requirements arising from both the Port and Harbor Law and the Fishing Port Law. The internal component buildings on the Mega-Float were regulated by both the Building Standard Law and the Fire Defenses Law. In addition, floating structures are regulated by the Ship Safety Law.

Given the interaction of these regulatory regimes, a Mega-Float Safety Evaluation Committee was proposed and accepted by the government. Experts and all government bureaus in charge of the approval gathered in the committee and evaluated the application. Once the plan was judged to be acceptable, each bureau approves the plan. Figure 21 shows the process of government approval.

The Safety Evaluation Guideline of Mega-Float, which forms the backbone of the committee, was investigated, proposed and accepted by the government bureaus. The contents of the guideline are shown in Table 5. Although each elemental structure is designed by a well-established design rule, it is necessary to confirm that the whole system satisfies expected safety levels. A risk-based safety evaluation of the total system was included in the guideline.

9 Semi-Submersible-Type Mega-Float

A Semi-Submersible-type Mega-Float (SSMF) was also investigated (Yoshida et al. 2001). SSMF is supported by many column footings or lower hulls and has good response characteristics in waves. A SSMF is expected to be an efficient form of floating structure that can be utilized in open seas, such as the Exclusive

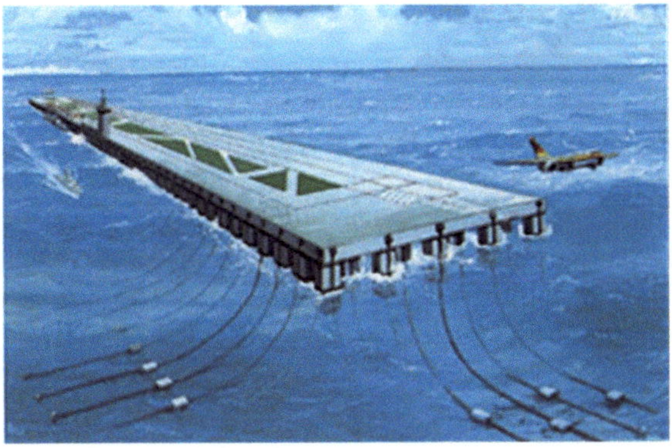

Fig. 22 Schematic illustration of semi-submersible type Mega-Float

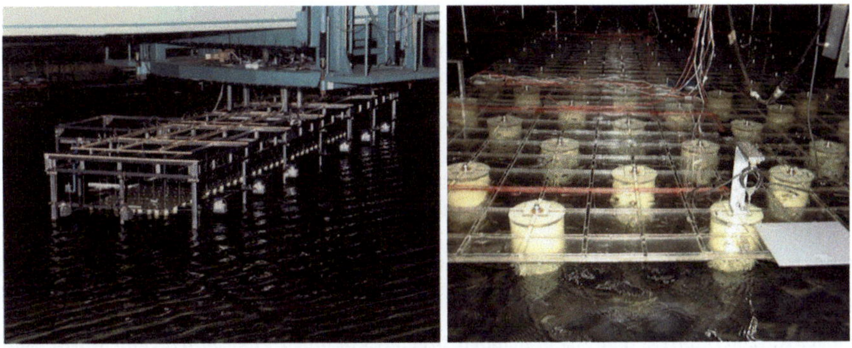

Fig. 23 Wave tank and wind tunnel experiment of SSMF

Economic Zone of Japan. A typical middle-sized SSMF, 2,200 m long, 300 m wide and 7-m deep, that could be used as a floating runway was used for this research. Figure 22 shows a schematic diagram of the SSMF. The deck is supported by 320 column footings with a draft of 24 m. The feasibility and optimization of the SSMF have been investigated and validity of the VLFS Oriented Dynamic Analysis Code (VODAC), an analysis code of hydroelastic responses of SSMF, was extensively tested. The computer code was verified by wave tank experiments with a dynamically similar model. Drag forces on multiple columns at critical Reynolds numbers were measured by wind-tunnel tests to study the design of the moorings. Photos of wave tank experiments and wind-tunnel experiments are shown in Fig. 23. The basic feasibility of the SSMF was confirmed.

10 Concluding Remarks

An overview of 6-year Mega-Float project in Japan and its major achievements have been presented. The fundamental design, construction and operational technologies were developed in Phase-I research and their effectiveness was successfully confirmed through a corroborative study on the use of the Mega-Float as an airport in Phase-II research. The achievements of the Mega-Float project have formed a basis of subsequent projects of large floating structures, e.g. a recent research and development of floating wind turbines. It is strongly expected that the Mega-Float type VLFS will be more extensively applied to various ocean space utilizations.

References

Bhattacharya, B., Basu, R., & Ma, K. T. (2001). Developing target reliability for novel structures: The case of the Mobile Offshore Base. *Marine Structures, 14*, 37–58.

Fujikubo, M., & Yao, T. (2001). Structural modeling for global response analysis of VLFS. *Marine Structures, 14*, 295–310.

Fujikubo, M. (2005). Structural analysis for the design of VLFS. *Marine Structures, 18*, 201–226.

Fujikubo, M., Xiao, T. Y., & Yamamura, K. (2005). Structural safety assessment of pontoon-type VLFS considering damage to the breakwater. *Journal of Marine Science and Technology, 7*, 119–127.

Hara, S., Ohmatsu, S., Yamakawa, K., Hoshino, K., Yukawa, K., & Toriumi, M. (2003). *Towing field test and critical towing operation on Mega-Float unit, Proceedings of 4th Very Large Floating Structures* (pp. 261–270).

Inoue, K. (2003). *Stress analysis of detailed structures of Mega-Float in irregular waves using entire and local structural models, Proceedings of the 4th International Workshop on Very Large Floating Structures, Tokyo*, Japan (pp. 219–228).

Isobe, E. (1999). *Research and development of Mega-Float, Proceedings of 3rd International Workshop on Very Large Floating Structures, Honolulu, Hawaii,* USA (Vol. 1, pp. 7–13).

ISSC2000. (2000). *Report of Special Committee V.1. Risk assessment, Proceedings of the 14th International Ships and Offshore Structures Congress* (Vol. 2, pp. 1–41).

Kato, S., Namba, Y., Masanobu, S., & Shimoyama, M. (2001). *Safety evaluation of dolphin mooring system of hybrid semisub Mega-Floats, Proceedings of 20th International Conference on Offshore Mechanics and Arctic Engineering*, OMAE2001/OSU-5016.

Ohmatsu, S. (1997). Numerical calculation of hydroelastic responses of pontoon type VLFS. *Journal of the Society of Naval Architects of Japan, 182*, 329–340 (in Japanese).

Ohmatsu, S. (2000). Numerical calculation for the hydroelastic response of a pontoon-type very large floating structure close to a breakwater. *Journal of Marine Science and Technology, 5*, 147–160.

Sasajima, H. (2002). *Mega-Float—results of phase-2 research, Proceedings of 12th International Offshore and Polar Engineering Conference, Kitakyushu*, Japan (Vol. 1, pp. 1–5).

Sato, C. (2003). *Results of 6 years research project of Mega-Float, Proceedings of 4th International Workshop on Very Large Floating Structures, Tokyo*, Japan (pp. 377–383).

Seto, H., Ochi, H., Ohta, M., & Kawakado, S. (2003). *Hydroelastic response analysis of real very large floating structures in regular waves in open/sheltered sea, Proceedings of 4th International Workshop on Very Large Floating Structures, VLFS'03, Tokyo*, Japan (pp. 65–73).

Suzuki, H., Yoshida, K., & Iijima, K. (1996). A consideration of the structural design of a large-scale floating structure. *Journal of Marine Science and Technology, 1*, 255–267.

Suzuki, H. (2001). Safety target of very large floating structure used as a floating airport. *Marine Structures, 14*, 103–113.

Torii, T., Ohkubo, H., Hayashi, N., Matsuoka, K., & Kanai, H. (2000). Development of a very large floating structure. *Nippon Steel Technical Report*, No. 82, pp. 23–34.

TRAM (Technical Research Association of Mega-Float). (1997). Analysis of strength against airplane collision (in Japanese).

TRAM (Technical Research Association of Mega-Float). (1999a). Summary of Practical Research on Mega-Float Airport in 1999 (in Japanese).

TRAM (Technical Research Association of Mega-Float). (1999b). Technical Guideline of Mega-Float (in Japanese).

TRAM (Technical Research Association of Mega-Float). (1999c) Safety Guideline of Mega-Float (in Japanese).

TRAM (Technical Research Association of Mega-Float). (2001). Summary of Practical Research on Mega-Float Airport in 2001 (in Japanese).

TRAM (Technical Research Association of Mega-Float). (2002). Summary of Practical Research on Mega-Float Airport in 2002 (in Japanese).

Utsunomiya, T., & Watanabe, E. (1998). *Wave response analysis of a box-like VLFS close to a breakwater, Proceedings of 17th International Conference on Offshore Mechanics and Arctic Engineering*, OMAE98-4331.

Yago, K., & Endo, H. (1996). *Model experiment and numerical calculation of the hydroelastic behavior of mat-like VLFS, Proceedings of 2nd Very Large Floating Structures* (pp. 209–216).

Yamashita, S., & Harada, T. (1997). *A practical method of hydroelastic analysis of a very large floating structure in head sea, Proceedings of 20th International Conference on Offshore Mechanics and Arctic Engineering, OMAE1997* (Vol. IV, pp. 99–106).

Yamashita, Y., Yonezawa, M., Shimamune, S., & Kinoshita, Y. (2003). *Joining technology for construction of very large floating structures, Proceedings of 4th Very Large Floating Structures* (pp. 229–236).

Yasuzawa, Y., Kagawa, K., Kawano, D., & Kitabayashi, K. (1997). Dynamic response of a large flexible floating structure in regular waves. *Proceedings of Offshore Mechanics and Arctic Engineering, 6*, 187–194.

Yoshida, K., Suzuki, H., Kato, S., Sumiyoshi, H., & Kado, M. (2001). *A Basic study for practical use of Semisub-Megafloat, Proceedings of 20th International Conference on Offshore Mechanics and Arctic Engineering, OMAE2001/OSU5015.*

Large Spar Drilling and Production Platforms for Deep Water Oil and Gas

J. Halkyard

Abstract A "spar" is a term for a long slender pole commonly used in sailing vessels as booms or masts. In the context of this book, spars are floating structures that are characterized by their deep drafts and long vertical hull forms. Spar platforms have been used for offshore oil and gas operations for a number of years. The first application was a storage spar, "Brent," installed in the North Sea in the 1960s. Spars were introduced for production, drilling, and workover operations in the 1990s. They remain one of the few hull forms that can support top-tensioned risers for drilling and production in ultra-deep water.

1 What Is a Spar?

A "spar" is a term for a long slender pole commonly used in sailing vessels as booms or masts. In the context of this book, spars are floating structures that are characterized by their deep drafts and long vertical hull forms. Their deep draft permits these structures to minimize heave better than semi-submersibles and ship-shaped floaters, making this hull form attractive to deep water oil and gas operations involving risers that are sensitive to heave motions.

This chapter investigates the use of Spars in deep water oil and gas production. It traces the history of Spars from its initial usage in marker buoys through to its use as offshore relay stations and oil storage facilities, and to its employment in oil and gas production. The advantages of the hull form to the oil and gas industry is then presented, before the structural elements of the two primary types of spars are explored. The potential scale of Spar projects is then discussed, and construction methodology is presented. Finally, the structural design considerations and current solutions are presented.

J. Halkyard (✉)
John Halkyard and Associates, Offshore Engineering Consultants,
14121 Cardinal Lane, Houston, TX 77079, USA
e-mail: jhalkyard@halkyard-associates.com

C.M. Wang and B.T. Wang (eds.), *Large Floating Structures*,
Ocean Engineering & Oceanography 3,
DOI 10.1007/978-981-287-137-4_9

2 History of Spars

Spars have been used for decades as marker buoys and for gathering oceanographic data. The first significant Spar for our purposes is Flip (see Fig. 1), a structure owned by the U.S. Navy and operated by the Scripps Institution of Oceanography in California (Fisher and Spiess 1963). Flip was put into service in 1965 and is used primarily for ocean acoustic measurements. Operating draft is about 82.3 m (270 ft). Its diameter is 3.66 m (12 ft) at the waterline, tapering to 6.1 m (20 ft) over most of the hull. Its heave natural period is 29 s. Although Flip on occasion has been tethered to the sea floor, it is more commonly allowed to drift in the ocean currents.

In the early 1960s, Nippon Telegraph installed a Spar off the coast of Japan to carry a microwave relay station. This Spar is 135.6 m (445 ft) long with a stepped hull ranging in diameter from 3.05 m (10 ft) to 6.1 m (20 ft). The topside structure is a cylinder, 15.2 m (50 ft) in diameter by 10.0 m (33 ft) high, with equipment, accommodations and a heliport on top. A four point catenary mooring of 76 mm (3 inch) chain connected to a 159 kN (175 ton) clump weights keeps the Spar in place. The operating draft is 100.5 m (330 ft).

In the mid 1970s, Shell installed an oil storage and offloading Spar at Brent Field, in the North Sea (van Santen and de Werk 1976). The hull is 29.0 m (95 ft) in diameter, necks down to 16.8 m (55 ft) at the water plane, and the operating draft is 108.8 m (357 ft) (see Fig. 2). This Spar was designed to store 35,772 cu m (300,000 bbl) of produced crude and to transfer it to bow loading tankers. The oil storage system used a water displacement principle which allowed the tanks to be designed for ambient pressures. The mooring system consists of 6 lines, each made up of a 9964 kN (1000 ton) concrete gravity anchor, 792.5 m (2,600 ft) of 89 mm (3.5 in) wire rope, and 285.0 m (935 ft) of 102 mm (4 in) chain. The top part of the superstructure rotates to allow the tanker to weathervane about the Spar. Accommodations, power plant, other equipment and pumps are in the superstructure and a heliport is fitted.

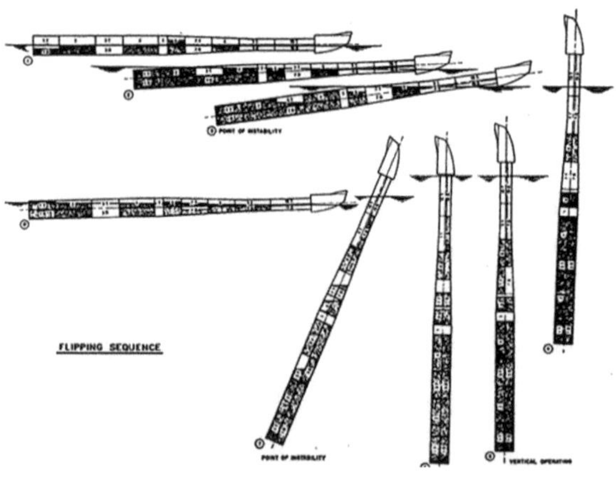

Fig. 1 FLIP was built in 1962 (Fisher and Spiess 1963)

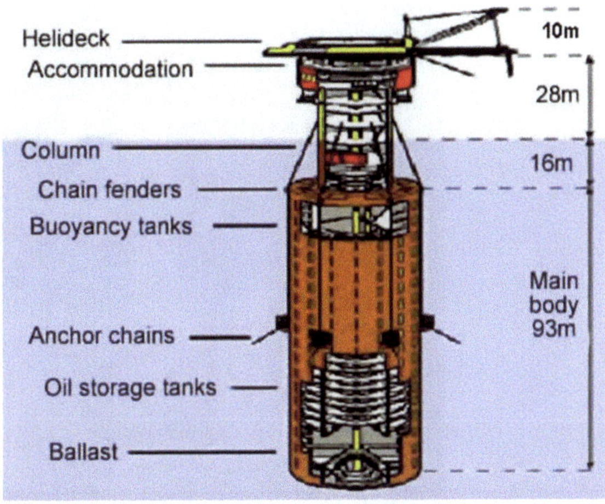

Fig. 2 Brent Spar (courtesy Royal Dutch Shell)

Agip installed a flare Spar off West Africa in 1992 designed to flare 2.8 MM cu m (100 million cu ft) of gas per day. The Spar is 71.0 m (233 ft) long, with a diameter of 2.3 m (7.5 ft) that tapers to 1.7 m (5.5 ft) through the water plane. With a draft of 51.8 m (170 ft), it is held in place with a 4 leg chain catenary mooring. In 1993 Shell installed a loading Spar at Draugen. The Spar topsides include a rigid boom to which the flexible loading hose is attached. The hull diameter is 8.5 m (28 ft) and the operating draft about 76.2 m (250 ft).

The world's first oil and gas production spar was the Neptune Spar installed in 1996 by Oryx Energy Company (now Kerr McGee) and CNG (see Fig. 3) (Vardeman et al. 1997). Design production rates were 25 mbod and 30 mmcfd, respectively. The maximum topsides weight with the workover rig is 54800 kN (5500 tons). Wells would be predrilled with a semisubmersible and completed with a platform workover rig placed temporarily on the Spar. The Neptune spar has a hull 214.9 m (705 ft) long with a 9.8 m × 9.8 m (32 ft × 32 ft) centerwell and a diameter of 21.9 m (72 ft). The six point mooring system consisted of driven pile anchors, 121 mm (4-3/4 in) spiral strand wire rope and chain for the section leading up to the fairleads and onto the hull.

As of this writing there are 18 spars in production and two under construction. Figure 4 shows the 18 spars. Williams Gulfstar 1 is not shown in the figure. The Perdido spar was the deepest water floating production platform when it was installed. As this goes the print, the Aasta Hansteen spar is being built in Korea, and will be the first spar installed on the Norwegian continental shelf.

Aside from the oil and gas industry, the spar concept has been considered a leading contender as a platform to support Ocean Thermal Energy Conversion (OTEC), see (Cohen 2009; Lockheed Martin 2011 and Ross 2012) for example. Figure 5 shows a concept from 1975 developed by Lockheed Martin. The spar is ideal for OTEC because there is a need for large volumes of heat exchangers to be placed in a protected environment below the water surface, as deep as possible. The spar provides this volume.

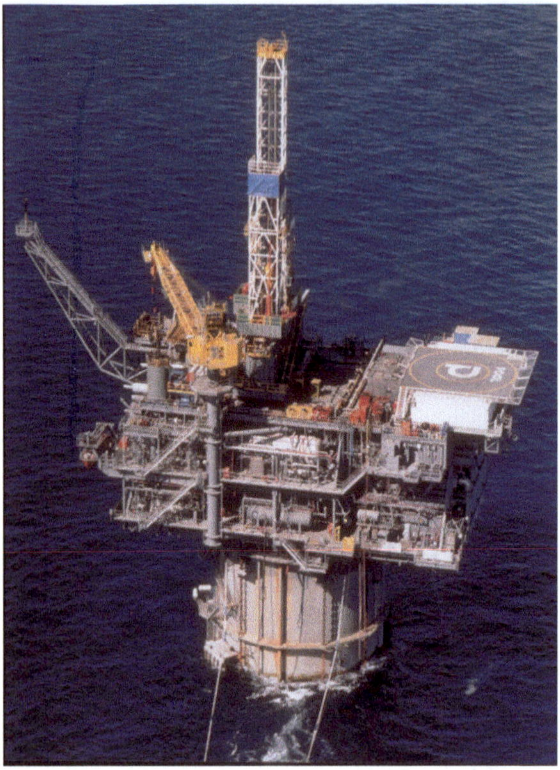

Fig. 3 Oryx/CNG Neptune Spar, now Kerr McGee (Vardeman et al. 1997)

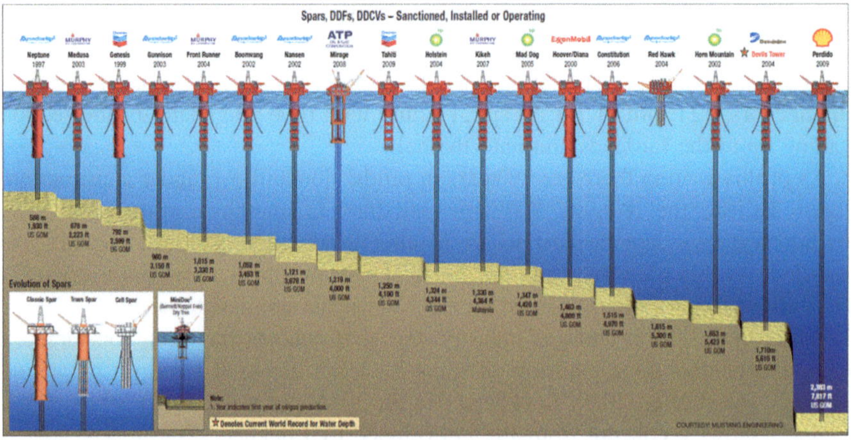

Fig. 4 Progression of spars (Offshore magazine/Mustang Engineering)

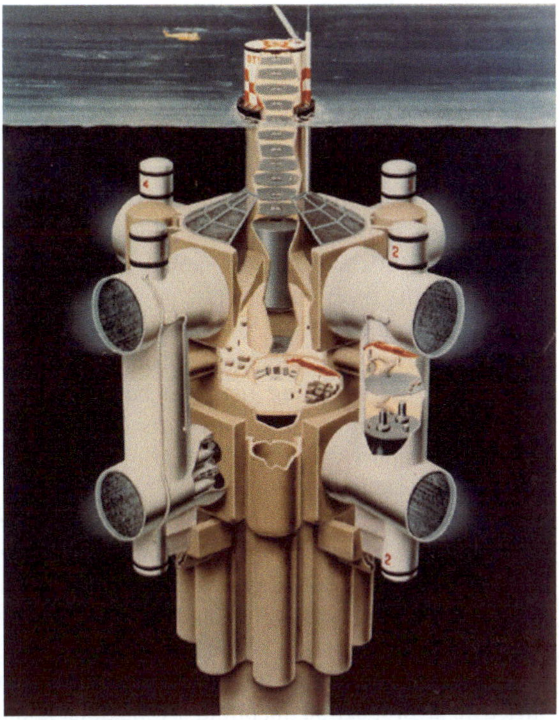

Fig. 5 Early concept for ocean thermal energy conversion (Cohen 2009)

3 Increasing Importance of Spars in Oil and Gas Extraction and Production

The unique feature of the drilling and production spar is its ability to support top tensioned steel risers. This enables the spar to connect a riser pipe from the surface directly to a seafloor wellhead. The favorable motions of the spar allow these "top tensioned risers", or "TTRs", to withstand the worst sea states expected over the life of the platform. This means that production controls, or "Christmas Trees", may be placed at the top of the risers at the spar deck rather than on the seafloor. This enables direct access to the wells for maintenance from the platform using an inexpensive platform workover rig. This ability to support "dry trees" in deep water is unique to the spar and the Tension Leg Platform, or "TLP". Semi-submersibles and ship shaped floaters experience too much heave motion, and in the case of ships roll motion, in the typical survival sea states to allow top tensioned risers to survive [design standards for drilling and production risers are described in offshore recommended practices such as API RP 2RP].

For drilling, a top tensioned riser is used as a conduit for the drill pipe and as a return for the drilling fluids. In this case, a blowout preventer (BOP) is placed at the top of the drilling riser, again at or above the spar deck. The riser itself must be designed to withstand the full shut-in pressure of the well, as well as the dynamic loads resulting from platform motions and wave forces.

Spar risers run through a centre well which extends the length of the hard tank (see below for description of the hull structure). The riser tension may be provided either by buoyancy cans placed in the centre well, or by use of hydraulic tensioners which are more typically used on floating semi-submersible type units (for drilling only) or on Tension Leg Platforms (TLPs). The buoyancy cans are located in guides which provide lateral constraint. As the spar heaves the buoyancy cans remain in place while the spar moves relative to them. As the spar offsets, the geometry of the risers causes draw down and again the guides allow the risers to slide relative to the hull. The buoyancy can/guide system allows stroke of the risers to be relatively large compared to that for typical hydraulic tensioners. This arrangement of supporting risers on buoyancy cans in the centre well enables larger relative motions between the risers and spar than would be practical with hydraulic tensioners. This was the primary inventive claim in the original production spar patent (Horton 1987).

A schematic of the riser arrangement in the centre well is shown in Fig. 6. The buoyancy cans have an outer shell and an inner stem. The stem has an internal diameter sufficient to allow passage of the riser tie back connector. When in place, the riser itself is centralized within the stem. The stem extends above the buoyancy cans to the tree elevation at the production deck level. The risers and the production tree are supported by the stem there. Production fluids from the risers are carried to a production manifold by flexible jumpers which accommodate the relative movement between the risers and the hull.

Figure 7 shows one of the buoyancy cans being installed. Figure 8 shows the upper stem, surface trees and flexible jumpers.

The “surface tree”, sometimes referred to as a “Christmas Tree” is a critical part of a hydrocarbon well system. It consists of a series of valves and connections which allow communication of fluids and electrical power and controls from the platform and the downhole components of the well. The spar and the Tension Leg Platform (http://en.wikipedia.org/wiki/Tension-leg_platform) are the only floating platforms currently able to support a riser with the tress at the surface. That is because the risers that support the surface tree must be capable of containing the well pressure in the most extreme conditions, and this is only possible for platforms with very low motions, particularly in heave. The other alternative is to place this critical component at the seafloor, attached directly to the wellhead. In this case the fluids pass from the tree to the platform through flowlines on the seafloor and relatively expensive flexible pipes that can withstand larger dynamic motions than the “rigid” vertical steel risers used with surface trees. When the tree is located at the wellhead it is commonly referred to as a “wet” or “subsea” tree. The surface tree well system has the advantage that the well may be directly accessed from the platform for well maintenance, whereas the wet tree wells require hiring an expensive mobile drill rig to service the well. This often leads to abandoning wells with subsea trees long before the reservoir is fully utilized, because the cost of well intervention is too large.

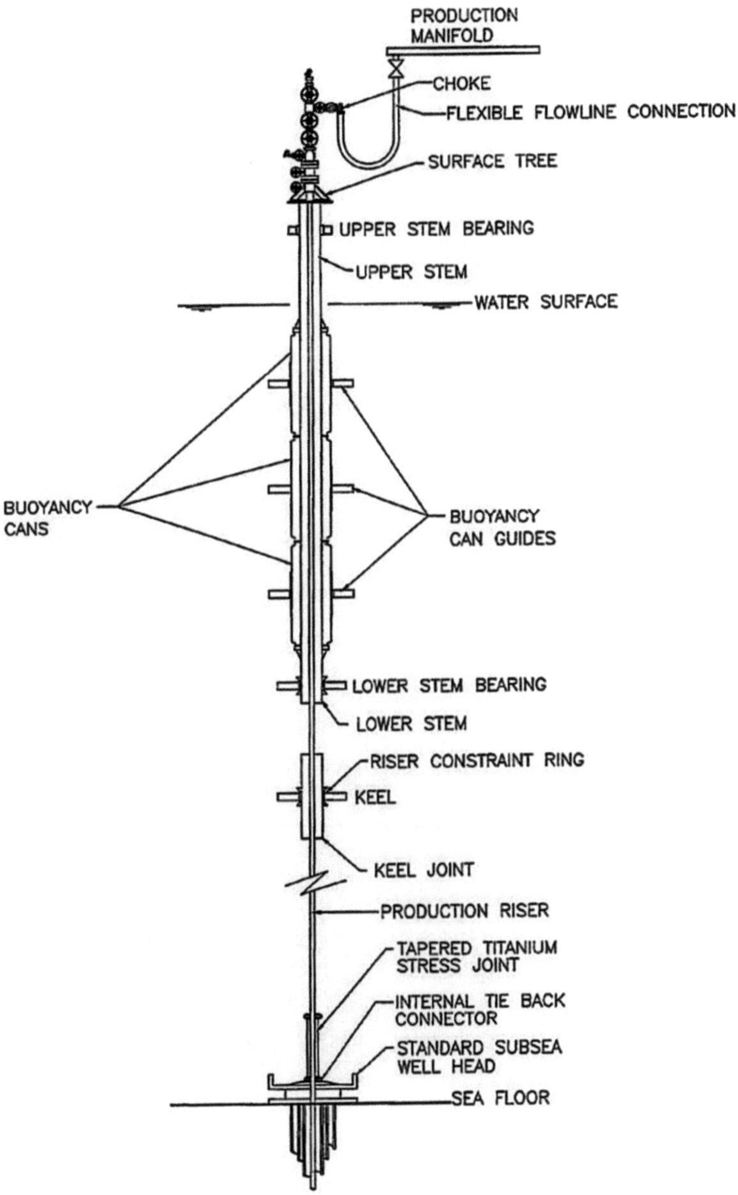

Fig. 6 Schematic of production riser and buoyancy cans (Glanville et al. 1997)

The Holstein Spar was the first spar with top tensioned risers that did not utilize the buoyancy can system for support. Instead, the risers were supported on RAM type or "push up" tensioners as illustrated in Fig. 9. Installation of the production riser tensioners in shown in Fig. 10. These utilize hydraulic cylinders oriented with

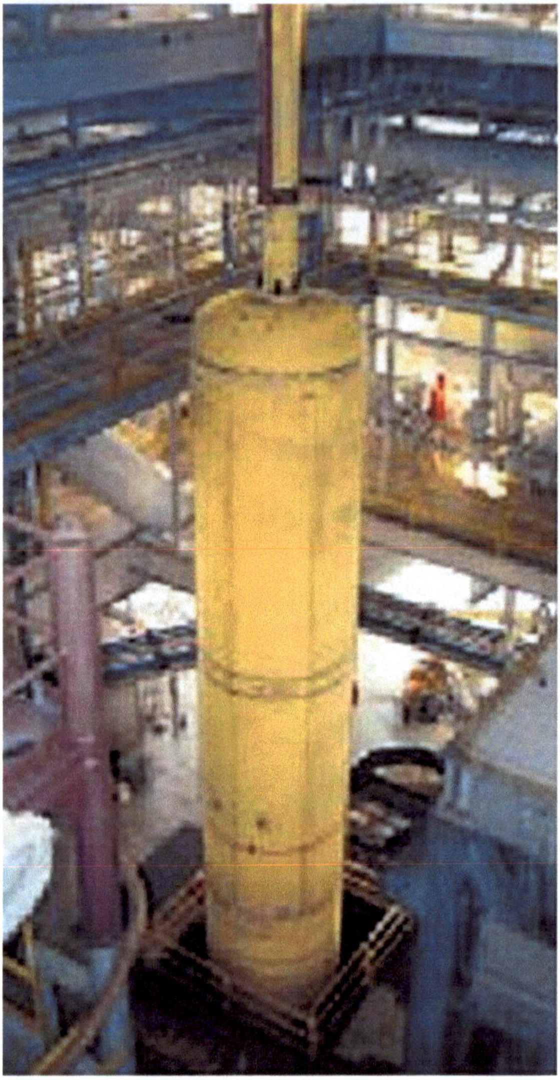

Fig. 7 Buoyancy can being installed (courtesy Kerr McGee)

the rods upwards. The riser load is supported by the rods in compression, as opposed to traditional riser tensioners which consist of a number of cylinders in tension supporting a riser cassette (see Fig. 11). The RAM tensioners take up less space and allow the risers to be more closely spaced, resulting in increased efficiency of space and a smaller centre well and spar overall. The drawback of the RAM style tensioners is that the riser needs to be aligned with the axis of the tensioners. The spar configuration allows this. Full drilling risers, used on the Genesis and Hoover spars, have also used RAM tensioners (in the case of Hoover).

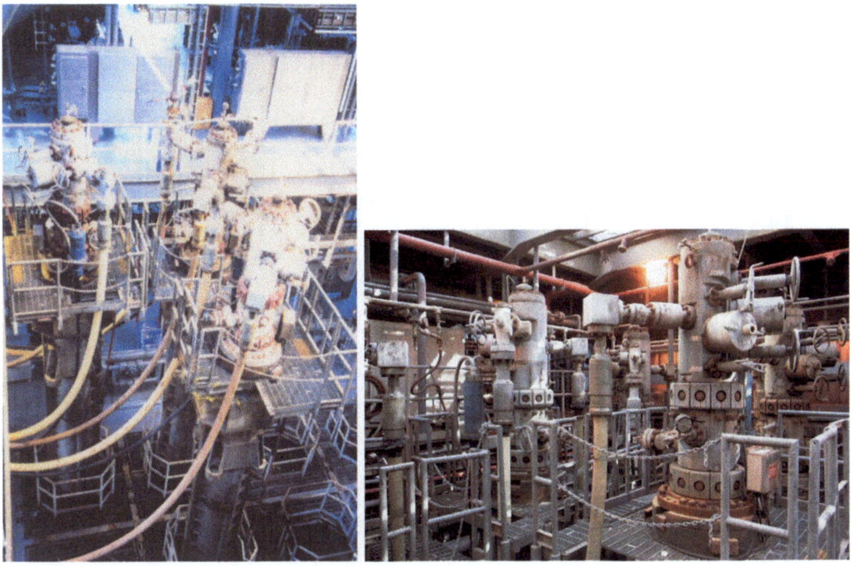

Fig. 8 Surface trees on a spar (*courtesy* Kerr McGee)

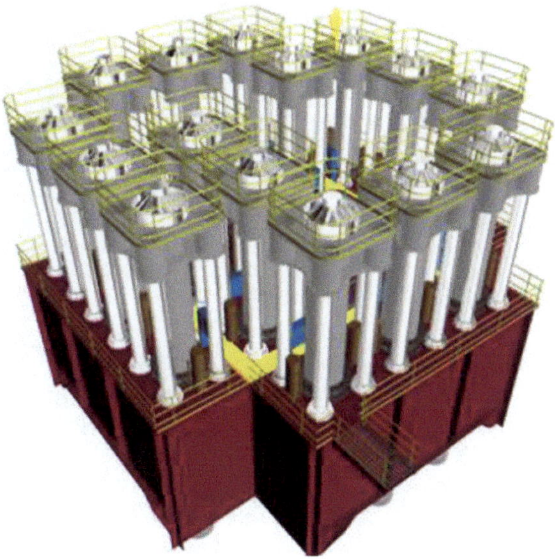

Fig. 9 Illustration of production riser tensioners for Holstein spar (Smith et al. 2005)

Spars have been used for both drilling and workover of wells (see Moyer et al. 2001; Bates et al. 2001; Glanville and Vardeman 1999). The Genesis, Diana and Holstein spars are equipped for full drilling using platform rigs. Drilling operations on the Genesis spar are illustrated in Fig. 12. This figure shows the arrangement of

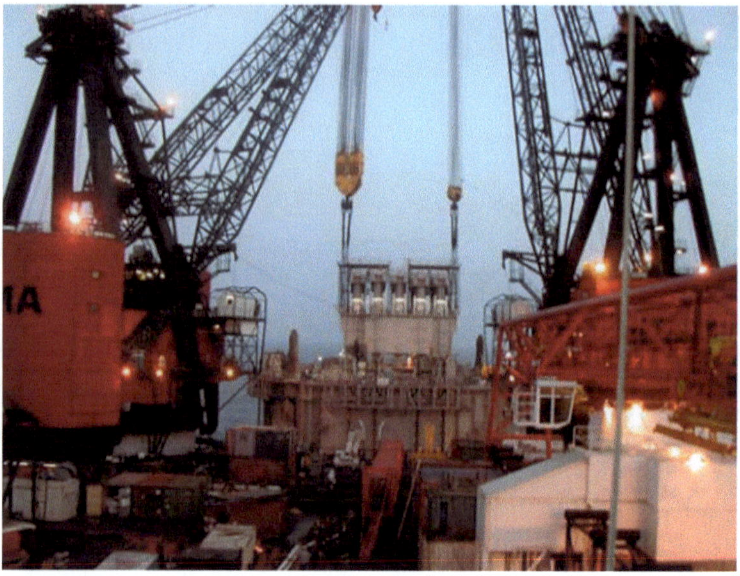

Fig. 10 Installation of Holstein PRT and substructure (Smith et al. 2005)

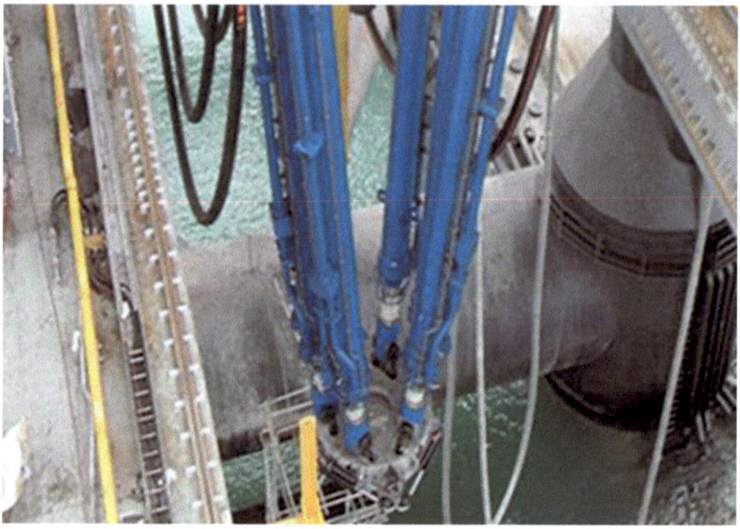

Fig. 11 Direct acting tensioners (National Oilwell Varco)

wells at the spar, and the corresponding well layout on the seabed. The drilling riser is first run through the central slot of the wellbay while the spar is positioned in the center of the seafloor pattern (neutral position). When the drilling riser is close to the seabed, the mooring lines are adjusted on the spar to position the

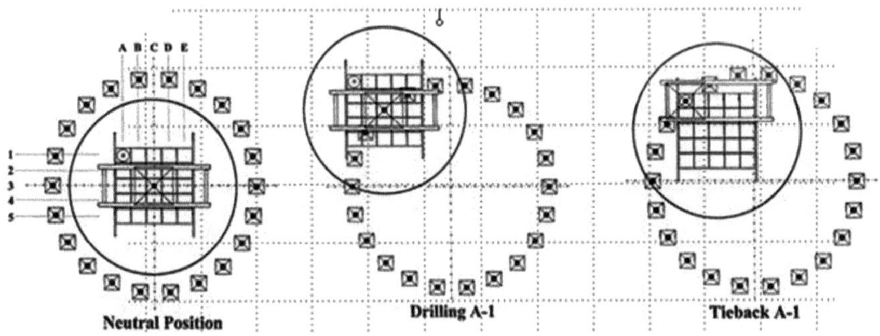

Fig. 12 Drilling operations from a spar (courtesy Technip USA Inc.)

bottom of the drilling riser next to the seafloor wellhead (Drilling A-1 position). The active mooring on the spar allows the connection of the drilling riser to the wellhead without guidelines or special seafloor guidance. The connection process is controlled by the chain jack operator while monitoring a television camera on the riser and an ROV camera. Positioning is controlled manually through a joystick connected to the master computer. When the drilling riser connector is stabbed into the wellhead, it is latched in place by the ROV actuating a lockdown mechanism in the tieback connector. The well is now sealed and drilling may proceed. If there is a large current the connection may still be made by maneuvering the spar up current. As long as the entry angle is less than about two degrees the riser can be stabbed and connected.

Once the well is drilled and ready for completion, the drilling riser is unlatched and moved to another well where it is "parked" while the well is completed. The drilling riser BOP is removed. The drilling riser is then hung off the platform in the center slot while the production riser is run and the well is completed. In order to complete the well, the drill rig is skidded to the appropriate well slot on the platform (tieback A-1 position). The production riser is run and the spar mooring system is again used to position the bottom of the riser over the wellhead where it is latched to the wellhead with a tieback connector using the same methods as the drilling riser. Once latched, a surface tree and BOP is installed and the well is completed. Subsequent to the completion of the well, production commences and the rig may be moved to the central drilling slot and drilling of another well may commence.

Most of the spars in service do not have full drilling; instead they have workover and completion rigs. These operations are conducted through the production risers and don't require a special drilling riser. Nevertheless, the unique form of the spar allows for simultaneous drilling and production even with this reduced capability. This is accomplished by using "pullover drilling", a procedure pioneered on the Neptune spar (Glanville and Vardeman 1999). The procedure is illustrated in Fig. 13.

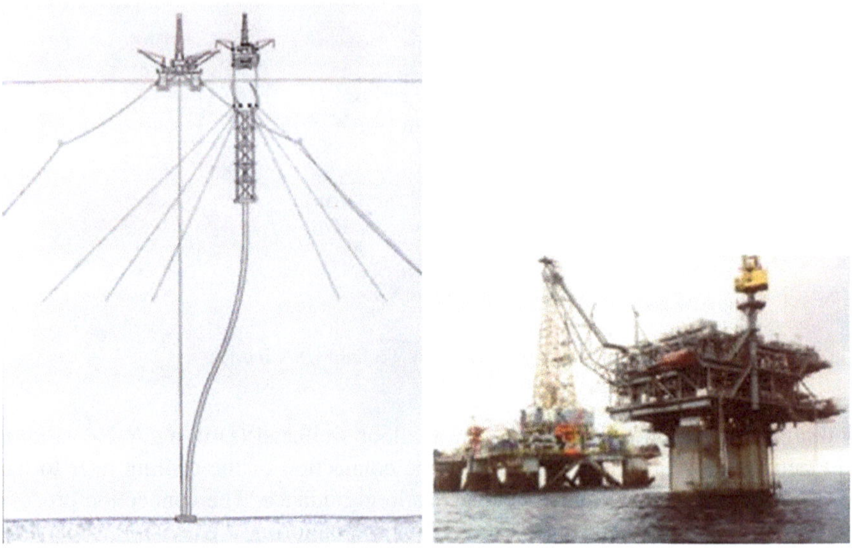

Fig. 13 Pullover drilling (courtesy Technip USA Inc.)

Pullover drilling involves using the mooring system of the spar to move far enough from the location of the wells that a leased Mobile Offshore Drilling Unit (MODU) may be moored alongside to drill the wells. In the case of the Neptune spar this involved moving about 250 ft from its neutral position. This large offset in 1930 ft of water necessitated using a titanium stress joint at the base of the production risers (Berner et al. 1997). Once this maneuver is done, a MODU is moored in place over the wells alongside the spar. This requires a preinstalled mooring with buoys to prevent interference between the spar and the MODU moorings. The MODU may spud in new wells, or drill wells already batch set to an intermediate casing point. Figure 14 shows the seafloor well pattern for Neptune as of 1999. The original wells and new wells are shown. New wells were drilled with the spar offset to the southwest. Future wells will be drilled with the spar moved to the northeast.

Once the MODU drills the wells, it may be removed and the spar can be moved back over the wells where the production riser is run and the wells are completed and/or tied back for production. In this design, the offset mooring system with the MODU is designed for a 10-year hurricane survival condition.

Kerr McGee found that pullover drilling helped improve their project in two ways. First, by delaying drilling some of the wells the production of the first wells could be examined to determine reservoir characteristics prior to committing to a specific drilling program. Secondly, the costs of drilling could be deferred until revenue was being produced.

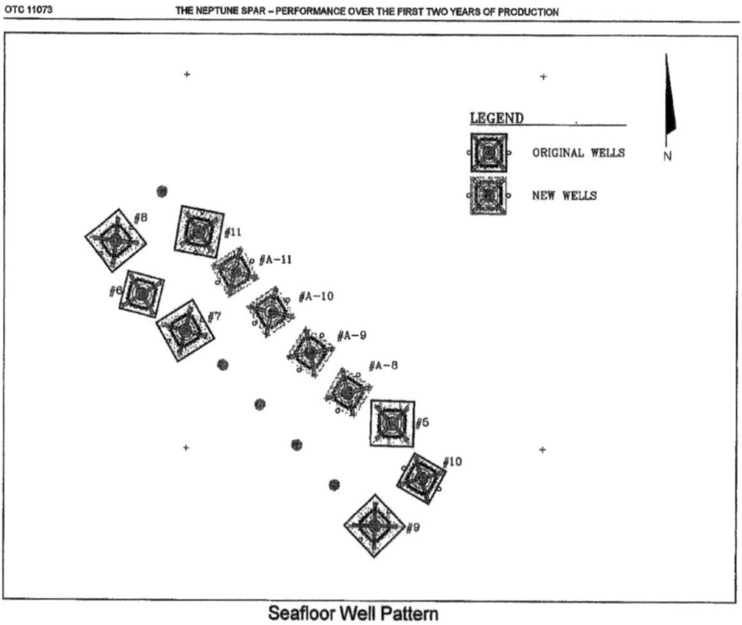

Fig. 14 Seafloor well pattern for neptune (Glanville and Vardeman 1999)

4 Structural Components of a Spar

The two primary types of spars built to date have been the "Classic" and the "Truss" spar as shown in Fig. 15. The basic parts of either spar include:

1. Deck.
2. Hard Tank.
3. Midsection (steel shell or truss structure).
4. Soft Tank.

The Topside Deck is typically a multi-level structure in order to minimize the cantilever requirement. For decks up to about 18,000 tons the deck weight is supported on four columns which join hard tank at the intersection of a radial bulkhead with the outer shell. Additional columns are added for heavier decks. Figure 16 shows the arrangement of the top of a spar with the deck supports. Decks up to about 100,000 kN (10,000 tons) may be installed offshore with a single lift. Larger decks require multiple lifts.

The Hard Tank provides the buoyancy to support the deck, hull, ballast and vertical tensions (except the risers). The term "Hard Tank" means that its compartments are designed to withstand the full hydrostatic pressure (see section of Hull Structure). Figure 17 illustrates a typical structural arrangement. There are five to six tank levels between the spar deck and the bottom of the hard tank, each

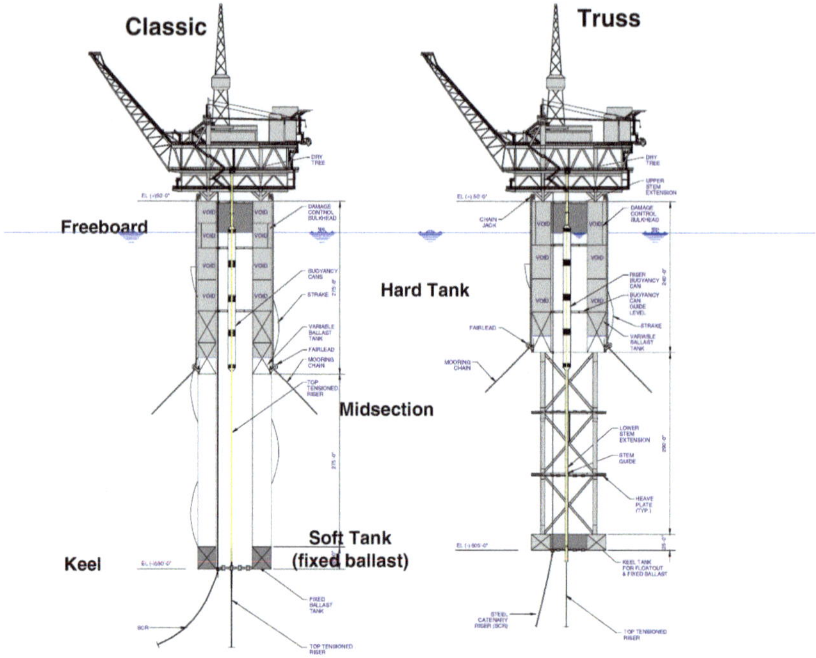

Fig. 15 Classic and truss spar (courtesy Technip USA Inc.)

level separated by a watertight deck. Each level is further divided into four compartments by radial bulkheads emanating from the corner of the centre well. The tank level at the waterline includes additional cofferdam tanks to reduce the flooded volume in the event of a penetration of the outer hull from a ship collision. Thus there are up to 28 separate compartments in the hard tank. Typically, only the bottom level is use for variable ballast, the other levels being void spaces.

The Midsection extends below the hard tank to give the spar its deep draft. The selection of the spar draft is discussed below. In the early "classic" spars the midsection was simply an extension of the outer shell of the hard tanks. There was no internal structure. Longitudinal stringers are added to compensate for global bending. The scantlings for the midsection were determined by construction loads and bending moments during upending. Alignment of adjacent sections during construction requires a tight tolerance on the diameter and roundness of the outer shell. Cable stays are added to control the deformations while the section is resting on its side. Later spars replaced the midsection with a space frame truss structure. This "truss spar" arrangement resulted in a lower weight, less expensive hull structure. Also, the truss has less drag and reduces overall mooring loads in high current environments.

The Soft Tank at the bottom of the spar is designed to provide floatation during the installation stages when the spar is floating horizontally. It also provides compartments for the placement of fixed ballast once the spar is upended. The soft tank has a centerwell and a keel guide which centralizes the risers at that point.

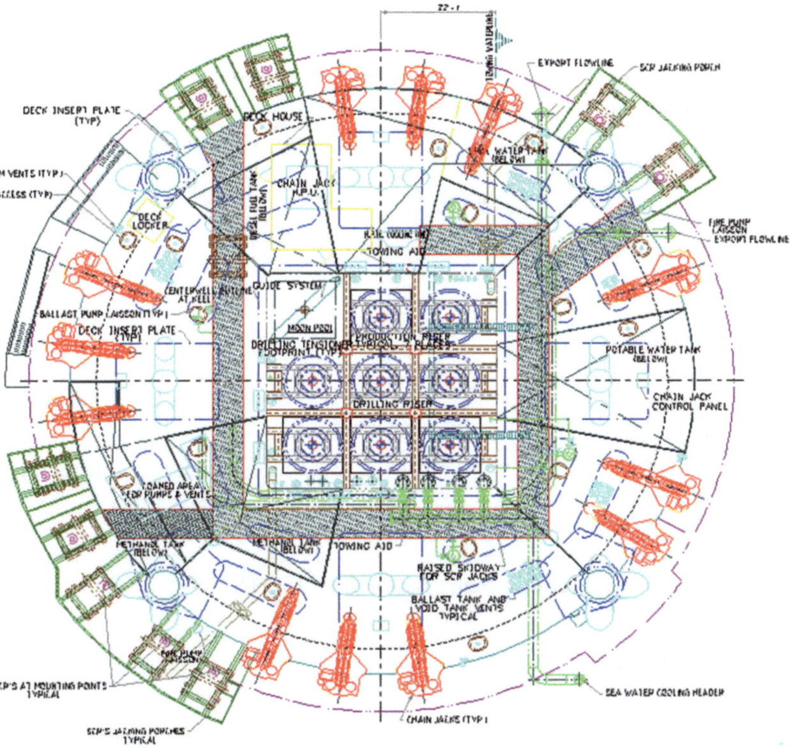

Fig. 16 Spar deck arrangement (courtesy Technip USA Inc.)

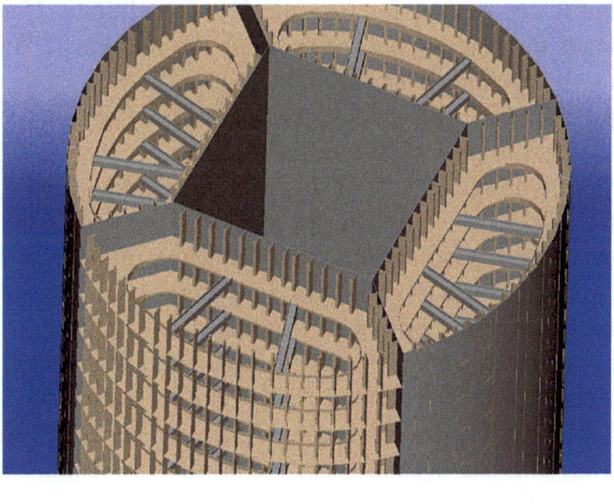

Fig. 17 Hard tank structural arrangement (courtesy Technip USA Inc.)

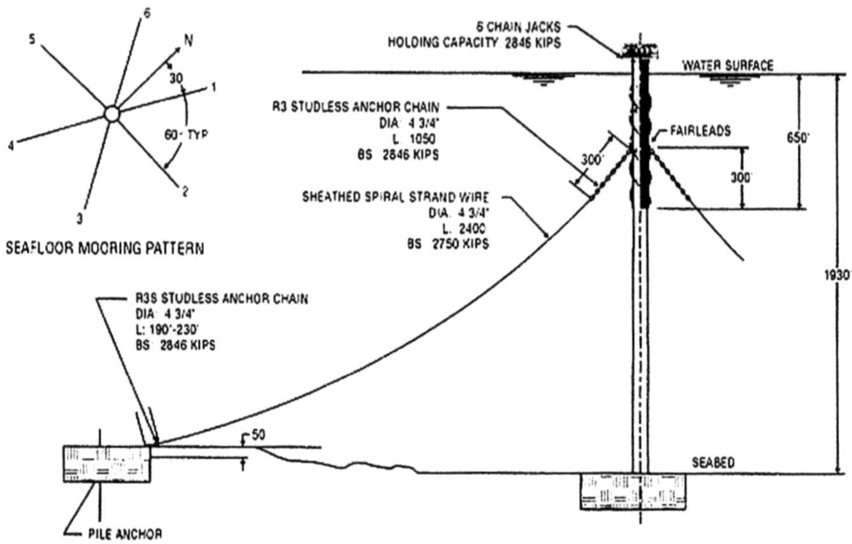

Fig. 18 Neptune spar mooring system (Vardeman et al. 1997)

5 Spar Mooring

The mooring system for the spars built to date consists of two types of moorings: a "semi-taut" chain-wire-chain catenary system similar to the one shown in Fig. 18 for the Neptune Spar, and a "taut" system consisting of chain-polyester and chain. A "semi-taut" mooring is defined here as one in which the anchor loads have an uplift component for extreme load conditions. Under nominal and calm water loads the anchor chain may lie on the bottom. A "taut" system always has uplift on the anchors. The semi-taut systems rely on chain and wire weight and suitable pre-tension to provide restoring forces. The taut system relies on synthetic line stiffness to provide the restoring force.

The design of a semi-taut system requires careful selection of pretension such that, under the mean conditions of a survival conditions, the catenary still has sufficient sag to allow first order wave motions at the fairlead to occur without over-stretching the line. Survival metocean conditions can range from hurricanes and tropical cyclones with maximum wave heights of up to 35 m (115 ft) and wind speeds of 50 m/sec (98 kts) to extreme currents and solitons having velocities of 2 m/sec (4 kts) or more. Spar motions are small enough, even in the 100 year hurricane, that the semi-taut system can be used without synthetic mooring lines. The semi-taut system saves a considerable length of wire and chain needed for a conventional catenary mooring.

Beyond a certain water depth the chain-wire systems' self-weight limits their efficiency. The lighter synthetic lines are more cost effective beyond about 1,200 m (3,900 ft) water depth.

The platform chain is tensioned using chain jacks or windlasses which are installed on the periphery of the upper deck of the hull (the "spar deck"). Figure 16 shows an example of the arrangement of this deck. This type of tensioning is easier and more controllable than using a winch if, for example, the mooring system consisted only of wire rope or strand.

The chain runs from the chain jack to a fairlead which is located from up to 107 m (350 ft) below the mean waterline. The length of platform chain is determined by the amount of chain that needs to be pulled in or paid out to maneuver the spar.

The midsection of the mooring system consists of spiral strand wire rope or polyester line. For longevity, the steel strand is typically sheathed with a urethane coating. The lower end is attached to a length of anchor chain. The length of chain and the mooring tension is selected so that the wire will not make contact with the sea bottom except under the most extreme conditions.

The anchor chain is connected to a piled anchor which can sustain uplift and lateral loads. The pile padeye is usually about 15 m (50 ft) below the mudline so that the bending moment from the mooring forces is minimized.

The total scope of the taut moor from the fairlead to the anchor is typically about 1.5 times the water depth.

6 Potential Scale of Spar Installations

Given there is a pattern of expanding spar sizes, there is a strong likelihood for the trend to continue with installations increasing in size. This section studies the design considerations that affect the potential scale of a spar installation.

The largest spar built to date is the Holstein spar: 45.4 m in diameter and 227.4 m long. From its base to the top of the derrick is over 300 m. Figure 19 shows the dimensions of the Perdido spar in perspective.

The main requirements for sizing a spar are:

1. Maximum weight of topsides and risers supported by the spar that needs to be accommodated,
2. Eccentricity of the deck which needs to be trimmed by variable ballast,
3. Area required in the centerwell to enclose risers and buoyancy cans,
4. Maximum pitch motions in a 100-year event less than about 10 degrees,
5. Elevation of the deck above the sea to prevent the wave crest from impacting the deck (positive airgap),
6. Maximum riser stroke less than about ± 15 ft.
7. Ability to be single piece transported on a heavy lift vessel,
8. Ability to float-off heavy lift vessel (maximum draft less than 10 m in horizontal position).

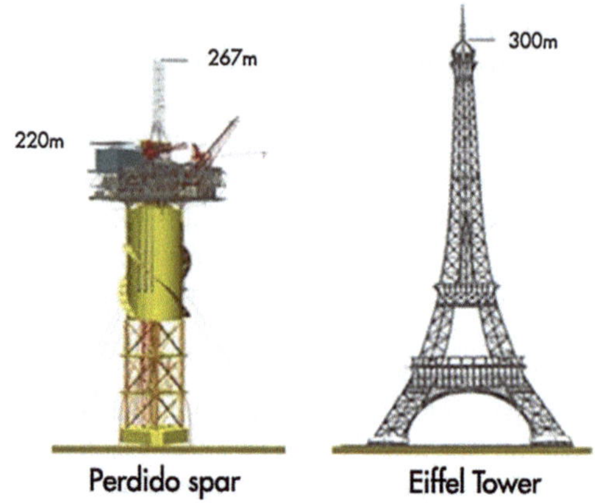

Fig. 19 Perdido spar size in perspective (shell). http://www.shell.com/global/aboutshell/major-projects-2/perdido/assembling-deepest-platform.html

Topsides weight and other payload weights are obviously key drivers. Aside from this, the most critical 1st decision is the size of the centerwell, because this sets the minimum diameter for the hull.

6.1 Centerwell Sizing

All spar centerwells to date have been square, lending themselves to 4 × 4 (16), 5 × 5 (25) or 6 × 6 (36) well slots. Sometimes the four slots in the center of the pattern are allocated to a drilling riser. Also, four slots may be opened to provide a moonpool for running of tools, ROVs, etc. Space may also be allocated for SCR J-tubes, or even access trunks for personnel entry to lower tanks. Figure 20 shows a typical layout with two slots dedicated to running of flowlines, umbilicals and SCRs.

Slot spacing is dictated by the diameter of the buoyancy cans. It is best to perform sufficient riser analysis early in the design in order to determine the maximum top tensions required. Once this is known the buoyancy cans may be sized. Spacing for existing spar centerwell slots has ranged from 8 (for the Neptune Spar) to 14 ft. The size increases with water depth because of the higher tensions required.

6.2 Spar Hull Sizing Parameters

Aside from the centerwell size, the key basic design parameters for the spar hull include:

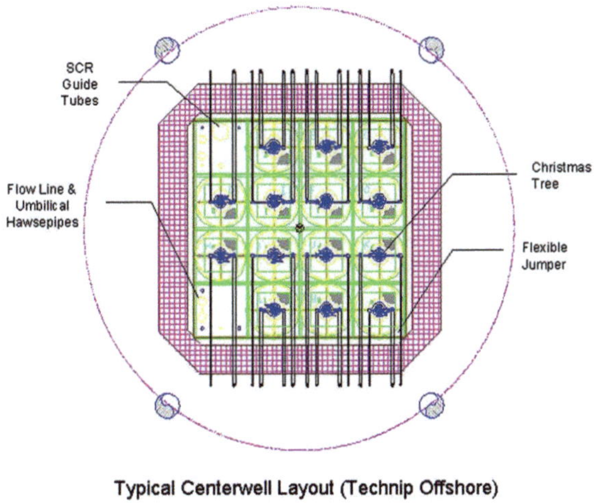

Fig. 20 Typical centerwell layout (courtesy Technip USA Inc.)

1. Diameter.
2. Hard Tank Depth.
3. Fixed Ballast.
4. Draft.
5. Fairlead Elevation.

Initial classic spar designs were based on a draft of 198 m (650) This draft was chosen based on providing adequate separation between the wave energy and the natural heave period of the spar so that resonant heave would not be excited (Glanville et al. 1991). Heave predictions based on linear theory (Newman 1963) suggested this was a point of diminishing returns. The heave response in the 100 year hurricane, maximum wave height of 22.5 m (74 ft), was predicted to be less than ±1.2 m (4 ft) Experience with full-scale spars has proven that the heave response is less than predicted, and also that the spar heave was not controlling the amount of riser stroke, hence more recent spars have had shallower drafts, around 152 m (500 ft).

Spar sizing is primarily determined by its heel and pitch response. “Heel” is the static inclination in response to a ballast shift or steady environmental loads. “Pitch” is the total inclination including transient or dynamic responses. For the Gulf of Mexico environment, hulls the maximum pitch response is achieved by designing for a static heel angle in the 100 year hurricane of less than 5 degrees. Figure 21 shows a free body diagram of a truss spar. The static heel angle is determined by the resultant moment caused by the couple between the steady environmental forces (wind, current and wave drift loads) and the resisting force of the mooring lines.

A third generation spar, the “cell spar”, was installed for in 2004 (Finn et al. 2003). This design embodies a new construction technique using ring stiffened tubulars assembled in a hexagonal formation to form a spar. Figure 22 shows an illustration of this concept. The first cell spar is designed for wet trees only.

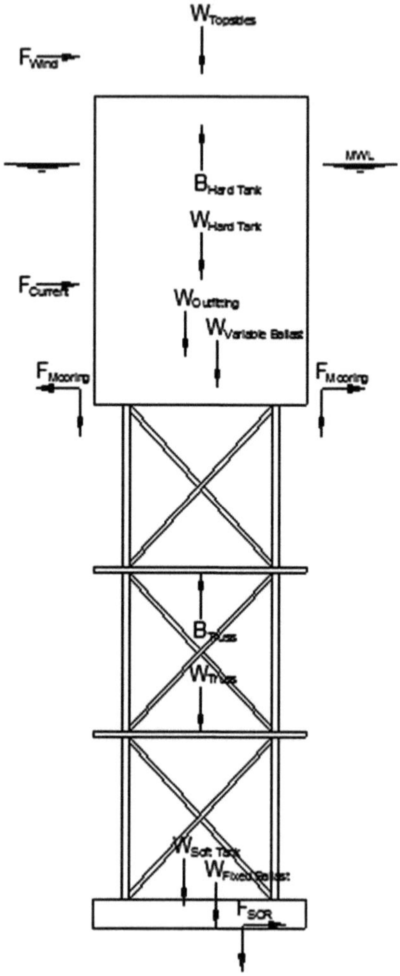

Fig. 21 Free body diagram for spar hull sizing (courtesy Technip USA Inc.)

7 Spar Construction and Installation

7.1 Construction

Spars are traditionally built and transported horizontally. They are upended while afloat near the installation site. This facilitates the use of conventional shipyard and fabrication yard equipment to fabricate and assemble the hull.

The particular method of fabrication and assembly will be unique to each yard and is dependent on the shop space and equipment available. Over 70 % of the spars built to date have been constructed at the Technip Mantiluoto yard in Pori, Finland. Figures 23, 24, 25, 26 and 27 show the sequence of construction used there.

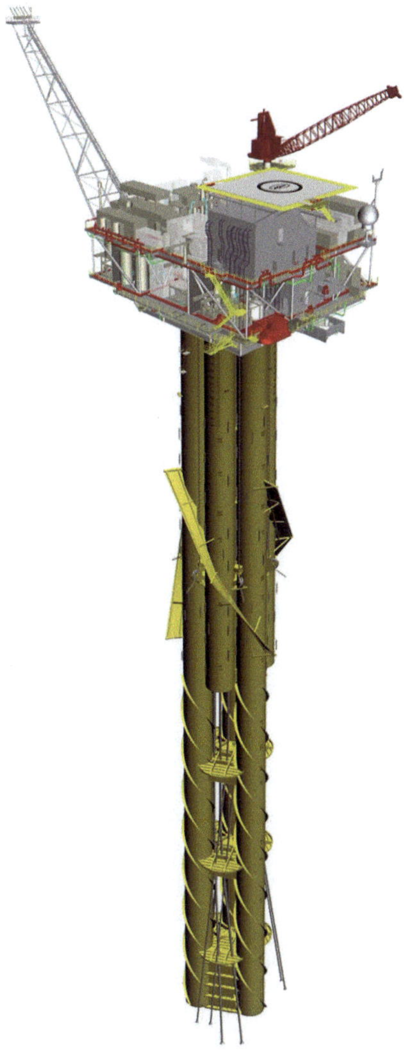

Fig. 22 Cell spar (courtesy Technip USA Inc.)

Subassemblies (Fig. 23) are fabricated in a shop equipped with automatic plate cutting and welding equipment. The size of the subassemblies is limited by the lifting and transporting capacity available.

The subassemblies are joined to form half round ring sections of the spar (Fig. 24). If the spar is very large, the half section may be too large for the fabrication shop and instead smaller sections may be constructed. The upper partial ring section is transported to a lifting tower and raised. The lower section is transported below this section and the two sections are welded together (Figs. 25 and 26). As each ring section is assembled, the entire section is moved to a skidway and connected to other sections to make the entire spar (Fig. 27). This procedure would

Fig. 23 Fabrication of subassemblies (courtesy Technip USA Inc.)

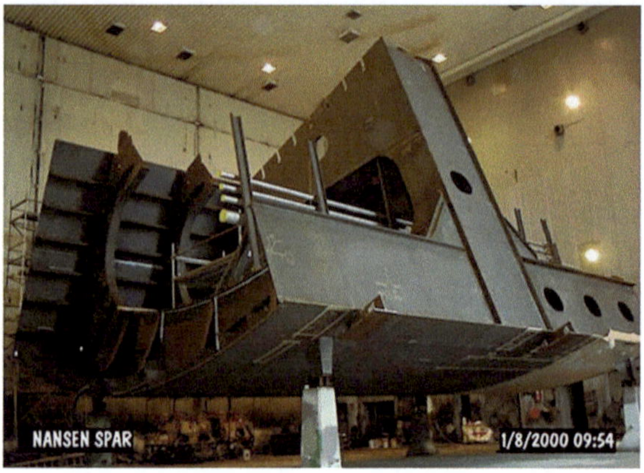

Fig. 24 Rollout of half-section (courtesy Technip USA Inc.)

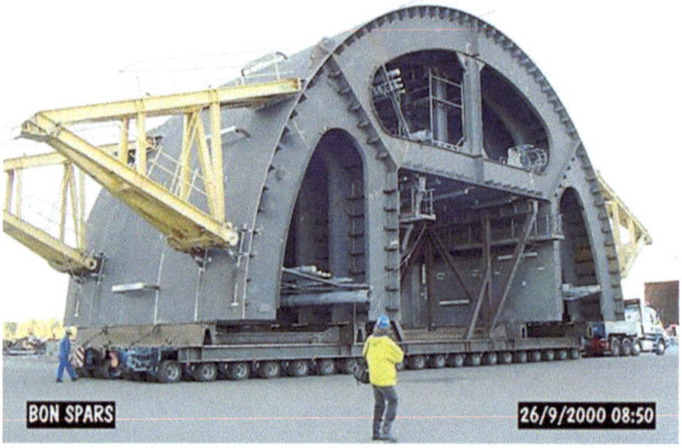

Fig. 25 Lifting of upper half-section (courtesy Technip USA Inc.)

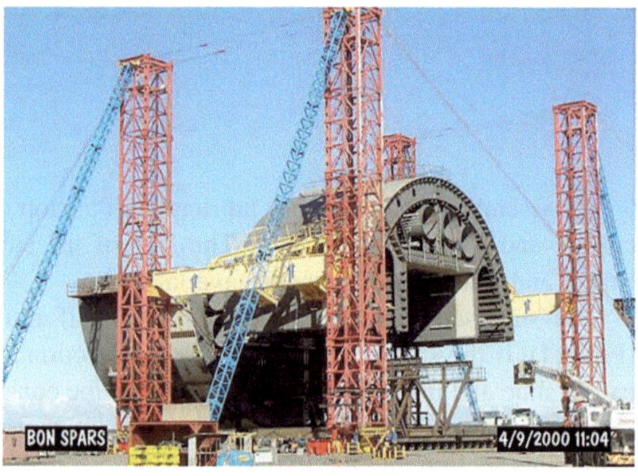

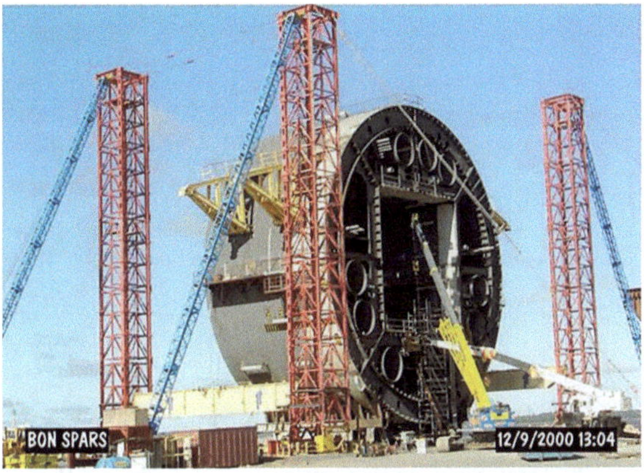

Fig. 26 Connection of two half-sections (courtesy Technip USA Inc.)

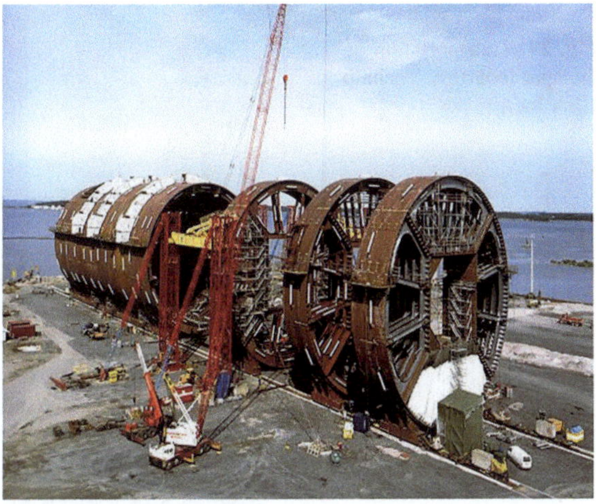

Fig. 27 Construction of the spar (courtesy Technip USA Inc.)

apply to the hard tank of a truss spar. The truss section would be fabricated separately and joined to the hard tank on the skid ways.

After construction, the spar is loaded out onto a heavy lift vessel for transportation using winched and skid beams to move the hull. The hull may then be transported to a staging area in protected waters near the installation site, or transported to a facility near the installation site (Fig. 29). The Gulf of Mexico is a 21-day transit from Finland. This requires careful planning and analysis of the vessel responses for any potential storms during transit (Fig. 28).

The construction phase begins with the completion of Front End Engineering Design (FEED) which includes appropriate class approvals for the design. Construction includes detailed design and fabrication. One facet of the spar construction that aids in faster fabrication time is concurrent engineering. This means that detailed design proceeds from section to section in the order in which they will be fabricated. For example, Fig. 26 shows one section being assembled and Fig. 27 shows the fabricated sections lined up on the skidway ready for joining.

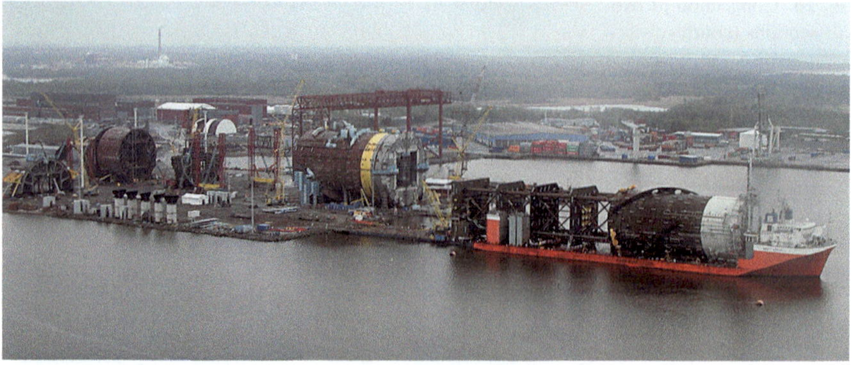

Fig. 28 Three Spars under Construction at Technip's Pori, Finland yard (Technip USA Inc.)

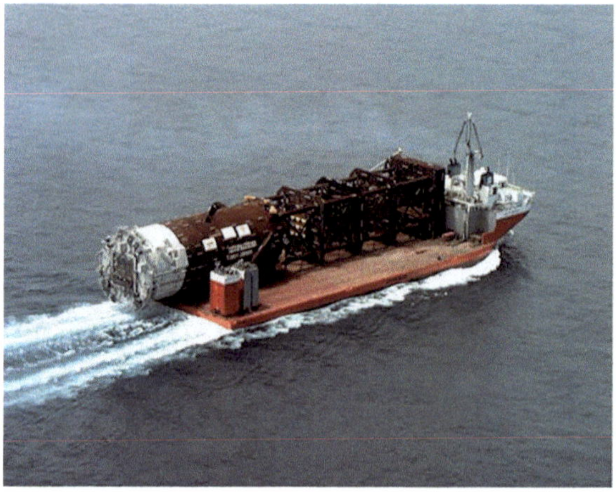

Fig. 29 Transportation of the spar (courtesy Technip USA Inc.)

The entire spar does not have to be designed before construction begins because the pieces are relatively independent. In fact, on occasion, it has been necessary to lengthen the hard tank (e.g. when topsides weight increases) AFTER the hull construction has started. As long as the assembly is from the top section down this change could be accommodated without a major schedule impact. The total construction time is typically 17–19 months.

The cell spar (Fig. 22) was designed specifically for ease of fabrication in a typical offshore fabrication yard. The cell spar consists of rolled and ring stiffened tubes that are joined on a pipe rack and assembled on a skidway. A challenge of build the spar this way is getting it in the water without the use of an expensive heavy lift transport ship. The spar may be skidded onto a barge, but launching introduces many design challenges. The Red Hawk spar, weighing 7,000 tonne, was lifted of the barge by a large stiff leg crane at the Kiewit yard in Corpus Christi, Texas (Fig. 30).

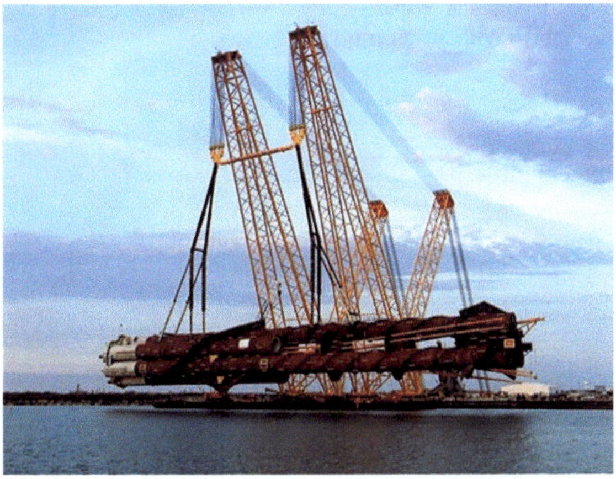

Fig. 30 7,000 tonne *red* hawk cell spar being lifted off of launch barge (Kiewit offshore)

Williams Corp. has introduced a new generation of a classic spar which is also designed for fabrication yard construction. The first one, Gulfstar 1, was built in a graving dock and floated out by itself (Fig. 31).

7.2 Mooring and Hull Installation

The installation of a permanent mooring requires that the anchor piles and mooring lines be preinstalled several months before the spar is installed. Anchors usually consist of steel piles which are either driven into the soil with hydraulic hammers, or steel piles which are forced into the soil using suction (Suction Piles). Figure 32 shows a pile driving installation using a derrick barge. A wire to support the hammer and pile is run through the main block of the derrick. Separate windlasses and winches are used to pay out the main anchor chain, spiral strand and an umbilical for the hammer.

When the pile is driven to its design penetration, the chain and spiral strand are laid on the sea floor in a pattern that will facilitate retrieval and connection to the spar upon its arrival. Figure 33 shows the lay down pattern for pre-installed mooring.

Once the spar is offloaded from the transportation vessel and rigged for installation, it is wet towed to the installation site. The spar is potentially vulnerable to large bending moments and high loads on heave plates (in a truss spar), so this wet tow is limited to smaller sea states, typically an annual storm event. If the tow is long, a 10-year design storm criterion may be invoked to reduce risk. When the spar arrives at the installation location, it is upended by flooding the soft tank and midsection as shown in Fig. 34.

Once the hull is upended it is made ready for attachment of the mooring lines. This has typically involved these steps:

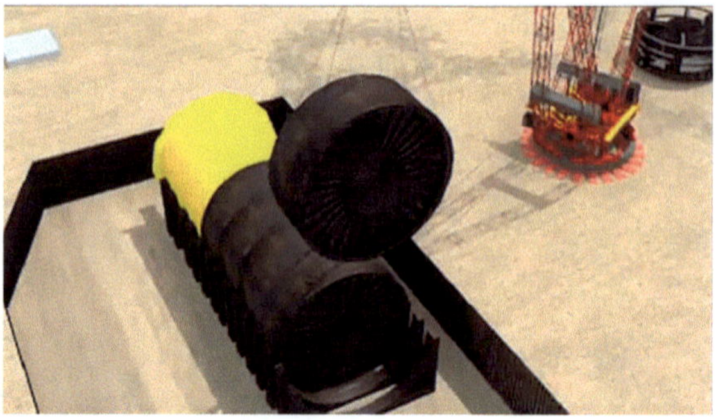

Fig. 31 Gulfstar 1 spar assembly (Williams energy)

1. Ballasting of the spar for an even trim and acceptable freeboard,
2. Lifting of a temporary deck containing power supply, messenger wire and winches,
3. Paying out of the messenger wire to the derrick barge,
4. Retrieving the bitter end of the pre-laid wire rope and chain using an ROV (Fig. 35),
5. Connecting the platform chain to the messenger wire,
6. Pull-in of the chain.

This procedure is repeated for each mooring line. When two opposite lines are installed, chain jacks are activated to pretension the lines. Once all the lines are attached and the mooring is secure, the temporary deck is removed.

Once the spar is upended and moored, and prior to the installation of the deck, fixed ballast must be added to the lower soft tank. The fixed ballast is typically iron ore in the form of magnetite shipped bulk from an iron ore producing country. It arrives in a barge where it is mixed with water to form slurry that can be pumped through pre-installed piping to the appropriate compartment in the soft tank (see Fig. 36). The fixed ballast may be distributed to achieve a permanent offset for the center of gravity to counter an offset on the topsides. This reduces the amount of variable ballast required.

7.3 Deck Installation

Once the mooring system is installed, the spar is ready to receive the topsides deck. This requires a derrick vessel to lift the deck. Decks weighing from 30,000 kN (3,000 ton) to more than 200,000 kN (20,000 ton) have been installed on spars. Figure 37 shows the 32,880 kN (3,300 ton) lift of the Neptune spar deck using the McDermott DB 50 derrick barge.

Larger decks require multiple lifts. Figure 38 shows a module support frame being installed using the McDermott DB 30 derrick barge on the Chevron Genesis spar in preparation for setting additional production modules. This entails a protracted period for offshore hookup and commissioning of the deck. An alternative

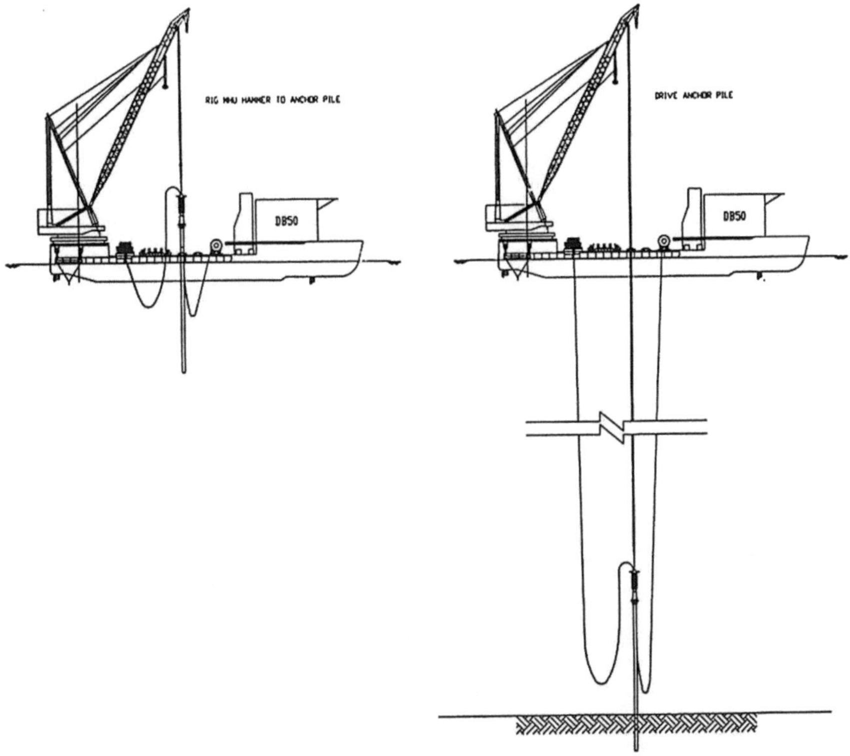

Fig. 32 Running pile from derrick barge (Kocaman et al. 1997)

to deck lifting is the floatover deck method which has been used extensively for fixed platforms (O'Neill et al. 2000). A spar floatover requires a catamaran transfer in order for the deck to pass over the spar. This method allows for the installation of very large, fully integrated decks on the spar (Maher et al. 2001). Figure 39 shows the first deck floatover on a spar platform. This was done on the Kikeh spar in Malaysia where derrick barges are not available. The sea state limits for float over installation are similar to those for a derrick barge installation.

A challenge with floatover deck operations on the spar is the relative motions between the deck and the platform. This is not as critical for lifted decks, as the load transfer may be accomplished in a matter of seconds. The spar floatover operation requires minutes or hours of ballast pumping to effectively transfer the load. In the meantime the two bodies are moving relative to each other and an effective means of absorbing these motions in the form of shock cells is required. A concept for mitigating this developed by Deep Oil Technology and Technip in the late 1990s and 2000s is illustrated in Fig. 40. Here the barges carrying the deck have a small waterplane area; thereby reducing the motions in swell significantly. Model tests of this concept indicated a large weather window for performing the operation (Tahar et al. 2004).

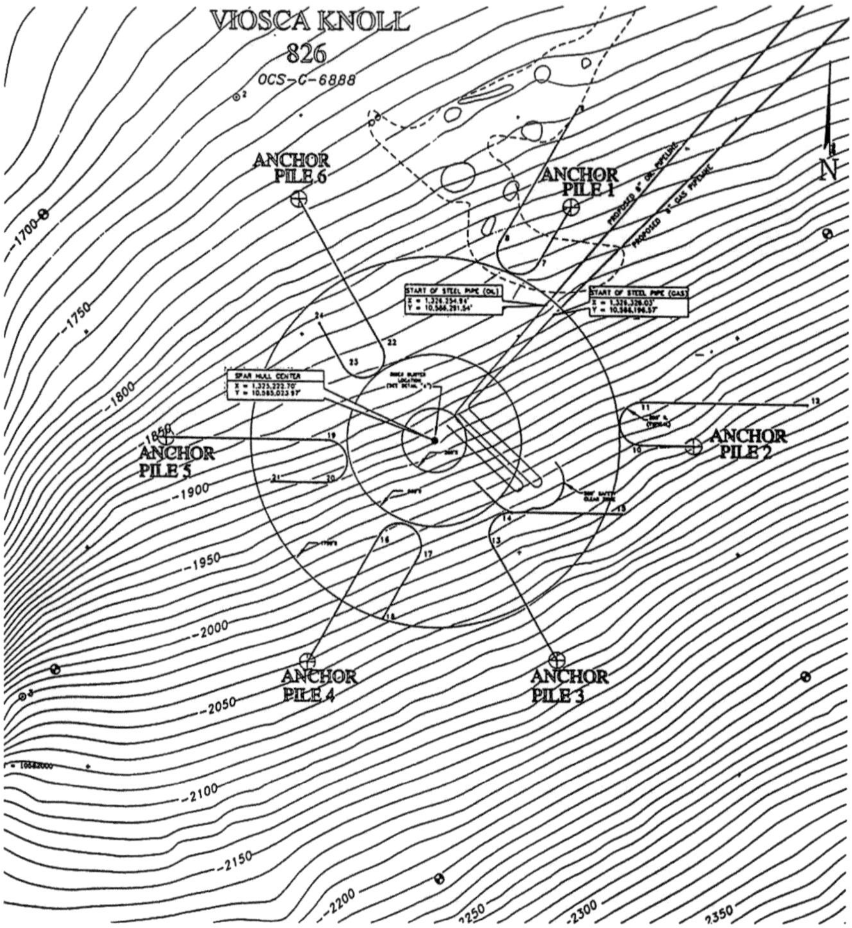

Fig. 33 Lay down pattern for pre-installed mooring (Kocaman et al. 1997)

8 Spar Responses

8.1 Waves and Wind

Spars are able to support top tensioned risers because the spars' heave motions are minimal. Figure 41 illustrates the wave motions. Because the wave pressure and motions deck with depth, the heave producing dynamic pressure fluctuations at the keel, typically at a depth of 160–180 m, are negligible except for the very longest waves. The horizontal motion of the spar is accentuated at the surface, where the spar moves essentially with the wave particles. Because of the decay of wave motions with depth, the bottom of the spar has negligible surge motion. Hence the

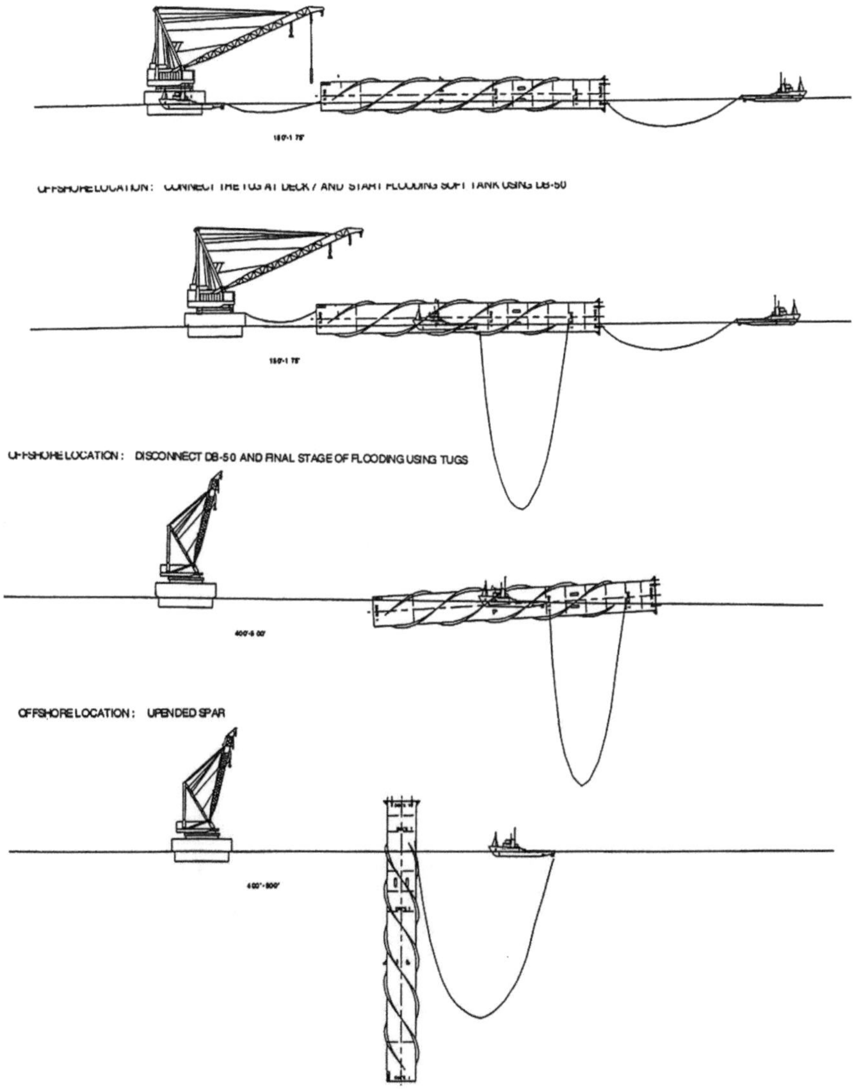

Fig. 34 Spar upending sequence (Kocaman et al. 1997)

wave frequency pitch angle (in radians) is approximately equal to the amplitude of the waves divided by the spar draft (about 4° amplitude in a 100 year hurricane).

Originally the draft of spars was selected at 180 m, which appeared to be a point of diminishing returns for the heave responses. Shortly after the first spar was installed, and fully instrumented, a 50-year hurricane passed directly over (Fig. 42). The measured heave responses, Fig. 43, were about 1/3rd of the predicted values. The reason for this was that the predicted values, based on linear theory, over-predicted the response *at resonance* due to an underestimation of damping.

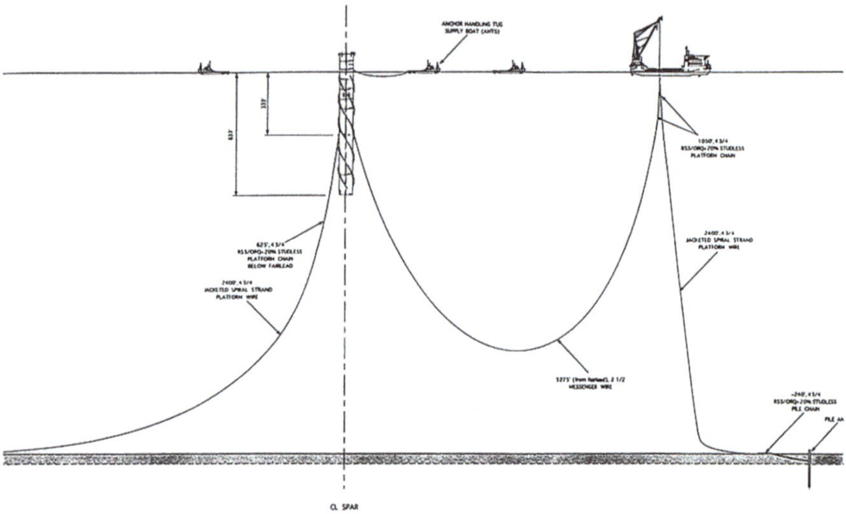

Fig. 35 Spar mooring line hookup (Kocaman et al. 1997)

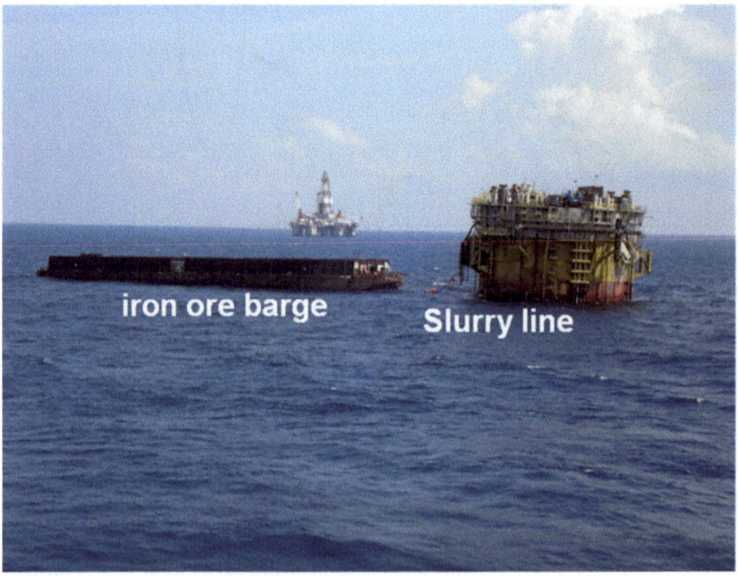

Fig. 36 Fixed ballast (iron ore) installation (courtesy Technip USA Inc.)

Subsequent to this more attention has been paid to damping effects, especially those due to riser interactions with the hull, i.e. Coulomb friction. As a results, second generation spars and the more recent Gulfstar spars by Williams have been built at some savings at a shallower draft, around 150 m.

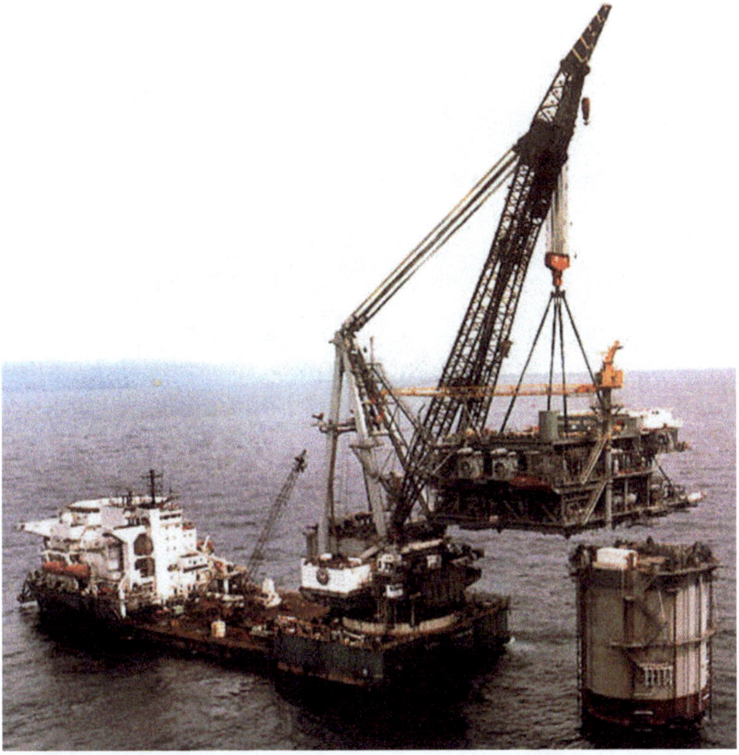

Fig. 37 Derrick barge setting deck on spar platform (Anadarko)

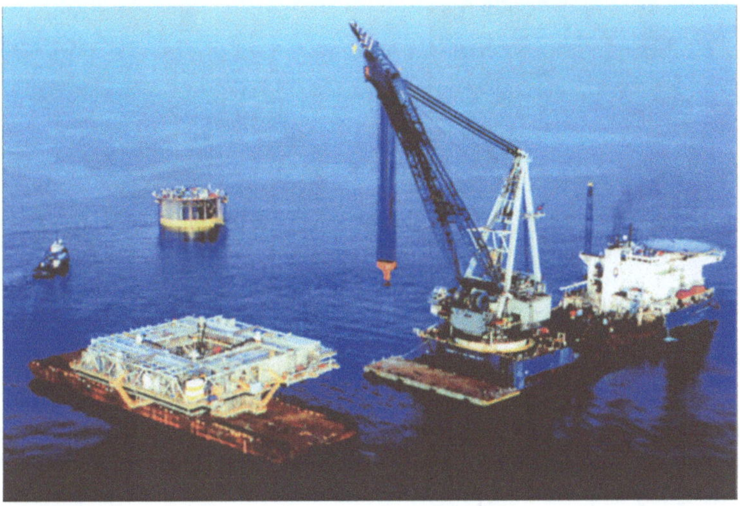

Fig. 38 Module support frame for a multi-deck lift installation (Spars International)

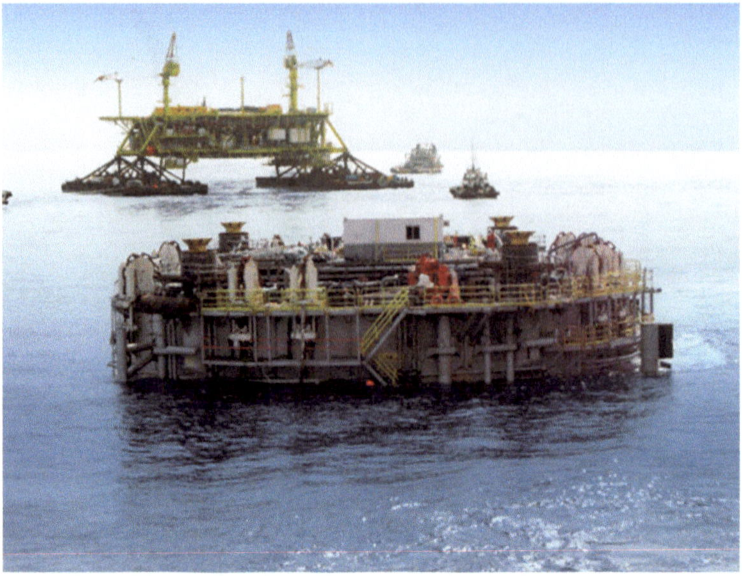

Fig. 39 Kikeh spar deck floatover (Halkyard)

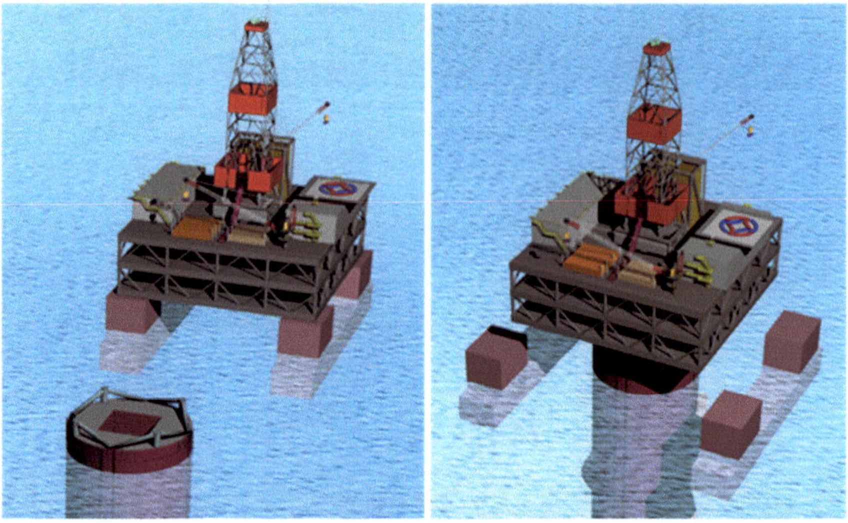

Fig. 40 Small water-plane area pontoons for spar floatover (Tahar et al. 2004)

Spars typically have very long natural periods in surge and pitch. This gives rise to resonant responses, called "slow drift" or "2nd order motions". Again, damping is very important in predicting the magnitude of these responses. Once again field data indicated that these responses are over predicted by normal numerical tools. Figure 44 shows the response spectra for spar pitch motion in a

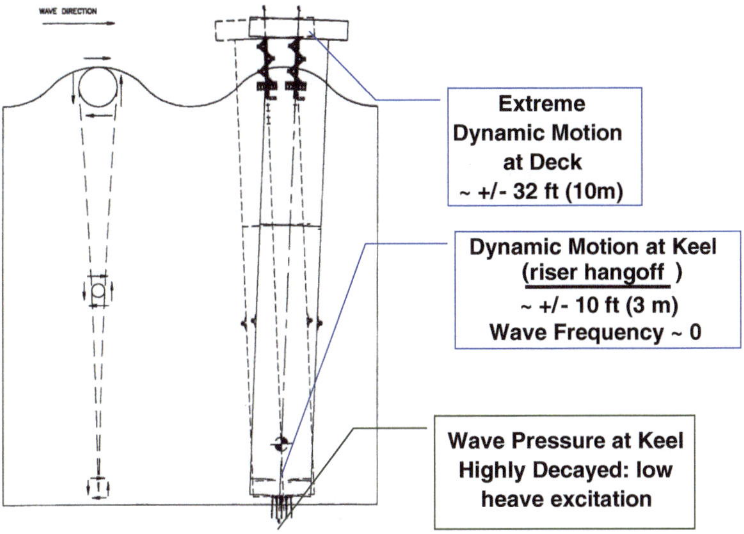

Fig. 41 Illustration of spar motions (Glanville et al. 1991)

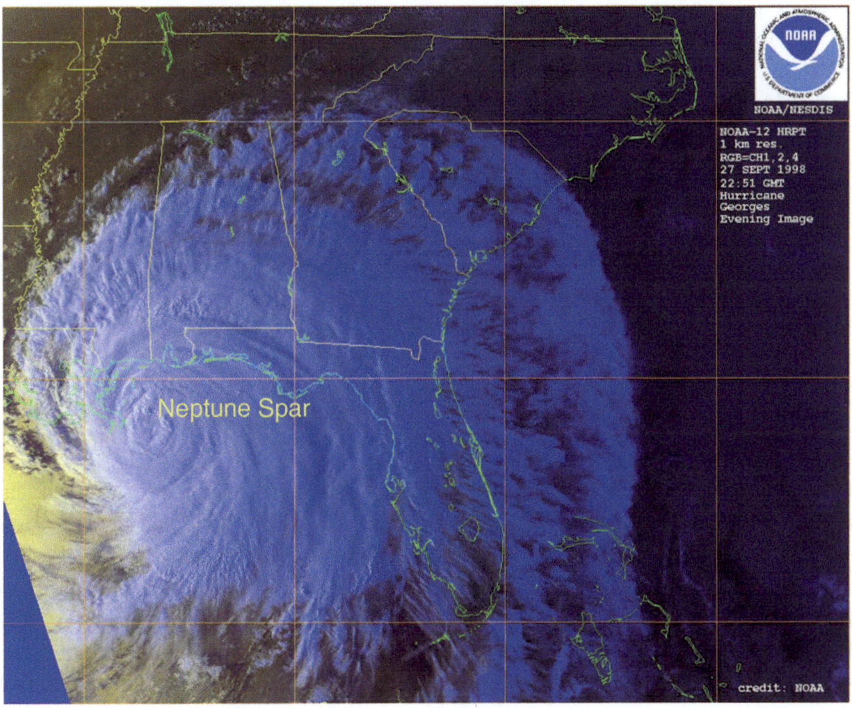

Fig. 42 Position of neptune spar for hurricane georges (Prislin et al. 1999)

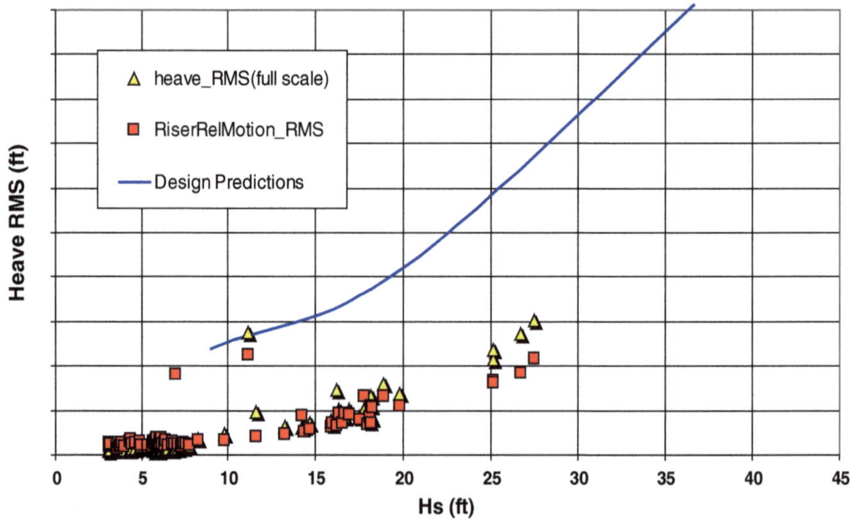

Fig. 43 Measured heave motions of neptune spare (Prislin et al. 1999)

Fig. 44 Measured pitch response spectra for Horn Mountain spar (Tahar et al. 2004)

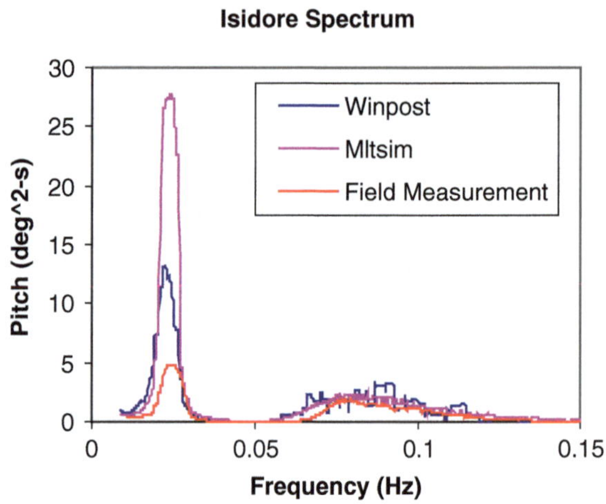

Gulf of Mexico hurricane. The design predictions were made with a non-linear time domain analysis tool, MLTSIM (Paulling and Shin 1985). The wave frequency responses, between .05 and .1 Hz, follow the theory almost exactly (see above discussion of wave frequency pitch). The slow drift pitch responses at resonance, around 0.023 Hz, were highly over-predicted by MLTSIM. It was believed part of the reason for this was that MLTSIM is an uncoupled program, i.e. the risers and moorings are treated quasi-statically. Dynamics and the effect of drag on these members are not considered. Subsequently a coupled program, WINTPOST (Kim 2012), which includes these effects was used to analyze the motions. The results show that the responses were less, but still not as low as measured. The accurate estimation of full scale damping, and wave loads, remains an area of active interest in the research community. Fortunately, prediction methods have proved conservative.

8.2 Vortex Induced Motions

A response issue unique to deep water floating bluff platforms like spars, semi-submersibles and TLPs is the vortex induced response to current. It is well known that bluff bodies shed a vortex wake of regularly spaced vortices. These cause a periodic forcing function on the body, primarily in the direction normal to the current. When the period of these vortices is close to the resonant period of the body (if it is allowed to oscillate), then the body responds with a periodic motion. For slender bodies like risers and pipelines this is called vortex induced vibrations. When it occurs on a floating bluff body, the term "Vortex Induced Motions", or "VIM" is used.

For typical spar platforms this resonant "lock in" occurs at current velocities greater than about 1 m per second. When this occurs, a bare spar would oscillate at amplitude nearly equal to its diameter. Furthermore, at lock in the effective drag coefficient may be almost doubles. At the current velocities encountered in the Gulf of Mexico it would be very costly to design a mooring system for spars. Because of this, every spar so far installed has been outfitted with helical strakes, as shown in Fig. 45, that mitigate these effects. Figure 46 shows the potential effectiveness of

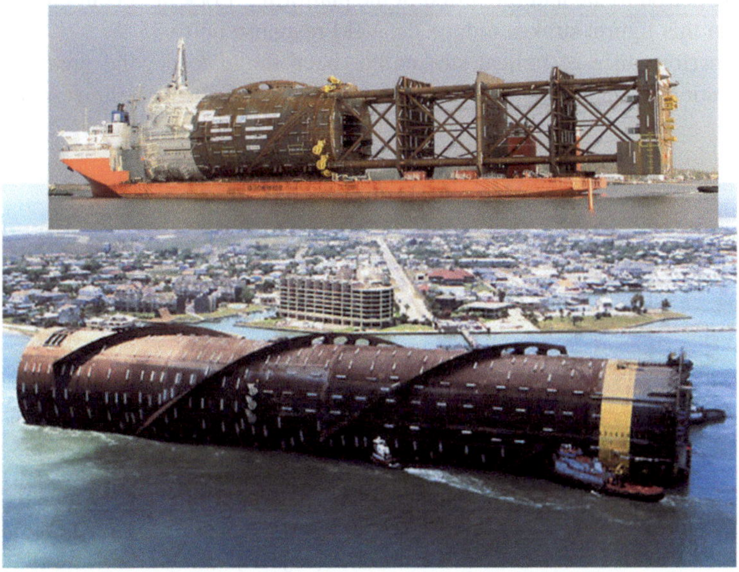

Fig. 45 Helical strakes are required on spars (courtesy Technip USA Inc.)

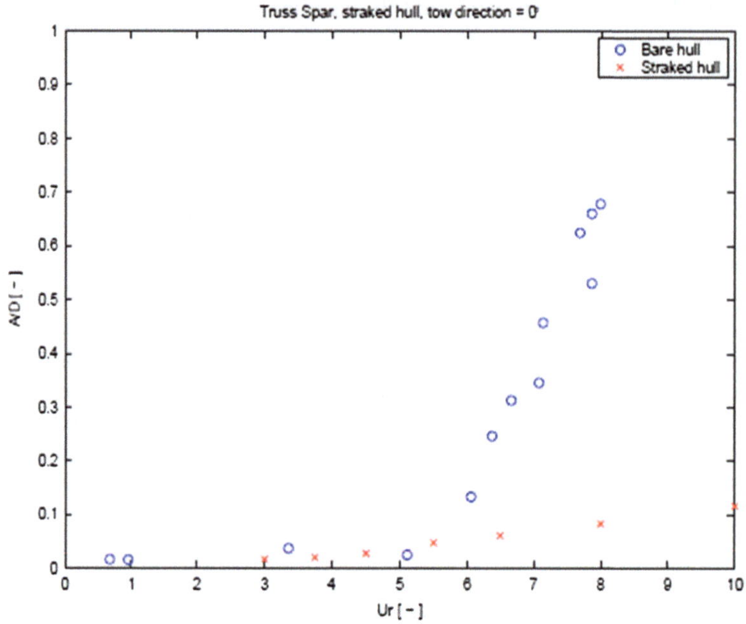

Fig. 46 A/D as a function of Ur for a truss spar with and without strakes (van Dijk et al. 2003)

strakes. "A/D" refers to the ratio of VIM amplitude to platform diameter, and "Ur" is the reduced velocity, a non-dimensional measure of the ratio of the vortex shedding period to the natural period of the platform. "Lock in" occurs for Ur $\geq$ 5. Based on this figure, strakes reduce the VIM response up to 85 %.

The performance of helical strakes has been found to depend greatly on the configuration of appurtenances on the outside of the hull. And the responses are

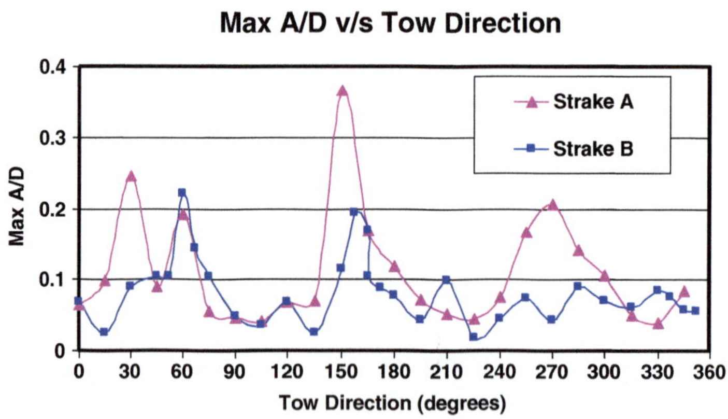

Fig. 47 Truss spar VIM amplitude versus direction (Halkyard et al. 2005)

Table 1 Spar project data

Project	Year	Operator	Type	Location	Water depth (ft)	Contractor	Oil (bpd)	Gas (mm)scfd	TTRs	Drilling	Topsides (st)	Hull (st)	Dia. (ft)	Length (ft)	Mooring
Neptune	1996	Oryx Energy	Classic	GOM	1960	J Ray McDermott/ Aker Maritime/Deep Oil Technology	25000	30	16	Workover	5500	13000	72	705	6–4.75″ Chain/ Strand
Genesis	1998	Chevron	Classic	GOM	2598	Spars International	55000	72	20	Full drilling	17000	28700	122	705	14–5″ Chain/ Strand
Hoover/ Diana	2000	ExxonMobil	Classic	GOM AC 25/ 26	4800	Aker Maritime (SII License)	100000	325	6	Full drilling	24000	35000	122	705	12-5.6″ Chain/ Strand
Medusa	2002	Murphy	Truss	GOM MC 582	2224	SparTec	40000	110	6	Workover	9000	13000	94	586	10
Nansen	2002	Kerr McGee	Truss	GOM	3678	Spars International	40000	200	9	Workover	8800	12000	90	543	9–4.9″ Chain/ Strand
Boomvang	2002	Kerr McGee	Truss	GOM	3451	Aker Maritime(Technip)	40000	200	5	workover	8800	12000	90	543	9–4.9″ Chain/ Strand
Horn mountain	2002	BP	Truss	GOM	5423	Aker Maritime (Technip)	65000	68	14	Workover	9000	14600	106	585	9–5″ Chain/ Strand
Gunnison	2003	Kerr McGee	Truss	GOM	3150	Aker Maritime (Technip)	40000	200	9	Workover	10700	13800	98	549	6–4.9″ Chain/ Strand
Front runner	2004	Murphy	Truss	GOM	3330	SparTec	60000	110	8	workover	11000	14000	94	586	9
Holstein	2004	BP	Truss	GOM GC 645	4344	Technip	110000	90	15	Full drilling	27000	35550	149	746	16–5.9″ Chain/ Strand
Mad dog	2004	BP	Truss	GOM	4419	Technip	80000	40	16	Full drilling	18500		128	555	11–5.8″ C/10.1″ P

(continued)

Table 1 (continued)

Project	Year	Operator	Type	Location	Water depth (ft)	Contractor	Oil (bpd)	Gas (mm)scfd	TTRs	Drilling	Topsides (st)	Hull (st)	Dia. (ft)	Length (ft)	Mooring
Red hawk	2004	Kerr McGee	Cell	GOM	5299	Technip	15	300	wet tree		4400	7000	64	560	6–4.5″ C/8.5″ P
Devils tower	2004	Dominion	Truss	GOM MC 773	5610	SparTec	60000	60	8	Workover			94	586	
Constitution	2006	Kerr McGee	Truss	GOM GC 680	4970	Technip	70000	200	6	Workover	10770	14800	98	554	
Kikeh DTU	2007	Murphy	Truss	Malaysia	3937	Technip			10	Tender assist			106	466	10
Tahiti	2007	Chevron	Truss	GOM GC 640	4000	Technip	125000	70	wet tree						
Perdido	2008	Shell	Truss	GOM AC 857	7817	Technip	130000	150							

highly directional. Figure 47 shows an example of the response of a spar with two different strake configurations at various angles. Deriving this kind of data for design remains a difficult challenge for both spars and other floating platforms. Currently, model tests are the only recognized method of securing design data, although much effort has been placed on CFD as an alternative.

Qualified field data to verify the VIM results from model tests is sparse; however there is evidence that our methods over predict the responses (Ma et al. 2013)

9 Spar Data

Table 1 tabulates available data on spars currently in service.

References

Bates, J. B., Kan, W. C., Allegra, A. P., & Yu, C. A. (2001). Dry tree and drilling riser system for hoover DDCV. *Proceedings of Offshore Technology Conference*, OTC 13084.

Berner, P. C., Gendron, K., Baugus, V. E., & Young, R. (1997). Neptune project: production riser system design and installation. *Proceedings of Offshore Technology Conference*, OTC 8386.

Cohen, R. (2009). An overview of ocean thermal energy technology, potential market applications, and technical challenges. *Proceedings of Offshore Technology Conference*, Paper 20176, Houston, USA.

Finn, L. D., Maher, J. V., Gupta, H. (2003). The cell spar and vortex induced vibrations. *Proceedings of Offshore Technology Conference*, OTC 15244.

Fisher, F. H., & Spiess, F. N. (1963). FLIP-floating instrument platform. *Journal of Acoustical Society of America, 35*, 1633–16544.

Glanville, R. S., Halkyard, J. E., Davies, R. L., Steen, A., Frimm, F. (1997). Neptune project: spar history and design considerations. *Proceedings of Offshore Technology Conference*, OTC 8382.

Glanville, R., Paulling, J. R. Jr., Halkyard, J. E., & Lehtinen, T. (1991). Analysis of the spar floating drilling, production and storage structure. *Proceedings of Offshore Technology Conference*, OTC 6701.

Glanville, R.S., Vardeman, R. Don (1999) The Neptune Spar - Performance Over the First Two Years of Production, *Proceedings of Offshore Technology Conference*, OTC 11073.

Halkyard, J. E., Sirnivas, S., Holmes, S., Constantinides, Y., Oakley, O., & Thiagarajan, K. (2005). Benchmarking of truss spar vortex induced motions derived from CFD with experiments. *Proceedings of 24rd International Conference on Offshore and Arctic Engineering* June 12–17, 2005. OMAE 2005-67252, Halkidiki, Greece.

Horton, E. E. (1987) Drilling, production and oil storage caisson for deep water. *US Patent Number: 4702321*, Filing Date: September 20, 1985, Issue Date: October 27, 1987.

Kim, M. H. (ed). (2012). SPAR PLATFORMS—Technology and Analysis Methods, ASCE Monograph ISBN 978-0784412091.

Kocaman, A., Verdin, E., & Toups, J. (1997). Neptune project: spar hull, mooring and topsides installation. *Proceedings of Offshore Technology Conference*, OTC 8385, Houston.

Lockheed Martin Corporation. (2011). NAVFAC ocean thermal energy conversion (OTEC) project contract number N62583-09-C-0083 CDRL A014 volume 4 of 4 SSOPP configuration report and CDRL A021 volume 4 of 4 SSOPP development plan for 2.5 MW OTEC mini-spar pilot plant. Report OTEC-2011-001-4 prepared for the U.S. Naval Facilities Engineering Service Center (NFESC). [http://hinmrec.hnei.hawaii.edu/wp-content/uploads/2010/01/LM-OTEC-Mini-Spar-Design_December-2011.pdf]

Ma, W., Igor P., Wu, G. Y., & Soma M. (2013). Vortex induced motions of a column stabilized floater. *Proceedings of the D.O.T International Conference*, October 22–24, 2013. Houston, Texas, USA.

Maher, J. V., Prislin, I., Chao, J. C., Halkyard, J. E., & Finn, L. D. (2001). Floatover deck installation for spars. *Proceedings of Offshore Technology Conference*, OTC 12971.

Moyer, M.C., Barry, M.D., Tears, N.C. (2001) Hoover-Diana Deepwater Drilling and Completions, *Proceedings Offshore Technology Conference*, OTC-13081.

Newman, J. N. (1963). The motions of a spar buoy in regular waves. U. S. Dept. of the Navy, David Taylor Model Basin, Report 1499

O'Neill, L. A., Fakas, E., Ronalds, B. F., & Christiansen, P. E. (2000). History, trends and evolution of float-over deck installation in open waters. *Proceedings of SPE Annual Technical Conference and Exhibition*, SPE 63037, Dallas, Texas.

Paulling, J. R., & Shin, Y. S. (1985). On the simulation of large-amplitude motions of floating ocean structures. *Proceedings of International Conference on Ocean Space Utilization* (Vol I, pp. 235–243), Nihon University. Tokyo: Springer

Prislin, I., Halkyard, J. E., DeBord, F. Jr., Collins, J. I., Lewis, J. M. (1999). Full-scale measurements of the Oryx neptune production spar platform performance. *Proceedings of Offshore Technology Conference*, OTC 10951.

Ross, J. M. (2012). Design considerations for a floating OTEC platform. *Proceedings of RINA Marine & Offshore Renewable Energy Conference*, London.

Smith, T., Allen, T., & Yu, A. (2005). Holstein spar supported vertical riser and dry tree system. *Proceedings of Offshore Technology Conference*, OTC 17254, Houston.

Tahar, A., Halkyard, J. E., Steen, A., & Finn, L. D. (2004). Float over installation method numerical and model test data. *Proceedings of 23rd International Conference on Offshore and Arctic Engineering*, OMAE 2004-51069, Halkidiki, Greece.

van Dijk, R., Magee, A., Perryman, S., & Gebara, J. (2003). Model test experience on vortex induced vibrations of truss spars. *Proceedings of Offshore Technology Conference*, OTC 15242, Houston.

van Santen, J. A., & de Werk, K. (1976). On the typical qualities of SPAR type structures for initial or permanent field development. *Proceedings of Offshore Technology Conference*, OTC 2716, Houston.

Vardeman, D., Richardson, S., & McCandless, C. R. (1997). Neptune project: overview and project management. *Proceedings of Offshore Technology Conference*, OTC 8381.

OTEC Platform

A.A. Yee

Abstract From our past experience, the most economical and durable structural material for the basic construction of floating OTEC platforms in a marine environment is a special mix of concrete reinforced with both conventional reinforcing steel and grouted prestressing steel tendons. When this material is molded into the shape of an integrated contiguous honeycomb pattern with cylindrical components in combination with straight walls, a robust structure of great stiffness and strength to provide high load capacity on exceptionally long spans uninterrupted by columns or vertical elements will result. In this manner, maximum spatial tolerance is made available for convenient access between pathways and equipment layouts for ease in maintenance work, replacement, rearrangements, etc. Special concrete mixes containing chemical and mineral additives can provide greatly increased durability so that the service life of the concrete structure can achieve a range of 100–150 years in a marine environment.

1 Introduction to Floating Platforms Supporting the Operation of OTEC Facilities

Ocean Thermal Energy Conversion (OTEC) facilities are generally supported on floating platforms on the sea surfaces where water depths are at least 1,000 m deep as this is the normal practical minimum water depth where adequate cold water temperatures can be obtained (4 °C at 1,000 m depth). For more details on

A.A. Yee (✉)
Precast Design Consultants Pte Ltd, 16 Jalan Kilang Timor, #06-02 Redhill Forum, Singapore 159308, Singapore
e-mail: alyee@precastdesign.com

A.A. Yee
Yee Precast Design Group Ltd., USA, 1217 Palolo Avenue, Honolulu, HI 96816, USA

C.M. Wang and B.T. Wang (eds.), *Large Floating Structures*,
Ocean Engineering & Oceanography 3,
DOI 10.1007/978-981-287-137-4_10

research and developments on OTEC, the reader may refer to the review paper by Wang et al. (2011).

Ocean experience with floating platforms of reinforced/prestressed concrete honeycomb framing systems over the past 30 years has proven that this framing concept is ideal for the support of the sophisticated equipment required for the OTEC facility. The integration of precast concrete cylindrical components in combination with a composite top slab, bottom slab and exterior side walls has proven that a structure of this framing concept will provide greatly increased robustness, resistance to large wave impact, dynamic forces, grounding, and wall collisions.

The 30-plus-year-old Concrete Island Drilling System (CIDS) platform, structurally framed with a reinforced prestressed concrete honeycomb structural system, has been actively exploring and drilling for oil in U.S.A. as well as Russian Territories in the Arctic Circle (VSL International Ltd 1992). The honeycomb module serves as the main ice impact resisting element of the CIDS platform (now renamed "The Orlan") after major additions in 2004 to convert the CIDS from an exploration to both exploration and production rig. It is in full service today and the 30-plus-year-old honeycomb module is showing no signs of damage or corrosion while actively operating to produce record profits. It has so far successfully withstood the pressures of large thick ice sheets, many intensive ice floes, frequent earthquakes of Richter 8.0 scale, 13 m high waves, and 6 m tall ice ridges without showing any signs of damage. We therefore conclude that the honeycomb framing is the most reliable proven structure for its purpose, adequately meeting all the structural, durability and special performance criteria.

2 Concrete Honeycomb Structural System

OTEC facilities generally involve a considerable array of sophisticated equipment and structural elements that must be integrated in a robust manner and supported by a floating ocean vessel constructed either in concrete, structural steel, or a combination of both materials in composite action designed to resist the powerful hydrodynamic and impact forces of wave action and sea storms of the open oceans. This floating vessel structure must be reliably water tight, extremely robust and durable. Ideally, the basic structural framing system in the lower compartment levels should provide large clear column free structural spans and spaces for uninterrupted internal areas where mechanical and electrical equipment layout for heat exchangers, pumps, valves, piping, wiring, control panels, etc. can be installed with maximum flexibility of efficiency of spacing and arrangement to achieve efficiency in functioning systems.

The top deck structure must also be designed to support significantly large gravity loads for machinery and heavy equipment of all sorts in different locations in a flexible manner to provide for maximum efficiency in the equipment functioning, maintenance, repair and replacement process. Design loads in these areas would be generally in the range of at least 48 kPa (1,000 psf).

The appropriate use of structural materials like mineral additives and chemical admixtures to the reinforced and prestressed concrete mixes will greatly increase strength, reduce cracking, minimize or totally eliminate surface spalling, deterioration and the need for maintenance activity including dry docking. These maintenance operations are costly and can result in major interruptions to the useful service life and revenue producing stream of the platform facilities and operations. Chemicals used in the maintenance activity may also be toxic and harmful to the environment and may increase the cost of maintenance work significantly due to environmental protection regulations in effect.

3 Concrete Island Drilling System

An example of one of the most impressive and proven concrete framing systems successfully applied to marine platforms and ocean going vessels is the Concrete Island Drilling System (CIDS) utilizing the author's invention of the concrete honeycomb framing system (see Figs. 3 and 4 and papers by Yee 1982,2009; Yee et al. 1984,). This framing concept provides an extremely robust concrete structure that can resist extremely high forces over long periods without structural deterioration or failure. It is a structure that has the greatest stiffness and strength achieved with the least amount of reinforcing, prestressing steel and concrete. Its capacity to resist grounding and collision forces and its inherent water tightness and damage stability cannot be matched by any other known marine structural frame designs. The basic strength of the concrete honeycomb framing system structure is derived from its favorable geometric features. Properly integrated concrete cylindrical shells used in this design will provide a disproportionate amount of stiffness and strength contribution to the structure in terms of its resistance to buckling, bending, shear and wracking forces encountered in ocean operations. Large unobstructed clear spans to accommodate flexible and efficient interior layouts for equipment locations and operations can be designed to provide unusual robustness, stiffness and strength to concrete hulls. The concrete honeycomb framed marine structures can withstand high magnitude hydrostatic and hydrodynamic forces as well as heavy impact grounding, collision, ice sheet and ice floe forces without incurring any significant damage partly due to the inherent redundancy in the basic structural framing system.

One of the earlier applications of this system was to build a mobile island to support a major exploratory drilling platform to operate in the midst of seas covered with heavy frozen ice sheets and high impact ice floe activity in the Arctic Circle (see Fig. 3). This drilling platform supported equipment, housing, repair shops, replacement parts, etc., for oil exploration activity and is moveable by floatation and tug boats. Drilling operations took place with the CIDS in a grounded mode. The platform included living quarters for the operating crew of 100 personnel, and drilling equipment and supplies for the oil exploration operation. It was literally a mobile concrete island that contains all the facilities required for the exploration of

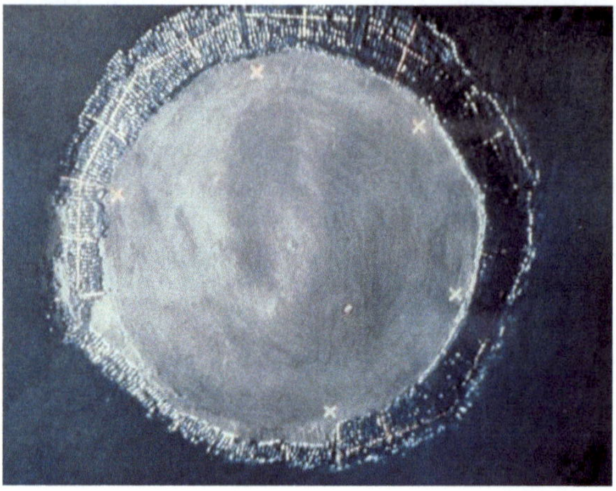

Fig. 1 Prior to the CIDS development, the usual method of supporting a drilling platform offshore in the Arctic Circle, North of Alaska, required constructing a US$100 million gravel island as the ice pressure was too severe for standard jack-up rigs. *Photo courtesy* Global Marine Development Inc

oil and gas deposits in the Arctic Circle where ice pressures are huge and destructive, air temperatures are extremely low (−35 °C) and the locations of drilling activity are extremely remote, difficult to access and challenging. Literally, the task was to build the moveable island "city" containing living quarters for the drilling crews, power plants for electricity, water treatment plants, sewage treatment plants, machine shops, repair-facilities, water and fuel and food storage, laundry facilities, heating systems and all other facilities to provide the necessary working and living comfort requirements.

Prior to the concrete honeycomb structure invention the only feasible and conventional method of oil exploration in the Arctic Circle was based on the construction of artificial islands man-made from gravel, sand and rocks mined from the nearby quarries. (See Fig. 1.) These aggregates were transported by barge to the drill supporting sites where they were used to construct new artificial islands to serve as large "stone pedestals" to support drilling equipment and housing for personnel. These islands were considered the only type of structure practical in cost and strong enough to resist the large ice pressures to be encountered. These islands were approximately 100 m in diameter across the top and with the sides sloping downward into the sea bottom the actual diameter of the island at the sea bottom would then be increased to as much as 150 m or more in diameter depending upon water depth and soil base profile. (See Fig. 1). These basic gravel islands cost about US$100 million each. The side slopes of these islands would then be lined with bags filled with concrete to form a protected revetment for shore protection against ice floes and sea wave erosion. These concrete filled bag linings would often be repaired by replacing units that get eroded and washed away by the ocean waves so they too create a maintenance problem. Figure 2 shows a gravel

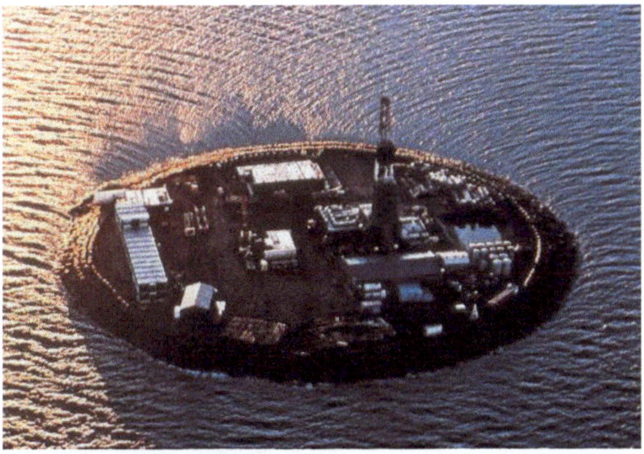

Fig. 2 Only after building the US$100 million gravel island can a drilling platform be mounted and operations begin. If no oil is found, the island is abandoned and removed, thus, incurring a loss of more than US$100 million. *Photo courtesy* Global Marine Development Inc

island that has been completed with all the living quarters, equipment, and infrastructure improvements installed ready for occupancy and drilling activity.

Only after drilling, one would know whether the island was built in the right location or not. If the drilling proves the island location to be a dry hole after several attempts the oil company will be forced to abandon this location and remove the island in order to protect the breeding grounds of the sea wild life (such as whales). The whole island is taken away bit by bit as the island is composed mainly of loose rocks, gravel and sand. Another drilling location is then selected and a new island of rocks, gravel and sand be built again in hope that oil will be found in the new location. Each gravel island generally costs about US$100 million to build.

Global Marine engineers, in their desire to develop the mobile island concept decided to adopt our honeycomb framing system for their main ice resistant concrete drilling module. We were then commissioned to design the basic hull of the CIDS in its entirety and participate in supervising its construction in a Japanese dry dock.

Figure 3 is a preliminary rendering of the CIDS as designed by the Global Marine Engineers. Note that a steel mud base resting on the sea bottom is supporting a concrete honeycomb module which serves as the principal ice resistant element of the platform. This honeycomb module is dimensioned to be effective for drilling and resisting ice sheets and ice floes in water depths between (9.15–18.3 m) 30 and 60 ft and is the key element to the success of the CIDS operations. Sitting above the CIDS honeycomb module are two main elements, the living quarters for the work crew and the drill rig with equipment shops and material storage for the drilling operations.

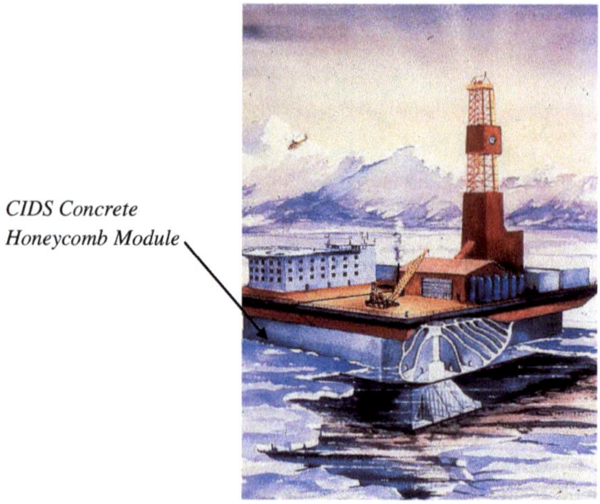

Fig. 3 Isometric of the US$75 million CIDS exploration drilling platform with concrete module to resist heavy ice pressures. The CIDS has been in service since 1984 in various locations with no sign of concrete cracking, spalling or rusting. *Photo courtesy* Global Marine Development Inc

Structural Design Criteria for the CIDS honeycomb module is as follows:

1. Hydrostatic Head

   Top slab 50 ft (seawater) (15.24 m)
   Bottom slab 70 ft (seawater) (21.34 m)
   Exterior sides 70 ft (seawater) (21.34 m)

2. Ice Forces

   a. Local ice pressure 900 psi on 5 ft × 5 ft footprint (632 tonnes/m^2)
   b. Global ice pressure 460 kips/ft (684 tonnes/m)

3. Temperature Gradient 52.8 °C on exposed areas
4. Live Load

$$\frac{\text{Top Slab 2 kips/sq.ft}}{9.767\ \text{Tonnes/m}^2}$$

$$\frac{\text{Bottom Slab 10 kips/sq.ft}}{48.8\ \text{Tonnes/m}^2}$$

The cost of this entire facility including the mud base, honeycomb concrete platform, ballasting systems, living quarters, drill rig, shops and equipment was US$75 million as compared to the old system where the gravel island alone costs US$100 million.

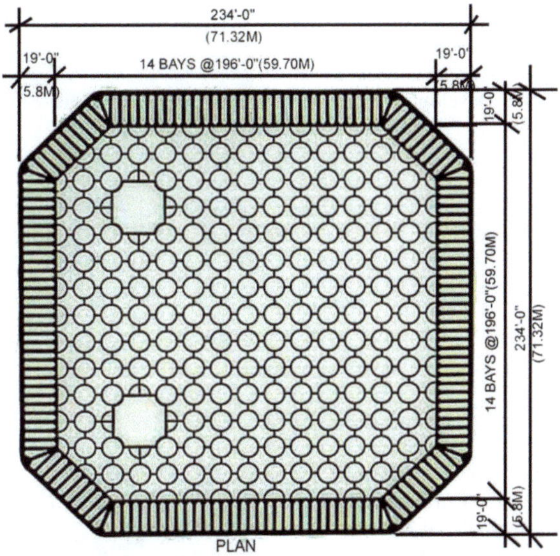

Fig. 4 Honeycomb concrete framing plan

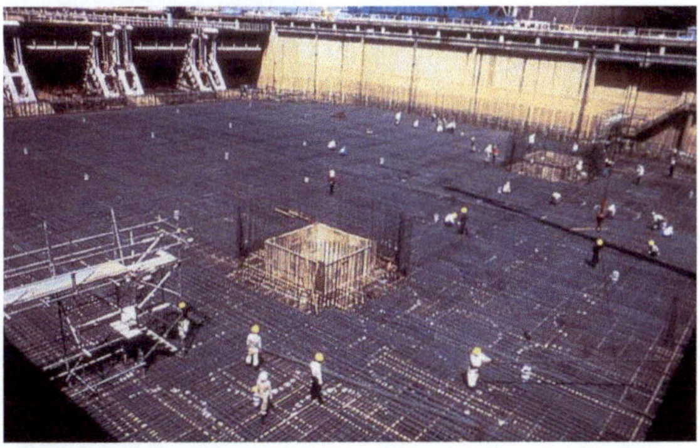

Fig. 5 Formwork, reinforcing and prestressing steel in concrete honeycomb structure. *Photo courtesy* Global Marine Development Inc

Figure 4 is a more detailed drawing of the concrete honeycomb structure as it was patterned for maximum resistance against the huge design ice pressures resulting from heavy ice sheets and large floating ice floe activity.

Figure 5 is an actual photo of the initial formwork, reinforcing and prestressing steel in preparation for pouring of the bottom slab of the concrete honeycomb module structure. Figures 6, 7 and 8 show the honeycomb construction module in progress.

Figure 9 shows the intricate 3-dimensional reinforcing required to resist the high combination of bending, shear and torsional forces imposed on the CIDS exterior concrete walls exposed to the unusually high ice pressure design criteria. The design ice pressure from the wall is based on 900 psi on a 5 ft $\times$ 5 ft footprint (632 tonnes/m^2), and the live load on the top deck is based on 2,000 lbs per square foot (9.767 tonnes/m^2). While on the bottom slab is 10,000 lbs per square foot

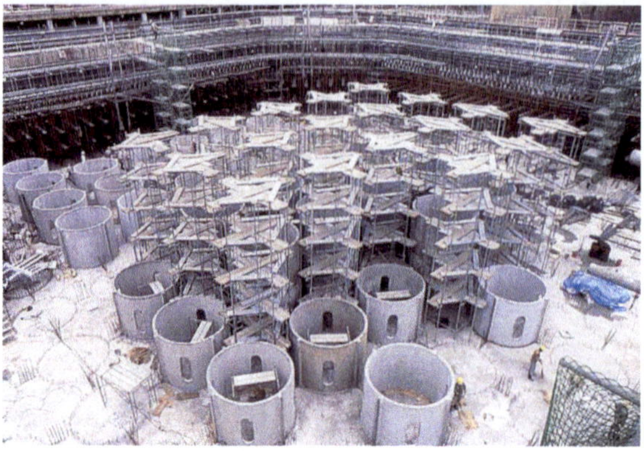

Fig. 6 Honeycomb construction at initial stage. *Photo courtesy* Global Marine Development Inc

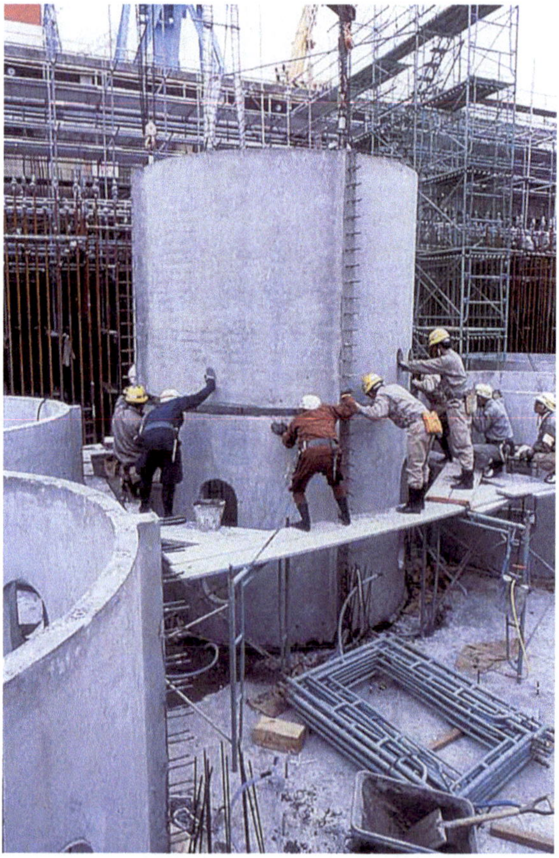

Fig. 7 Splicing of honeycomb structure. *Photo courtesy* Global Marine Development Inc

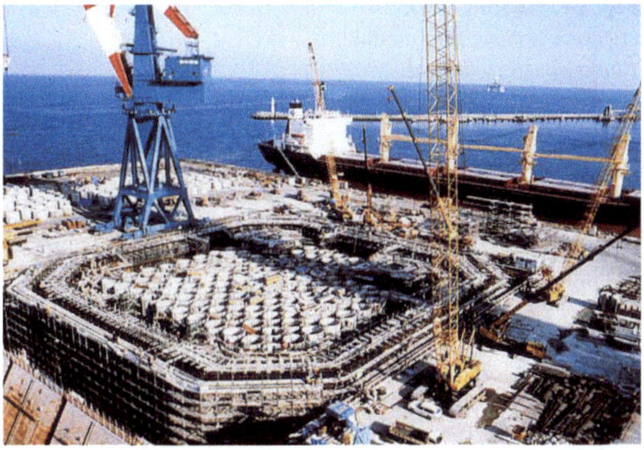

Fig. 8 Honeycomb construction at final stage. *Photo courtesy* Global Marine Development Inc

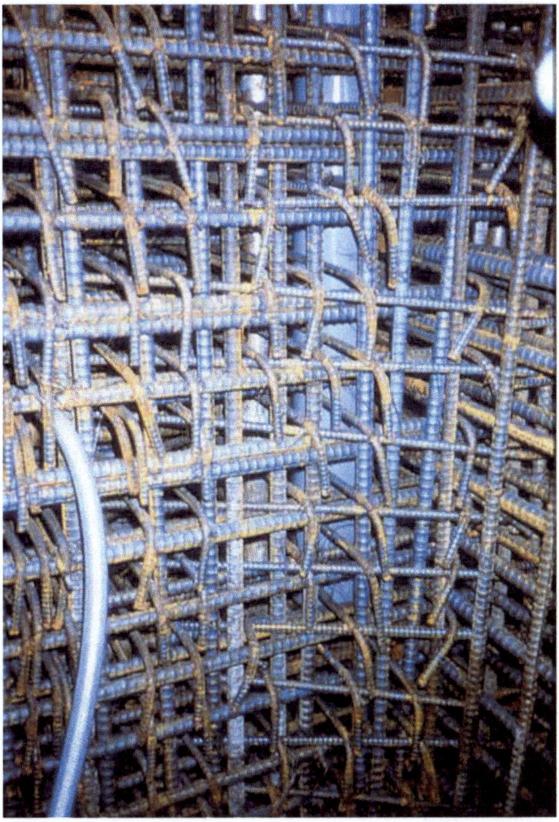

Fig. 9 Reinforcing steel arrangement in honeycomb structure

(48.8 tonnes/m^2). The honeycomb framing proved to be successful in forming a structure that is sufficiently robust to easily handle these very large loads. With over 30 years of operations so far no cracking, spalling or deterioration has ever been noticed on the concrete structure and we are satisfied with its performance.

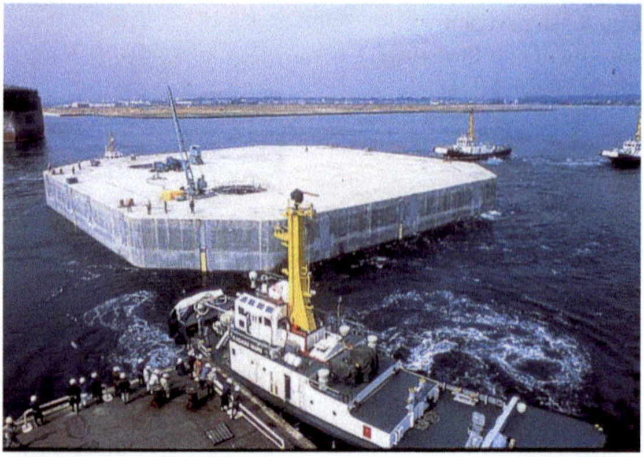

Fig. 10 Floating CIDS concrete honeycomb module. *Photo courtesy* Global Marine Development Inc

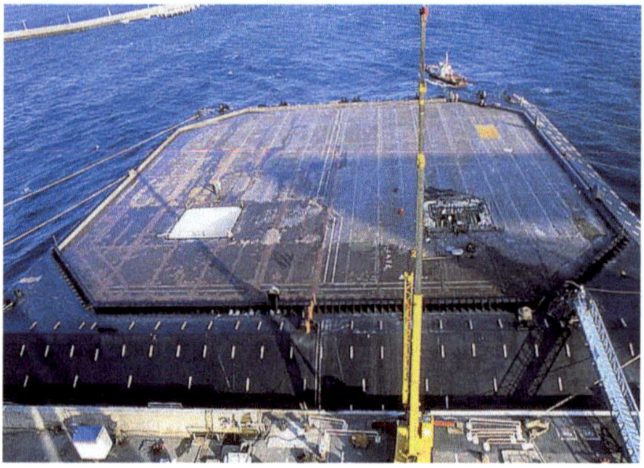

Fig. 11 Floating mud base. *Photo courtesy* Global Marine Development Inc

When the CIDS honeycomb module is completed in the dry dock it is then launched into the adjacent sea to begin the assembly procedure of the major CIDS components. Figure 10 shows the CIDS concrete honeycomb module afloat after launching from the dry dock.

Figure 11 shows the mud base also being floated to location in preparation for the mating procedure with the concrete honeycomb module.

Figure 12 shows the concrete module being positioned immediately above the temporarily grounded mud base for fitting together and permanent joining by welding and epoxy grouting.

Figure 13 is a close up photo showing the joint between the top of the mud base and the bottom of the concrete module executed by a combination of welding steel straps and epoxy grouting at the joint.

Figure 14 shows the final assembly of the CIDS principal components including the mud base, honeycomb concrete module, living quarter units, and drill rig unit.

Figure 15 shows the fully assembled CIDS being towed by two 22,000 horsepower tugs from Japan to the Beaufort Sea, North of Alaska.

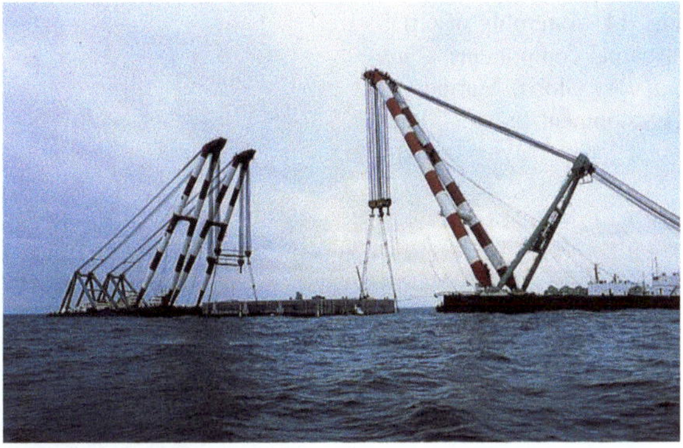

Fig. 12 Positioning of concrete module. *Photo courtesy* Global Marine Development Inc

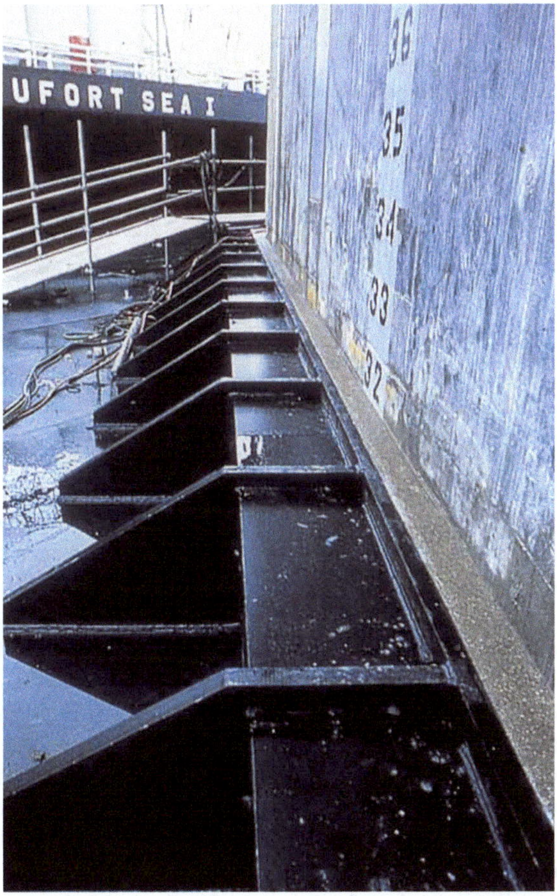

Fig. 13 Joint between top of mud base and bottom of concrete module. *Photo courtesy* Global Marine Development Inc

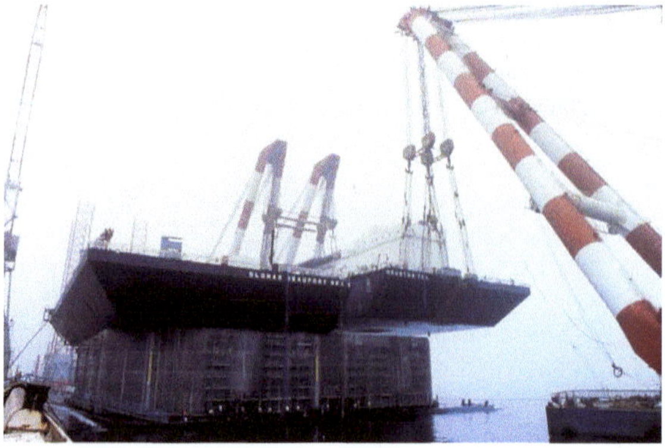

Fig. 14 Assembly of CIDS principal components. *Photo courtesy* Global Marine Development Inc

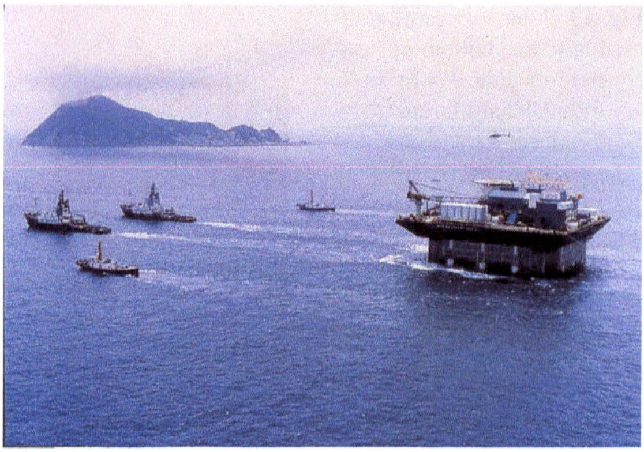

Fig. 15 Fully assembled CIDS being towed to site. *Photo courtesy* Global Marine Development Inc

Figure 16 shows the CIDS in parking position in the Beaufort Sea, North of Alaska, in the summer.

Figure 17 shows the CIDS in action in the middle of winter in the midst of a cracked ice sheet indicating its capable resistance to the high magnitude environmental pressures on this man-made structure.

The CIDS, originally owned by Global Marine Corporation, was initially operated as a facility leased to Exxon Mobil in 1984 for exploratory drilling in the U.S. territory of the Beaufort Sea, North of Alaska. Seventeen years later, when exploration of the entire U.S. territory in this region was completed, Exxon Mobil then purchased the CIDS from Global Marine in 2001 and modified the CIDS to be installed offshore the Russian Far East, Arctic Circle near Sakhalin Island. The CIDS was renamed The Orlan by the Russians after major renovations were made.

Some of these renovations included increasing the living quarters to accommodate 125 crew members instead of the original 100 crew members, the addition of oil production facilities to the then existing oil exploration facilities only, enclosing the entire topside of the CIDS with insulated cladding, and providing heating and lighting so that the crew can actively work 24 h a day, if necessary, regardless of the severe cold temperatures in the winter. The helicopter pad was

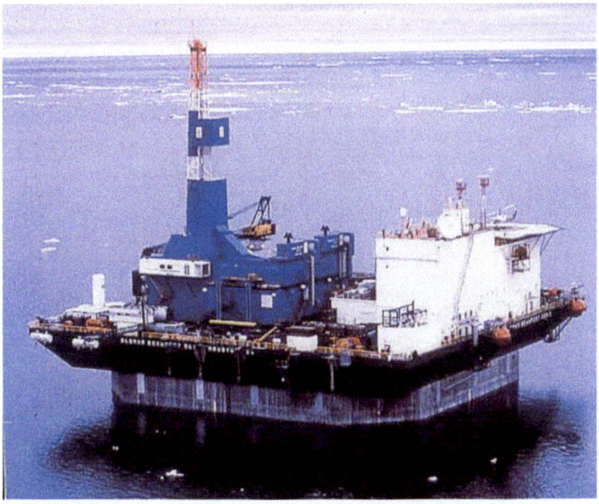

Fig. 16 CIDS in parking position in the Beaufort Sea, North of Alaska. *Photo courtesy* Global Marine Development Inc

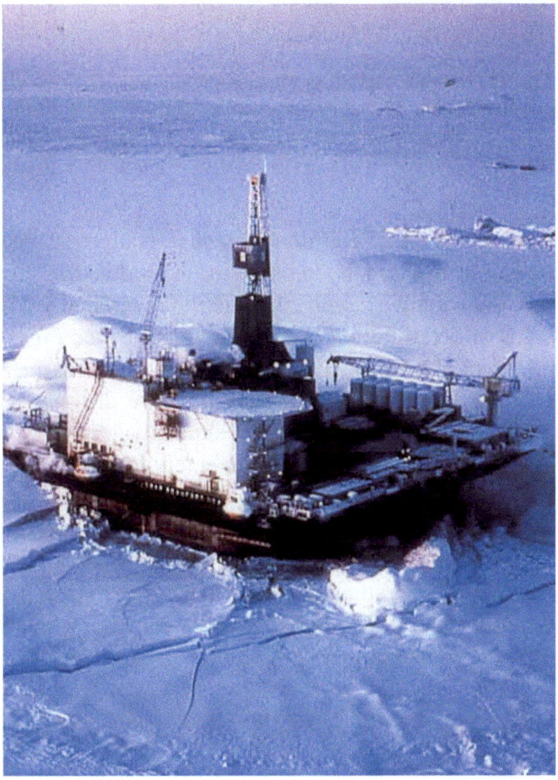

Fig. 17 CIDS under ice action. *Photo courtesy* Global Marine Development Inc

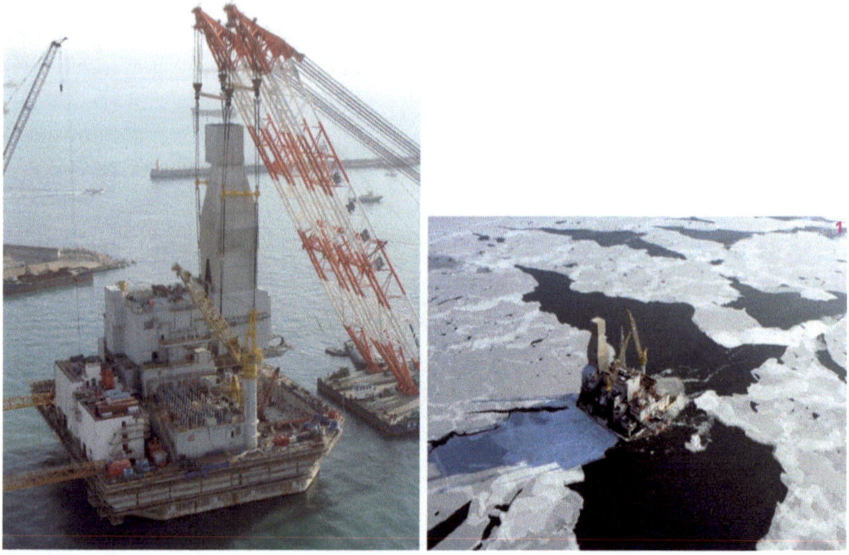

Fig. 18 In 2004, the CIDS was completely modified for oil exploration and production off the coast of Russia. It is now renamed as The Orlan Platform. The original honeycomb concrete module remains in full service as the main support and ice resisting element of the modified structure. *Photo courtesy* Global Marine Development Inc

also enlarged and additional cranage equipment installed. Except for some penetration work through the honeycomb module for utility and services modifications, the original honeycomb concrete module remained in use as is since there were no signs of deterioration or need for repairs (see Figs. 18 and 19).

According to the Russian publication, "*Russia Beyond The Headlines Asia*," July 19, 2013 issue, the Orlan offshore drilling platform is fully occupied producing oil in Sakhalin, Russia situated 11 km away from the coast in the Sea of Okhotsk. It is withstanding extremely low temperatures surviving under frequent 8.0 magnitude earthquakes, 13 m high waves and 6 m tall ice ridges. This platform has 14 MW available power and its 75 tonne drilling unit powered by a 2,300 horsepower drive can produce up to 23,000 tonnes of oil daily. The Orlan is capable of operating 20 wells, each with up to 13 km borehole deviation. The Orlan's capabilities and capacities are being constantly improved to handle a growing workload.

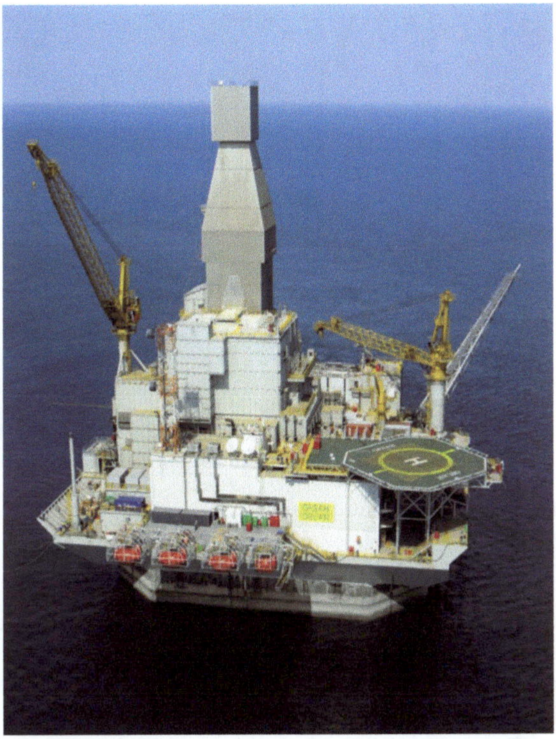

Fig. 19 Orlan Platform supported by original CIDS concrete module. *Photo courtesy* Global Marine Development Inc

4 Application of Concrete Honeycomb Module for OTEC Platforms

Our favorable experience with the CIDS concrete honeycomb module (now renamed The ORLAN) has encouraged us to consider this framing system for all types of permanent ocean and marine structures that require reliability, robustness, durability and long term service life. The oceans cover approximately 71 % of our planet's surface and are rich in minerals, sea life, and energy storage. The energy is stored in various forms and the largest of which is heat energy from the sun's exposure. The oceans serve as the largest heat sink in our planet and with OTEC technology we are able to convert this great heat energy into electrical energy that can be used in manufacturing processes such as to power the mechanical machinery and production equipment. With this electric power we can manufacture pure hydrogen and oxygen from sea water which can then be liquefied to greatly reduce the volume of material for convenient storage and transport when used for hydrogen fuel cell vehicles, alkaline fuel cell vehicles, and other equipment operations. The great heat sink provided by the oceans is a necessary operational facility in the process of liquefying these gases.

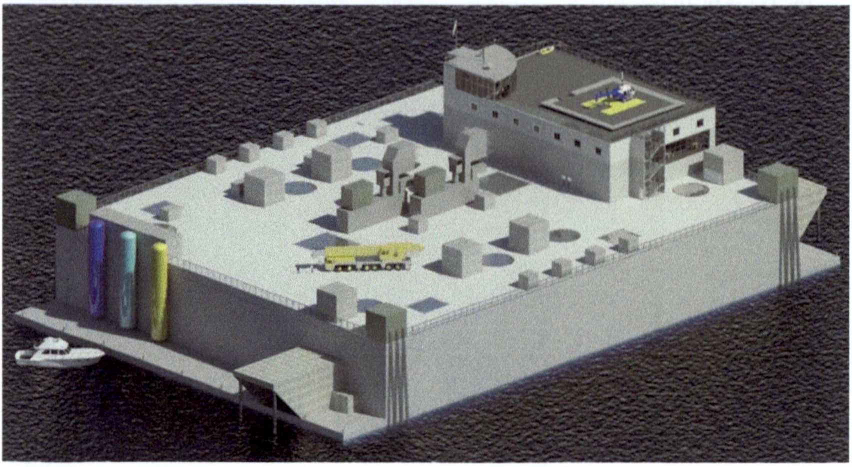

Fig. 20 20 MW OTEC plantship

It is obvious that for economic purposes, floating structures to support various equipment and materials for production activity, storage and transport will benefit from a long continuous service life with minimum maintenance and downtime in order to enjoy maximum production and use of equipment as well as return on investment.

The most reliable construction material with reasonable capital cost and proven long service life while requiring minimum maintenance and repair work is reinforced and/or prestressed high performance concrete for the hulls of these supporting ocean vessels. Well-designed concrete mixes will produce concrete of extremely high strength, water tightness and very long term maintenance free service life generally in the range of 100 years or more even under the worst marine exposures. Economically this works out to provide most favorable life cycle costs.

Among the most important structures that will require material of this type of performance are the hulls of the Ocean Thermal Energy Conversion (OTEC) plantships. These are best built to operate afloat on the seas with special equipment to station, over ocean areas to process the warm sea water and deep cold water through a system of pumps, heat exchangers and turbines to generate electric power. Figure 20 shows a rendering of a proposed OTEC plantship of 20 MW capacity. A rendering of a 60 MW capacity OTEC plantship is shown in Fig. 21, and a rendering of a 125 MW capacity OTEC plantship is shown in Fig. 22. Each plantship can produce electric power from which seawater desalination can be realized as well as the production of liquid hydrogen and oxygen can be produced by the electrolytic process.

With the honeycomb framing system, large clear column-free spans can be realized thus capability of providing wide flexible unobstructed spaces for convenient and efficient equipment layout can be realized. This will also allow

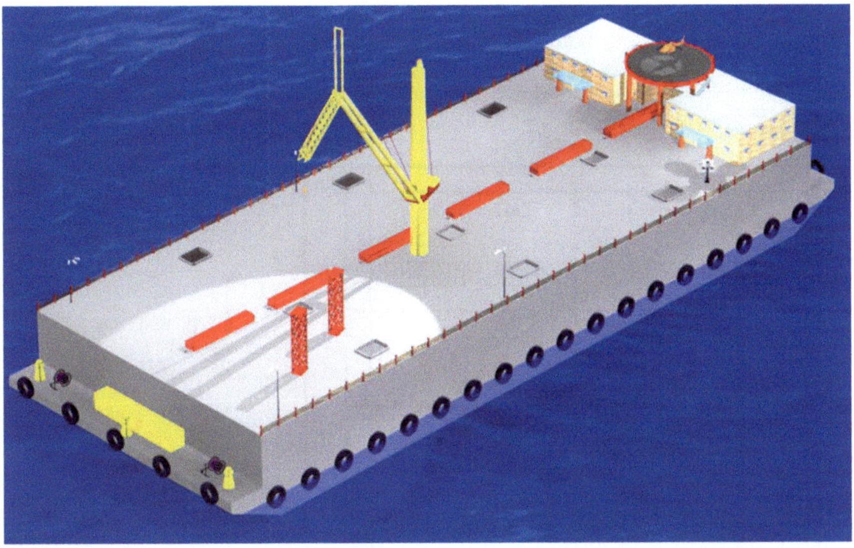

Fig. 21 60 MW OTEC plantship

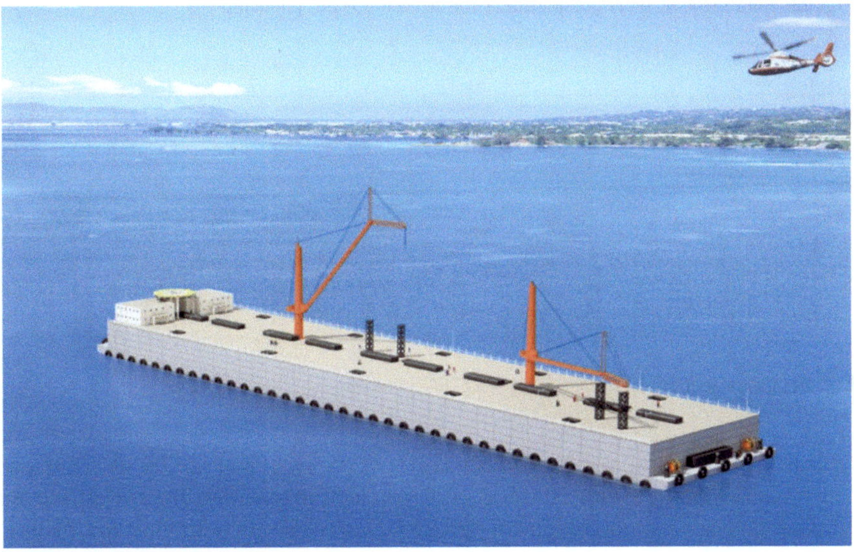

Fig. 22 125 MW OTEC plantship

convenience and freedom of access to equipment during maintenance, repair and replacement activity.

For example, if we review the dimensions of the smallest capacity OTEC plantship of 20 MW capacity (see Figs. 23 and 24) the clear space dimensions of a

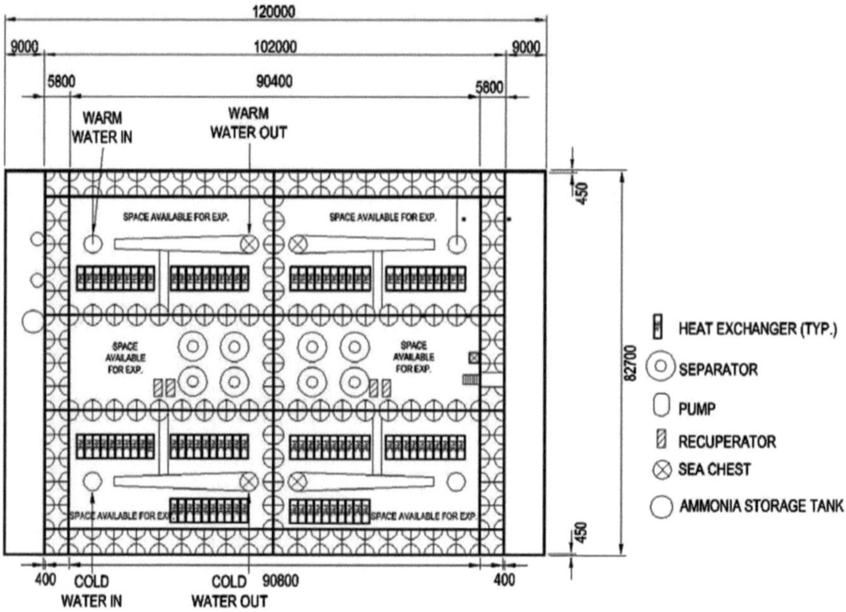

Fig. 23 Plan view of level compartment for 20 MW OTEC plantship

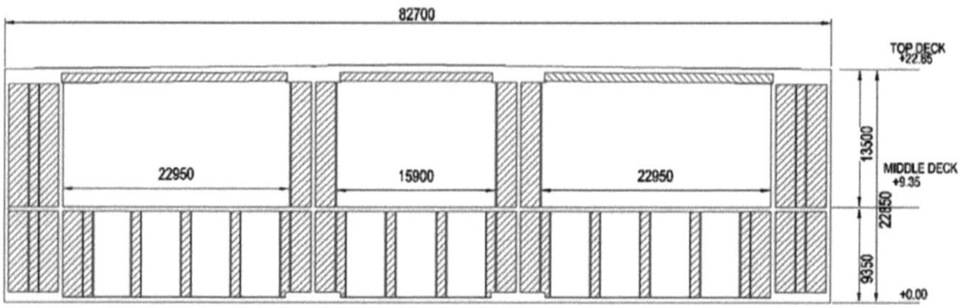

Fig. 24 Cross sectional view of 20 MW OTEC plantship

typical cold or warm water chamber are L = 45 m × W = 22.95 m. Thus one can visualize the generous confined unobstructed column free areas available for efficient and accessible equipment layout for convenient operations and maintenance economy.

Equipment is usually located within the lower hull sections of these platforms (see Figs. 23 and 24). In our engineering studies the honeycomb framing system has enabled us to divide these lower hull sections into large clear span cell spaces of areas of at least 1,032.75 m^2. A standard NBA basketball court total area is approximately 437 m^2 in plan 28.66 m × 15.24 m (94 ft × 50 ft). A championship tennis court is about 846 m^2 (39.63 m × 21.34 m) (130 ft × 70 ft).

Therefore, our proposed column free space is equivalent to approximately 2.36 NBA basketball courts in combined area or 1.22 times the area of a championship tennis court area size. The clear ceiling height varies from 11.2 to 12.3 m.

Our concrete mixes for these vessels involve ordinary Portland cement, silica fume, fly ash, granulated ground blast furnace slag, retarders, and superplasticizers. The W/C weight ratio (water/cementitious weight) is generally 0.25–0.3. The durability and long term service life of the concrete mix, estimated to be in excess of 100 years, play an important role in the preservation of our environment, ecosystem and the economic viability of the project.

5 Concluding Remarks

For ocean platforms exposed to severe marine environments requiring long term service life performance at minimum maintenance and repair cost, actual experience strongly indicate that the most ideal material for the construction of hull structures is a combination of high strength reinforced and prestressed concrete consisting of a rich mix of cementitious materials and mineral additives (Portland cement, fly ash, granulated ground blast furnace slag, and silica fume) and chemical admixtures. Concrete mixes with an extremely dense impervious watertight high strength matrix can be achieved regularly with very high slump in the range of self-compacting concrete producing compressive strengths in the consistency range of 110–130 Mpa. The characteristics of this concrete mix allow thorough uniform penetration into reinforcing steel cages with densities in the range of 500–600 kg/m^3 without resulting in congestion or honeycombing of the concrete mass in the dense reinforcing steel cages.

Owing to the flowability of the concrete mix individual reinforcing steel bars will be completely encapsulated with rich cementitious materials and thus will not require galvanizing or epoxy coating and the reinforcing steel will remain extremely durable even though uncoated and will show no sign of active corrosion. The rich concrete mix will actually serve as a protective coating to the reinforcing and prestressing steel.

References

VSL International Ltd. (1992). *Floating concrete structures: Examples from practice*. Berne, Switzerland: Second Printing.

Wang, C. M., Yee, A. A., Krock, H., & Tay, Z. Y. (2011). Research and developments on ocean thermal energy conversion. *IES Journal Part A: Civil and Structural Engineering, 4*(1), 41–52.

Yee, A.A. (1982). The superiority of concrete honeycomb construction for marine structures. *First symposium on arctic drilling platforms*, Los Angeles, California, October 7, 1982.

Yee, A.A. (2009). Design and construction of composite precast prestressed concrete structures on land and sea. Keynote Paper in the *11th Arab structural engineering conference*, KFUPM, Dhahran, Saudi Arabia, October 25–27, 2009.

Yee, A.A., Masuda, F.R., Kim, C.N. and Doi, D. (1984). Concrete module for the global marine concrete island drilling system. *FIP/CPCI Symposia*, Calgary, Alberta, Canada, August 25–31, 1984.

Mobile Offshore Base

Paul Palo

Abstract SeaBasing is one of three core operational concepts of the 21st century U.S. Navy. Because there were no precedents to build from regarding safe operations between large floating vessels in the open ocean, the Office of Naval Research tasked EXWC to advance technology as well as bound the upper threshold of practical platform sizes. For example, the most capable conceptual SeaBase, labeled the Mobile Offshore Base (MOB), would need to be an unprecedented 2-kilometers in length to accommodate conventional take-off and landing long-range cargo aircraft. Government, industry, and academia made significant advancements in: hydrodynamics, the large (km) scale structure of ocean waves, drafting of the first probabilistic- (and performance-) based design methodology for U.S. marine structures, and feasibility studies and physical demonstrations of multiple module dynamic positioning. Taken together, these advancements lead the participants to conclude that a 2-kilometer MOB platform could be economically built and safely operated.

1 Introduction

"SeaBasing" is one of the three operational cornerstones (along with "Sea Strike" and "Sea Shield") of the U.S. Navy (Clark 2003), and represents an entirely new engineering and doctrinal challenge. The U.S. Navy defines SeaBasing as an organizational concept wherein multiple vessels already within a fleet are deployed in a coordinated manner to collectively perform various missions during times of peace and regional conflicts. When deployed, the functional requirements of SeaBasing include: open ocean ship-to-SeaBase or SeaBase-to-connector cargo

P. Palo (✉)
Naval Facilities Engineering and Expeditionary Warfare Center, P.O. Box 511, Cataumet, MA 02534, USA
e-mail: paul.palo@comcast.net

C.M. Wang and B.T. Wang (eds.), *Large Floating Structures*, Ocean Engineering & Oceanography 3,
DOI 10.1007/978-981-287-137-4_11

transfer; accommodation of combat troops; storage and maintenance of military hardware; and support for either vertical take-off and landing (VTOL) or conventional (fixed wing) take-off and landing aircraft (CTOL) aircraft operations. Within this SeaBasing culture, the Mobile Offshore Base (MOB) specifically denotes only the most capable multiple-module SeaBase platform capable of supporting CTOL cargo aircraft operations.

When first addressed in the 1990s, the functional requirements of a CTOL mission were totally unprecedented: first, depending on the aircraft, the runway length required could fall within an estimated range of 700 (2,300 ft) and 1,500 m (4,900 ft) (hereafter simplified as a nominal 1 km dimension). Second, such a MOB would need structural integrity in wind, waves, and current while on-station in the open ocean. Third, to be practical, it would need to be economical to construct, transit, operate, and maintain.

Not surprisingly, there was and still is no industry equivalent to this combination of platform attributes. Offshore platforms are generally no larger than 100 m (325 ft) in length and are used mostly for gas and oil exploratory drilling. So while the Navy could hope to extrapolate from the industry's extensive experience with design, construction and operation of these platforms, clearly a project focusing on significant advancements was needed. The first engineering look at MOB was sponsored by the Defense Advanced Research Program Agency (DARPA) in the early 1990s. The Office of Naval Research (ONR) assumed responsibility for and greatly accelerated MOB technologies between 1997 and 2000. This Chapter focuses on that 4-year, $35 M ONR Science and Technology (S&T) project which was managed at the Naval Facilities Engineering Service Center (NFESC, now the Naval Facilities Engineering and Expeditionary Warfare Center (EXWC)). Ultimately, 53 organizations (26 commercial, 16 academic, and 11 governmental) were selected and contributed.

Right from the start, NFESC and all participants recognized that there were serious deficiencies in the state-of-the-art tools, methods, and data for designing such a long structure to satisfy its unique mission. Furthermore, there was immediate consensus that such a long platform would need to consist of multiple modules (yet undefined), most likely semisubmersibles, which in turn introduced the need for some type of connection (again, undefined).

The MOB S&T project did deliver a broad spectrum of pioneering advancements. First, three candidate conceptual platform approaches were advanced to a prototype design stage to establish basic feasibility and in the process uncover any gaps in the methodology. Secondly, a pioneering probabilistic-based framework was created to guide the design of future MOBs (and is actually applicable to all large, multi-body floating platforms as addressed in this book). Lastly, these advancements were supplemented by numerous supporting studies in topics such as hydrodynamic and hydroelastic design tools for multi-body platforms, spatial characteristics of ocean wave fields at large scale, aircraft operations, multi-body dynamic positioning, and cargo handling. All of these developments advanced the state-of-the-art standard and are applicable to other offshore structures.

It was concluded that the shorter (700 m, C-130-capable) MOB consisting of semisubmersible modules was indeed economical with a high integrity for safe operations. *That is particularly noteworthy given that it was a consensus opinion amongst all 53 participants.* It was further concluded that there were no apparent "show stoppers" regarding feasibility of the longer 4,900 ft (1,500 m, C-17-capable) MOB, although further studies were necessary before any feasibility conclusions could be made.

The next section of this chapter captures the wide breadth of requirements that make MOB so unprecedented. It is followed by a general description of the three alternative MOB platforms, along with their advantages and disadvantages. The fourth section illustrates the process used to identify the S&T tasks. While the project objective—to advance the state-of-the-art sufficient to allow for confident design of MOB platforms—was deceptively simple, execution was a challenge. For example, *what was the full range of parties (designers, operators,* etc.*) associated with a MOB platform? Who needed to actively interface with whom? What were the major uncertainties and gaps in technology? Which were most important?* In that regard, *were ALL supporting technologies identified for advancement in a consistent manner that would be ultimately required in the engineering design phase? What were the risks and attainable goals associated with the various candidate concepts?* These questions are relevant to any unique maritime structure—namely, all the other large floating structures (LFS) in this book. The final section highlights some of the most significant individual technical accomplishments from this project—again, applicable to many other LFS systems.

2 MOB Requirements

Compared to a MOB, all of the other LFS in this book have one or more features that greatly simplify their design. Three common simplifications are:

- A location within a harbour or behind a breakwater that reduces or eliminates wave dynamics
- A fixed location, allowing for an extensive restraint or shore connection system, and eliminating concerns regarding a streamlined hull form for transiting
- An objective that can be readily divided into arbitrary sub-units, such as many moderate-sized facilities to satisfy a total requirement (e.g., house a given number of people)

The MOB allows for no such simplifications. Three absolute requirements were:

- Accommodate conventional (fixed-wing) take-off and landing of cargo aircraft operations
  - Nominal 1,500 m (4,900 ft) runway length for the largest C-17 aircraft
  - Nominal 700 m (2,300 ft) runway length for the smaller C-130 aircraft

  - Nominal 122 m (400 ft) runway width for simultaneous runway and off-loading/taxiway
  - Negligible or zero risk of connectivity failure (runway gaps) during operations

- Long-term station keeping in deep water sites anywhere in the world

  - Maintain alignment with prevailing wind direction for air operations
  - Limit runway slope discontinuities at module interfaces
  - Survive extreme events

- Moderate transit speed for timely arrival to theatre

Additional desirable features were:

- Large volume of climate-controlled storage to perform its logistic mission

  - House up to 3,000 troops
  - 800,000 m^2 (9 million ft^2) of environmentally controlled dry cargo storage space, arranged with maximum access (as opposed to dense packing)
  - 40,000 m^3 (10 million gal) of fuel storage

- High-throughput cargo transfer from large arriving cargo vessels and to local landing craft
- Economical acquisition cost, long-life maintenance, and ease of repair

In other words, the MOB required a dramatically unprecedented length and width, safe open ocean operations and survival, and the ability to transit long distances at reasonable speed. That combination made MOB unparalleled compared to any very large floating structure ever built.

The concept of a large, floating, offshore military Seabase had been considered since the early 1900s. A few system platform concepts were developed within the previously-mentioned DARPA project. These concepts and the contractors who worked on them were:

- Very Large Mobile Offshore Base (VLMOB): Seaworthy Systems, Inc. The VLMOB was based on converting a very large crude oil carrier. About 335 m (1,100 ft) long, this single hull ship was not long enough to meet the requirements of fixed wing aircraft operations.
- Ultra-Large Monohull Offshore Base (ULMOB): Band, Lavis and Associates, Inc. The ULMOB utilized monohull (displacement hull) modules, each about 505 m (1,660 ft) long. By rigidly connecting modules end-to-end, long runways were created. The initial estimates showed very large loads and resulting motions, and the results were generally inconclusive.
- Rigidly Connected Semisubmersible Modules: Brown and Root, Inc. This concept consisted of up to six semisubmersible steel modules, each 180 m (600 ft) long, connected rigidly. The rigid connectors effectively created one long semisubmersible hull, resulting in very large connector forces which exceeded reasonable design limits.

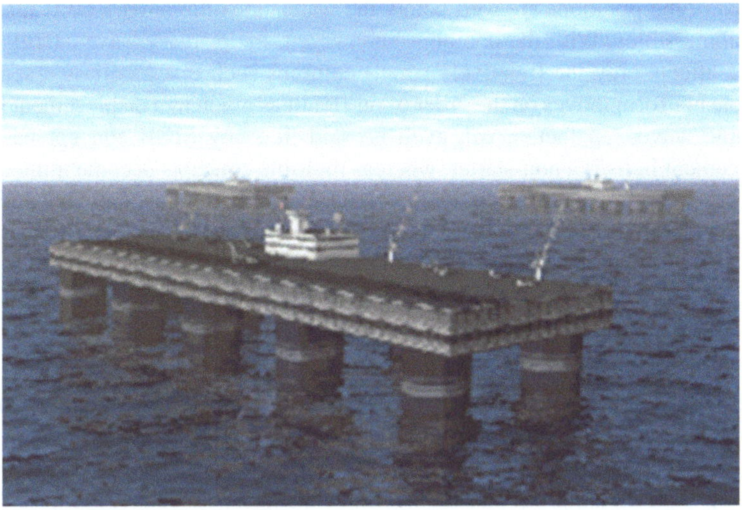

Fig. 1 Typical semisubmersible modules (unconnected)

It was concluded from these limited studies that any monolithic, full-length, *displacement* hull would have unmanageably-large structural stresses along the keel as it deflected in the wave field—not to mention difficulties navigating into almost any harbour. Conversely, allowing the structure to flex with the shape of the sea surface greatly reduces these stresses. Two remedies were adopted by the ONR program to alleviate these stresses. First, structural compliance was provided by dividing the monolithic hull into separate modules with some sort of connection or continuity. The second remedy was based on the observation that displacement hulls were non-optimum in that their buoyancy forced them to track the free surface and thereby induce large forces into any connector. This drove the ONR project to adopt the *semisubmersible* hull form for the module shape because of its inherent insensitivity to wave energy.

The use of semisubmersibles with some sort of connectivity /continuity thus formed the foundation for all MOB concepts. The project defined a notional set of mission requirements, and used them to pursue development and evaluation of four alternative MOB concepts. The main difference among the three system platform concepts was principally the method used for connecting the semisubmersible modules to form the assembled MOB platform.

For readers unfamiliar with semisubmersibles, they are extensively used by the offshore oil and gas industry. Figure 1 shows a typical semisubmersible hull module proposed for MOB.

The hull consists of two horizontally-oriented "pontoons", each of which has columns (here, five) that connect it to the upper deck unit. The term "semisubmersible" refers to the fact that the vessel operates as a catamaran during transit wherein the pontoons are at the water surface, but that once on station, water is deliberately flooded into ballast tanks such that the entire hull submerges to where the tops of the pontoons are well below the water line—nominally 60 ft for MOB semisubmersibles. With the pontoons submerged, the primary wave energy at the surface affects only the [relatively small] columns such that a ballasted semisubmersible platform moves on the order of one tenth as much as a displacement hull in the same waves. In fact, it can even be said that any MOB concept based on displacement hulls is a non-starter, whereas multiple concepts become immediately possible when semisubmersibles are involved. Thus, the semisubmersible hull form provides two platform requirements: on-site stability as well as rapid transit.

3 MOB Candidate Platform Concepts

Summarising, the aforementioned requirements define a very unique maritime platform. In one sense, this proposed Navy MOB platform can be thought of as incorporating "the most difficult" combination of all possible factors associated with any large floating structure. Performance factors associated with its operational phase include:

- 'Threshold' overall length of 700 m (2,300 ft) to accommodate the smallest cargo plane (C-130); this is at least four times longer than any existing platform
- 'Objective' overall length of 1,500 m (4,900 ft) m to accommodate the largest cargo plane (C-17)
- Negligible wave-induced runway motions for air operations
- Massive storage for personnel, cargo, and fuel/water
- Interfaces to offload arriving cargo vessels

For some phases, a MOB would be disassembled back into its component modules, with performance factors such as:

- Moderate speed to enable realistic transit times to sites of interest
- Safe on-site assembly and disassembly in moderate sea states
- Survival of semisubmersibles and/or connectors during extreme events

And finally, an optimum MOB design would feature

- Low acquisition cost
- Low lifecycle cost
- Note that in keeping with its logistics mission, military threats/vulnerability was not part of any conceptual design effort

Three different conceptual approaches were investigated and found to have potential as a MOB platform. These are summarised in Table 1.

Thus, these three approaches were defined by their connection scheme. A fourth design effort evaluated reinforced concrete modules for its potential life-cycle benefits.

All of the platforms described in this section were designed to the same basic set of functional requirements. For example, the 'objective' runway length was

Table 1 MOB system concepts

Connector type	Description
Discrete Five semisubmersible steel modules connected by compliant hinges	As 'hard connectors' between modules, this approach provides a sense of security to operators. The key is to allow some relative movements between the modules—mainly, pitch—to reduce structural stresses compared to a rigid connector
Flexible bridge Three semisubmersible steel modules connected by flexible steel truss bridges	This design inserts a nominal 300 m (1,000 ft) long 'flexible truss' between modules. This provides a smooth (no slope discontinuities) runway, yet through its flexibility it greatly reduces the structural stresses associated with a rigid connector
None (dynamic positioning) Three semisubmersible steel modules positioned dynamically relative to one another	This approach completely eliminates structural stresses due to inter-module shear loads by eliminating the physical connector altogether. Instead, dynamic positioning is used to keep modules aligned and adjacent. A non-structural, articulating bridge provides runway continuity

selected associated with the Boeing C-17 Globemaster operations. Designers then followed "best industry practice" to develop a prototype with reasonably-sized structural dimensions and preliminary hull mechanical outfitting.

Note that this section serves to give the reader the context as to what a massive platform like a MOB might actually look like. Here, the prototype system concepts are presented, complemented by a synopsis of their advantages and weaknesses. Further engineering and operational details and challenges are presented later in the chapter.

3.1 MOB Concept #1: Discrete Connectors

This category encompasses connectors that the typical reader probably visualizes for a multiple-module MOB. Example connectors in this family include hinges, universal, and ball joints. The commonality shared by all discrete connectors is that they are dimensionally small. The two major advantages of discrete connectors are: (1) their relative simplicity, and (2) the sense of security and peace of mind they provide to users who find comfort in a "simply-connected" platform. However, by their very nature they concentrate the large force and moment imbalances between modules into a small volume, leading to high stresses. A related structural characteristic is that their module foundations must be carefully designed to distribute the concentrated forces into the module. A final inherent disadvantage is the fact that, because of their small dimensions, they introduce the potential for runway slope discontinuities—which are a real concern.

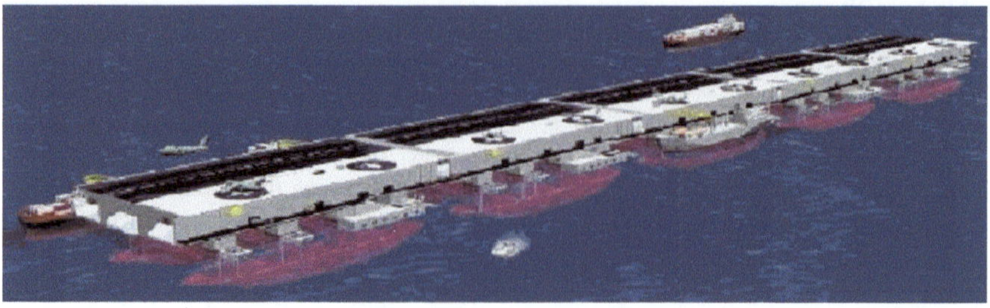

Fig. 2 Discrete connector MOB platform

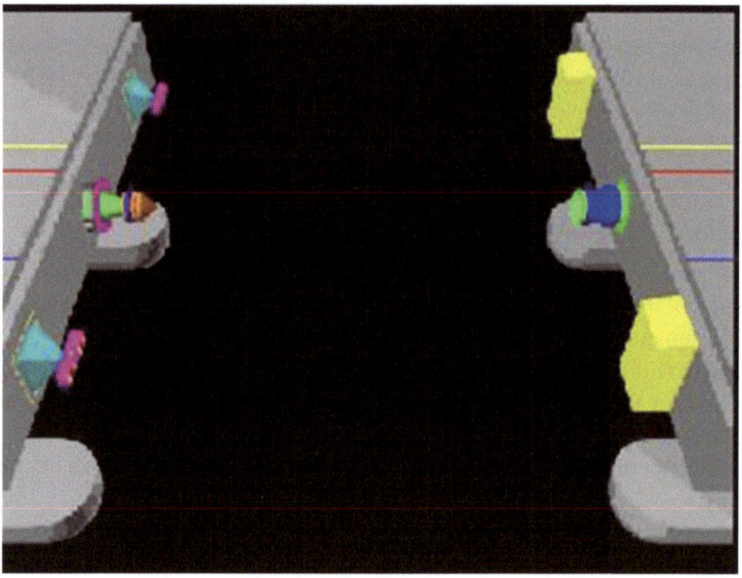

Fig. 3 Example discrete (compliant) connector

This subject concept study started with a simple, horizontally-aligned hinge connector that allowed for relative pitch between adjacent modules. This study selected five semisubmersible steel modules, each 300 m (985 ft) long as shown in Fig. 2.

The studies soon showed that the simple hinge connector concept could not be made viable for MOB. As a result, it was discarded in favour of a "compliant connector" and the system analysis was repeated. This compliant connector design is illustrated in Fig. 3, showing a docking centre ball-joint, and two outboard compliant hinges.

This connector change was important because with these new compliant connectors, the 'discrete connector MOB' concept was made viable for possible later refinement.

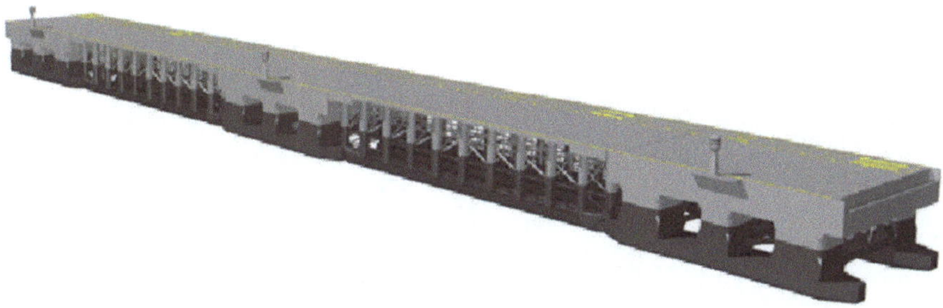

Fig. 4 MOB concept with flexible bridge connectors

3.2 MOB Concept #2: Flexible Bridge

This system concept began with three semisubmersible steel modules, each 235 m (770 ft) long. In contrast to the previous concept, a new type of connector is used—namely, two very long, flexible, floating, truss bridges, each about 410 m (1,350 ft) long as shown in Fig. 4.

These bridges bring a fundamentally different set of characteristics to the MOB platform. By far the most important characteristic is the fact that the long length and deliberate flexibility of each flexible bridge smoothly absorbs the semisubmersible relative motions and rotations without the stress concentrations found in a "discrete" connector. Secondly, the flexible bridges maintain a continuous flight deck with no slope discontinuities.

But these new features come at a price. Structurally, the bridge must be strong and compliant enough to resist the module dynamics and not deform significantly to the distributed wave action. The disadvantages of the bridge are that they must be designed as self-contained floating platforms which must individually transit (or be towed) across oceans to arrive at site, be manoeuvred during platform assembly and disassembly, and interestingly, survive storms (after the platform has been disconnected into individual units). Thus each bridge has dynamic positioning as well as ballasting subsystems.

3.3 MOB Concept #3: No Connectors

This represents one extreme limit of a connector: namely, none. This study settled on three extremely long semisubmersible steel modules, each 488 m (1,600 ft) long. The key to this third platform concept was that there were *no structural connections* between modules. Instead, they are dynamically positioned relative to one another as illustrated in Fig. 5.

A central dynamic positioning processor controls two functions: absolute positioning and orientation of the *platform* on site and relative positioning of each

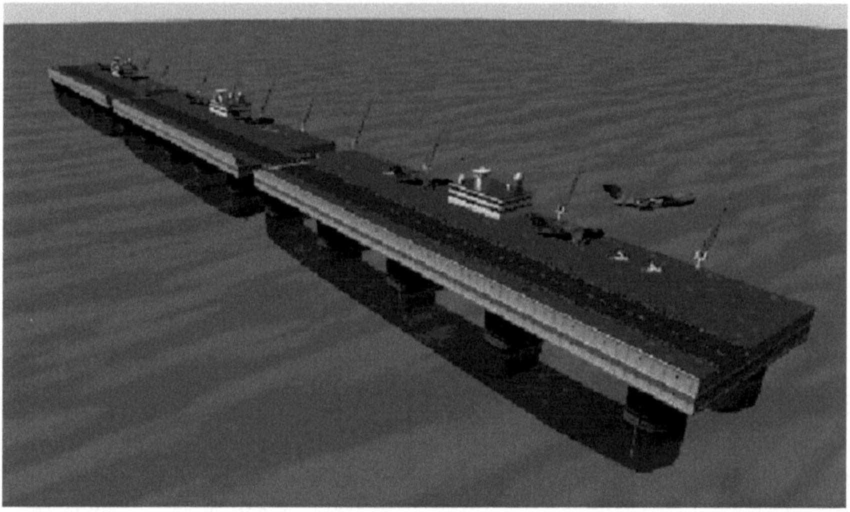

Fig. 5 Dynamically positioned MOB concept

module. This was an unprecedented maritime subsystem that required significant analytical and experimental investigations. Runway continuity between modules is provided via an articulated short deck which spans the nominal 7 m (25 ft) gap between modules. It includes a sliding drawbridge/docking mechanism to accommodate zero-mean wave surge motions—but most importantly, this deck does not provide any structural resistance. Thus, the main advantage of this concept is that it completely eliminates (problematic) connectors.

3.4 Module Alternative: Reinforced Concrete Modules

Simply referred to as the concrete MOB, this system concept was based on four semisubmersible modules, each 381 m (1,250 ft) long, consisting of a steel deck (dark grey) placed on a concrete lower hull. Modules were nominally connected by elastomer bearings and post-tensioned cables ("discrete" connector category). Finite element analyses showed that the required concrete thickness and the densities of reinforcing steel were well within the state-of-practice for existing offshore concrete structures.

The main advantage of concrete over steel is a lower lifecycle cost. With an intended service life of 40 years without docking for periodic maintenance, fatigue can be a critical issue for steel-hulled MOB concepts. By introducing post-tensioned concrete as the material for the lower hull, fatigue becomes essentially a non-issue. The acquisition cost for four hull units, i.e., a total MOB, is at the same level for steel and concrete. The life-cycle cost is about 17 % less for concrete compared to steel.

3.5 Summary: MOB Conceptual Feasibility

The system studies did progress far enough to establish that each of the three system approaches were indeed *feasible*—thus satisfying the first objective of this S&T program. However, the subsequent question as to which system was most suited for deployment could not be determined for two reasons. First, the U.S. Department of Defense (DoD) never formally defined the requirements for a MOB (and in fact have withdrawn it from consideration at this time). Secondly, the "best" qualifier can be interpreted multiple ways so is undefined as well (e.g., "lowest acquisition cost", "highest operational availability", "accommodates widest range of cargo aircraft").

4 MOB S&T Prioritization

The three system preliminary designs above continued throughout the 4 years of this S&T program. Thus the designers relied on a mixture of state-of-the-art procedures, in some cases supplemented with specially-focus studies. Simultaneous with that, the $35 M MOB S&T project proceeded on to its second objective: to advance all engineering design capabilities sufficient to allow for a future, credible MOB design by industry should the DoD choose to go forward with this vision of a CTOL Seabase.

While this exercise appeared straightforward, it was in fact quite challenging for MOB because of its unprecedented nature and size. Accomplishing this consisted of four distinct phases:

1. Identification of ALL engineering design tasks, as well as associated design tools. This phase was also required to ensure that all three of the MOB conceptual platform types were accommodated;
2. Identification of the industry state-of-the-art standard for all the tools in Phase 1;
3. Contrasting of those two phases to quantify all of the gaps and deficiencies in the engineering design; and
4. Identification of attainable advancement goals and performers.

It was recognized right from the start that certain technologies would require substantial and in some cases pioneering advances in the state-of-the-art technology to satisfy MOB design requirements. Such advances carried some obvious uncertainties, and perhaps some subtle, not-so-obvious uncertainties—each with developmental risks. To maximise success, this Navy S&T project created a number of special-focus committees and recruited renowned experts from industry, regulatory agencies, and academia to help "bound the problem". These groups were instrumental in advising the Navy project team regarding the eventual distribution of S&T funds.

4.1 MOB S&T Process

This first phase was arguably the most important of the entire S&T project because it set the framework for all subsequent deliverables (as well as missing deliverables). The two most important factors here were *thoroughness* and *consistency*.

- Thoroughness means that all necessary aspects of an engineering developmental goal were recognised. The objective was to reduce or hopefully avoid overlooking some necessary aspect of the design. For example, where a new numerical model was to be developed to address an entirely new scenario, such as multiple-module dynamic positioning, then in addition to development, that model in all likelihood must also be validated to establish confidence in its predictions. However, this validation exercise was problematic as the scenario was new and there were no existing, comparable data sets. Thus, an experimental test program, most likely at a laboratory scale, needed to be conceived, funded, and scheduled in parallel with the model software development.
- The second factor, consistency, addresses how the tools interact. For example, if a particular connector exhibited or even relied on highly nonlinear stiffness or damping characteristics, then that in turn could only be accommodated by a time domain, nonlinear platform dynamics model.

The reverse was also true: in the latter case, if a nonlinear dynamic model was not available, then there was no need to invest funds accurately quantifying highly nonlinear behaviour.

For some new scenarios it was quickly realised that development and delivery of a validated design tool was unachievable. Qualification and quantification of operational and survival (extreme) ocean wave behaviour at the kilometre scales corresponding to the MOB platform length best illustrates this situation. That information would be necessary for any final design, but was well beyond answering with the resources and schedule of this one S&T project. In such cases the MOB S&T program typically decided to make pioneering advancements to at least frame the level of effort necessary in subsequent design efforts.

These principles were applied to all three conceptual platform approaches described in Sect. 3.

4.2 Brief Synopsis of Engineering Design State-of-the-Art

This section very briefly outlines the state of marine hardware and analysis technology in 1997 as they pertained to a MOB. This information provides the baseline from which to appreciate the true scope and breadth of the MOB S&T challenge. Table 2 illustrates how MOB design exceeded industry in a significant number of categories.

Table 2 MOB-industry capability comparison

MOB design feature	MOB requirement	Industry capabilities
Semisubmersible modules	Between 235 m (770 ft) and 488 m (1,600 ft) long	Nominal 100 m (325 ft)
Connectors	Dynamic environment	No
Dynamic positioning	Multiple modules	No; one module only
	Absolute position and orientation	Yes
	Relative positioning (proximity)	No
	Dynamic environment	Yes
Transit	Moderate speed	No; low speed or towed
	Dynamic environment	No; deliberately scheduled only for calm season
Aircraft operations	Rotary wing	Yes
	Fixed wing	No
Hydrodynamic analysis, moderate seas. Ballasted semisubmersible(s)	Single rigid	Yes
	Single elastic	Yes
	Multiple rigid	CPU limited
	Multiple elastic	CPU limited
	Wave pressures over semi hull	No
Hydrodynamic analysis, large seas. Deballasted semisubmersible(s)	Transit mode	Approximate (LAMP)
Ocean wave characteristics at 1 km scale	Necessary	None
Comprehensive classification guide	Insures thorough and consistent engineering design effort	None

4.3 Highlights of MOB S&T Process

Some of the most significant disconnects between MOB and industry state-of-the-art in Table 2 are described in further detail in this section to illustrate the scope and breadth of the MOB S&T project.

- Multiple-Module Dynamic Positioning (DP) System
  - There are two alternative mathematical approaches for the controller algorithm: relative and global. The relative approach would use GPS to position and orient one 'master' module, say the most center one. An independent system would measure the relative separation (gap) between that center module to any adjacent'slaved' module and transmit those measurements to the central module's DP computer. That computer would

process the information and in turn direct the slaved module's DP computer on how to maintain position. In the global approach, each module is treated as a 'master' in that it would be assigned an exact latitude/longitude position and orientation to ensure runway continuity.
It was not a priori apparent which of those two was best (which is itself a vague term: 'best' could mean most reliable, or most economical to operate). Therefore, both controller approaches were advanced using numerical scenarios and experimental test data to reliably establish their benefits and disadvantages.

Note that this multiple module nature of MOB introduced an entirely unprecedented set of possible runway failure modes that needed to be identified along with DP remedies

- Validation data. Laboratory-scale experiments were conducted using 2 m (6 ft) long semisubmersibles with carefully scaled miniature thruster motors. As yet another example of how the unprecedented size of MOB cascaded throughout the project, the model scale was chosen to provide for correct thruster scaling—meaning a scale of at least 1/66 was needed. This scale produced very large MOB 2-module test platforms that were 4.1 m (13.6 ft) long and weighed 1,040 kg (2,290 lbs).

- Hydrodynamic numerical modelling during operations (ballasted mode)
  - The three conceptual platform studies all used semisubmersibles whose lengths were well beyond industry practice. Thus, it was believed necessary that an investigation should be undertaken into how to properly use state-of-the-art panel models to investigate their basic dynamic properties in waves.
  - Laboratory-scale experiments were conducted to validate the numerical predictions for such long individual modules.
  - It was also immediately recognized that state-of-the-art panel (diffraction) models were incapable of processing the total number panels associated with the three to five such long semisubmersibles that comprise a MOB platform. Therefore, two new treatments were advanced:

    A new numerical solution solver was implemented and evaluated to reduce processing times.
    A first-generation "higher-order panel" method was matured and evaluated that replaced large numbers of constant-potential panels with a much smaller number of panels that allowed for variable potentials in space.

  - The wave surface underneath an operating semisubmersible was of great interest. This surface is quite complicated because it is the superposition of the incident waves, the radiated waves from each of the semisubmersible columns, and finally reflected waves primarily associated with the pontoons.

It was important to understand this surface to assess the viability and seaworthiness of smaller craft engaged in cargo transfer.

A unique hydrodynamic model named LAMP (Large Amplitude Motion Program) was mathematically calibrated and software advanced to integrate up to the instantaneous instead of the incident free surface (where the former was approximated as the incident plus refracted wave components).
Here again, special laboratory-scale experiments were conducted using one module in the ballasted draft.

- Note that in the vast majority of cases the size of a MOB created uncertainty or complicated the study. In some cases however its tremendous size is instead an advantage. For example, as berthed cargo vessels are reaching their maximum allowable dynamics at a particular design wave height, the motion of the MOB is still generally negligible.
- It was felt that no studies were necessary for the dynamics of a cargo vessel berthed alongside a MOB semisubmersible module.
- No studies were initiated or data collected of the viscous flow around semisubmersible pontoons or DP thrusters because computational fluid dynamics (CFD) tools were not considered mature enough even with a moderate investment.

- Hydrodynamic numerical modeling during transit (deballasted mode)

  - Given the existing industry strives to transit their semisubmersibles at slow speed and only during seasons with low wave energy, the MOB's requirement for moderate speed transits through any arbitrary sea condition was unprecedented. The issue was the roll stability of these modules while at their unballasted drafts.

    Semisubmersible buoyancy, and hence its overall dynamics, is highly nonlinear when the semisubmersible was deballasted and waves regularly wash over the pontoons. Numerical modelling of semisubmersible dynamics was performed using the nonlinear LAMP model, and
    Laboratory-scale experiments were conducted which included measurements of the highly nonlinear wave "run-up" phenomenon on the columns

- Wave field characteristics (spatial coherence) at the 1 kilometre scale

  - It is as vital to know the correlation of wave crests at this unprecedented scale as it was to know the spatial correlation of wind gusts at a similar scale for the Golden Gate Bridge. This is necessary information for the hydrodynamic modelling and was an entirely new topic.

    Multiple pioneering mathematical studies (e.g., wavelets) were initiated to develop and evaluate various statistical signal processing tools and new metrics. Some limited applications to existing data sets were completed.

Partial financial support was provided to an ongoing NASA S&T project that mapped ocean wave fields during hurricanes. This collaboration did in fact obtain the very first measurements ever of actual waves in the vicinity of a hurricane eye.

- Air and other logistics operations
  - The MOB mission and constraints were tighter than typical FAA regulations for a terrestrial airfield, yet not as tight as for a Navy aircraft carrier. Studies with Boeing and McDonald Douglas and NAVAIR were sponsored to evaluate various taxiway 'set-backs' and nominal aircraft failure modes within the context of the intermediate-sized MOB flight deck to develop a draft guideline.
  - It is seductive to simply claim that a MOB would simply be disconnected into individual units prior to the arrival of an extreme event such as a hurricane. This would in fact likely avoid fabricating extremely large connectors for some concepts. However, it introduces the need to reconnect as the storm ebbs in order to resume operations, which could be a problem if assembly/disassembly was limited to low sea states. Waiting for such seas would in turn increase the time the MOB was unavailable to fulfil its mission before and after any extreme event. Thresholds defining 'acceptable durations' remain unknown; this a clear example of a type of unanticipated and non-engineering consequence associated with the design of a new system such as a MOB.
- Structural Stresses
  - "Hydrodynamic" models idealise the hull as infinitely rigid. Such an assumption may be adequate for short to moderate length hulls but are not suitable for MOB applications. Conversely, "hydroelastic" models allow for small hull deflections—say, for the first vertical, horizontal, and torquing bending modes. Note that even very small elastic hull deformations (e.g., slight curvature of the hull as a long wave passes under it) at the module scale can result in surprisingly large displacements at the local (e.g., column) scale. Linear hydroelasticity was already an active topic of research worldwide. So other than insuring that some MOB models accounted for it, there were only limited investments in advancing this technology.
  - A novel 'loads generator' was recognized as essential and therefore developed for these very long semisubmersible modules. This software mapped the wave potentials over the entire submerged surface of a semisubmersible as defined by its arbitrary higher-order panel representation into corresponding local pressures over its independent finite element surface representation. In other words, this mapped the time-varying wave excitation forces relative to their local position into the corresponding structural capacity of the module.

- MOB Classification Guide
  - The MOB is unprecedented due to both its unique size and combination of missions. This had two major impacts with respect to engineering design:

    It puts MOB well outside the scope of any existing design manuals.
    The absence of any 'lessons learned' from previous similar structures meant there was significant risk associated with unforeseen responses and failure modes. This risk was compounded by the variety of candidate platform hardware and characteristic responses as outlined in Sect. 3.

  A comprehensive MOB design guide was therefore considered as arguably the most important deliverable from this S&T project. In one sense it echoed the S&T project plan in that it was the very embodiment of the thoroughness and consistency demanded in a MOB design. This fundamentals-based approach follows advice given for all innovative marine structures (National Research Council 1991). Accordingly, major funding was assigned to this task. For example a large committee was organized and met regularly for the entire 4 years of the project to draft a preliminary guideline.
- Materials
  - Fenders are a commonly used piece of marine hardware. MOB was no exception. Here, as so many times before, the key technology driver was size; fenders between MOB modules would have to be mammoth to absorb the large forces. The simple requirement to cure such large objects was far beyond the current state of technology. A related challenge was the elastomeric material composition (e.g., graphite-fiber-reinforced, urethane-matrix). A limited number of fundamental studies were performed.

5 MOB S&T Deliverables

The previous sections have illustrated the process used by the Navy-industry-regulatory-academia MOB project team to advance the state-of-the-art standard. Detailed justifications were deliberately omitted for clarity. This section provides closure to this S&T program by itemizing many of the major technical deliverables. Again, context is omitted for clarity in most cases.

The challenges associated with the MOB unprecedented semisubmersible and platform sizes were reviewed for these specific design requirements:

- Runway length associated with the largest specified cargo aircraft (C-17)
- Open ocean air and sea logistics operations
- Open ocean survival (either connected or disconnected platform)
- Transit to/from theater in deballasted mode at 'reasonable speeds' through 'moderate sea states'

This was followed with an assessment of the three unprecedentedly large conceptual connector schemes to connect the individual semisubmersible modules:

- Discrete connector
- Flexible bridge
- None (dynamic positioning)

Major S&T contributions to state-of-the-art engineering design from the MOB S&T project associated with those criteria included:

- MOB Classification Guide and accompanying Commentary. This is the first performance and reliability-based standard for floating ocean structures published in the United States. It uses a reliability-based framework that assigns system and subsystem lifetime target reliability levels for key limit states (operational, fatigue, extreme) that are consistent with the required probability of mission performance. Structural adequacy (including weapons-related survivability), dynamic positioning, life-cycle maintenance, and environmental compliance are also addressed.
- WAMIT (linear, low-order hydroelastic model) was numerically accelerated with a new "pre-corrected FFT" equation solver and interfaced with the interactive geometry pre-processor MULTISURFTM. The new version reduces analysis times for 100,000 panel hulls from days to minutes.
- HIPAN (linear, higher-order, hydrodynamic model) was advanced with a b-spline panelisation scheme and MULTISURFTM interface. Also, HIPAN is the hydrodynamic engine that enabled development of the "universal loads generator" described next.

  2013 update: since the MOB program ended, the hydrodynamic community has favoured using HIPAN over WAMIT.

- Universal load generator (ULG) for HIPAN. This pairing is very significant because it exactly maps hydroelastic and structural models with independent panelisations. The ULG can transfer wave pressures time histories to arbitrary finite element meshes, and [linearly] scale them by an incident wave spectrum. The ULG accounts for the instantaneous flexure of the module (hydroelastic) including memory (convolution).
- HYDROMOB: this model efficiently solves for the motions of a multiple module, arbitrarily configured and connected platform assuming rigid bodies. It assumes that the hydrodynamic interaction can be neglected. It is used to facilitate preliminary, parametric studies.
- A second simplified model that does account for hydrodynamic interactions. A multiple scattering technique is used to model the interactions among any number of modules in any arbitrary arrangement. The approach is computationally very efficient. Used to facilitate preliminary, parametric studies.
- LAMP (nonlinear, time domain, hydrodynamic model): improved accuracy by integrating up to the (approximated) instantaneous vice incident surface. Upgrade was partially validated against MOB model scale test data. This model

was used for air gap in survival seas and the motions of a semisubmersible at the transit draft.

- Hydrodynamic Stability Model allowed investigation of dynamic motions that occurred while transiting at the deballasted draft. The model accommodates pontoon nonlinear buoyancy as the waterplane changes with pontoon immersion due to semisubmersible rotations and/or large incident waves. It is based on a dynamic phase-plane approach.
- Exercised five commercially-available hydrodynamic models with a variety of panelizations for semisubmersibles up to 700 m long. While the findings are more practical than innovative, this study did uncover some interesting traps that a SeaBase designer would easily fall into in the absence of such a coordinated study.
- Large-scale laboratory test for hydroelastic validation data. The model(s) for this test was one, two, and four elastically scaled (1:60 scale, 6 m long, 0.5 m column diameters, 2,347 kg) semisubmersibles.
- Laboratory-scale tests for dynamics at transit draft and air gap at survival draft. Used 1:70 scale (3.9 m long model) of a 274 m prototype semisubmersible.
- Environmental Specification. A global metocean database of joint wind/wave/current hindcast vector descriptors at 23 sites worldwide, averaged over and catalogued every 6 h, for 20 years. Includes an interactive MATLAB™ shell program to extract many useful environmental statistics for the chosen site (e.g., distribution of wave heights associated with all winds greater than 40-knots, or incident only from the North, or 20 years of wave statistics only for the month of January at a site of interest). Used for operating and fatigue studies.
- Typhoon Database. This second database provides wind, waves, and currents for 25 large typhoons from the Western Pacific. Each storm is described using a moving grid with 1-mile and 1-h discretizations. Used for survival studies.
- Pioneering numerical studies investigating the spatial coherence (crest lengths) of ocean waves at scales up to 2 km.

  - This saw the investigation of candidate signal processing techniques for quantifying coherence and wave crest lengths for analyzing three-dimensional ocean waves.
  - There was limited advancement of physics-based modelling techniques for the synthesis of three-dimensional ocean wave fields. This, in turn, was used for generating artificial wave fields with respect to a given wind speed and fetch condition.
  - Scanning Radar Altimeter data set (NASA) of instantaneous surface measurements in seven of eight octants in the Hurricane Bonnie wave field; this was the first ever measurement of a complete representative wave field ever for a tropical storm. This was significant to MOB because it measured a very large 18 m high wave with a *straight* 1.5 km long crest (the expectation was that the transient nature of hurricanes would naturally lead to only short crest lengths which were not a serious threat to a connected MOB platform).

- Two studies demonstrating that available data sets from Synthetic Aperture Radar (orbital scanners) are sufficiently accurate and extensive (30 m resolution for up to 100 sq km) for use in wave coherence studies.

- Regarding multi-body dynamic positioning:
  - A complete, representative multiple body dynamic positioning system was assembled including instrumentation, PID controller, thrusters, and power system. This allowed for generic parametric studies of DP performance.
  - Rigorous mathematical examinations of a number of mathematical controller approaches (global and relative) and real-time computer simulations. These simulations were extremely valuable for gaining insight into how the candidate controllers performed versus various measures (most robust alignment, most responsive, least fuel, etc.).
  - Dynamic Positioning Experimental Data. Three, 2 m long manoeuvrable semisubmersible modules were used to directly investigate/demonstrate the actual performance of candidate controller schemes.

See the MOB Final Report or any of the three Remmers conference articles for further details (Remmers et al. 1998, 1999; Remmers and Taylor 1998).

6 Other Navy VLFSs

In the 1990s, besides floating bridges, proposed airfields like MOB and MegaFloat were the exclusive focus of "Very Large Floating Structures" (VLFS) technology. Interestingly, depending on the definition of "VLFS", one other Navy VLFS structure already existed, and a second Navy VLFS has been conceived and fielded since the MOB Program ended.

- Navy Lighterage Floating Causeway (FC). The FC is an anchored, non-powered, floating pier up to 366 m long. It is assembled by flexibly connecting up to 19 small barge modules, each 24.3 m long by 7.3 m wide by 0.4 m draft. FCs have existed for decades for cargo discharge operations at undeveloped sites, and are designed for survival through Sea State 5. It can be easily classified as a "VLFS" due to (1) its extreme length-to-beam ratio (366/7.3 = 50), and (2) the associated elastic responses in its open ocean setting.
- Waterside Security Barriers (WSB). These are floating "fences" that are used to protect Navy assets from waterside threats. They were first fielded in 2002, and are steadily being installed in U.S. Naval harbour installations worldwide. The basic module is one 15.2 m long box beam, supported by 3-pontoons arranged orthogonally, and a nylon net assembly above the beam. The beams are structurally connected to form floating fences of arbitrary lengths; the longest to-date is 2,600 m. These innovative marine structures can also be loosely considered to be "VLFS" for the same two reasons as the FCs: (1) their

extreme length-to-beam and elasticity, and (2) the fact that their [sheltered] harbour wave excitations are actually relatively large compared to the pontoon dimensions.

7 2014 US Navy Seabasing Update

While not VLFS platforms, the US Navy took a giant first step forward in Seabasing with the delivery in May 2013 of the first dedicated ship, the USNS Montford Point (Mobile Landing Platform, or MLP 1). This is the first of three funded vessels including the USNS John Glenn (MLP 2) launched in September 2013 and the USNS Lewis B. Puller (MLP 3) under construction. These vessels have a length of 239.3 m, beam of 50 m, draft of 9 (fully loaded) and 12 (load line) m, and displacement of 78,000 long tons (fully loaded).

From Reference US Navy official web citation (http://www.navy.mil/navydata/fact_display.asp?cid=4600&tid=675&ct=4), the MLP ship class leverages on an existing commercial design to ensure design stability and low development costs. These ships will operate within the Maritime Prepositioning Ship Squadrons as mobile sea bases providing the U.S. Navy Fleet with a critical access infrastructure that supports the flexible deployment of forces and supplies. The approximately 80,000 ton, 785-foot ship will leverage float-on/float-off technology and a reconfigurable mission deck to maximize capability. Additionally, the ship's size allows for 25,000 ft^2 of vehicle and equipment stowage space and 380,000 gal of JP-5 fuel storage. The platform in its basic form possesses add-on modules that support a vehicle staging area, vehicle transfer ramp, large mooring fenders and up to three LCAC vessel lanes to support its core requirements.

Certainly the MLP will be a major player in future logistics missions. Might positive experiences with these MLPs over the coming years inspire the US Navy to someday upgrade to the vastly larger capabilities of a MOB Seabasing platform? The fact is that a MOB is the only viable candidate for logistics missions where extensive deliveries of cargo are critical and urgent - for example for humanitarian relief applications such as the Haitian earthquake. Such deliveries are only possible with large, fixed wing aircraft—which neither a MLP nor a drone can ever satisfy. That continual need, coupled with the regional political instabilities ever present around the world, means that even with the MLPs, the MOB will remain a valid candidate for decades to come.

8 Summary

While the sole purpose of the advances described in this chapter was to advance design tools for MOB design, the reader is encouraged to interpret them in the greater context as an extreme example of the thought process necessary for the proper treatment of any innovative large floating structure. The design challenge for all of these LFS is highly congruent given the common lack of formal design methodology or historical precedents to build from. This requires rigorous evaluation of engineering fundamentals and for these principles to be applied in a thorough and consistent manner.

Acknowledgments The MOB team of 53 commercial, academic, and government organizations spans too many performers to individually acknowledge. Mr. Gene Remmers at the Office of Naval Research administratively managed the project. At NFESC, Mr. Robert Taylor was the project manager, Mr. Ron Brackett was responsible for the financial and contracting, and the author technically managed the S&T. Other Navy technical team members included Dr. Ray Chiou, Dr. Robert Zueck, Dr. Theodore Shugar, and Ms. Michele Murdoch.

References

The following references are recommended for the interested reader. The second reference is the most comprehensive description of the S&T project and is available at www.dtic.mil

Clark, A. V. (2003). *Sea power* 21: Projecting decisive joint capabilities, *U.S. Naval Institute Proceedings*.

Mobile Offshore Base Science and Technology Program Final Report, Technical Report TR-2125-OCN, Naval Facilities Engineering and Expeditionary Warfare Center, Port Hueneme CA, December 2000.

National Research Council. (1991). *Assuring the safety of innovative marine structures*. National Research Council.

Remmers, G., & Taylor, R. (1998). Mobile offshore base technologies. In *Proceedings of the Offshore Mechanics and Arctic Engineering Conference (OMAE98)*, Lisbon, Portugal.

Remmers, G., Zueck, R., Palo, P., & Taylor, R. (1998). Mobile offshore base. In *Proceedings of the International Offshore and Polar Engineering Conference (ISOPE'98)*, (Vol. I, pp. 1–5). Montreal, Canada.

Remmers, G., Taylor, R., Palo, P., & Brackett, R. (1999). Mobile offshore base: A Seabasing option. In *Proceedings of the 3rd International Workshop on Very Large Floating Structures (VLFS '99)*, Honolulu, Hawaii, September 22–24, 1999.

Lilypad: Floating Ecopolis for Climatical Refugees

Vincent Callebaut

Abstract This chapter describes the Lilypad, a self-sufficient floating ecopolis model for future climatic refugees. This model serves as a long-term solution to rising water level as forecasted by the *Intergovernmental Group on the Evolution of the Climate*.

1 Editors' Note on the Lilypad

We present the Lilypad project as a means of continuing this conversation on how we might conceptualise the ocean as our new home. In many ways, Vincent Callebaut's *Lilypad* project falls squarely within the body of speculative architecture that seeks to tell a story about the way we may have to live in the future. It continues the architectural fiction vividly brought to life in the works of Archigram, Lebbeus Woods and the exceptional works of the Japanese Metabolist Movement.

C.M. Wang and B.T. Wang

V. Callebaut (✉)
Vincent Callebaut Architectures, Paris, France
e-mail: vincent@callebaut.org

C.M. Wang and B.T. Wang (eds.), *Large Floating Structures*,
Ocean Engineering & Oceanography 3,
DOI 10.1007/978-981-287-137-4_12

2 2100 A.D.: A Large Crowd of Ecological Refugees

As a consequence of anthropogenic activity, the climate will continue to warm up exponentially and ocean levels will increase. According to the principle of Archimedes and contrary to preconceived notions, the melting of the arctic ice-floe will not raise sea levels in much the same way as an ice cube melting in a glass of water will not make its level rise. Instead, the two significant ice reservoirs at Antarctica and Greenland, as well as the continental glaciers that are not on the water, will transfer their volumes towards the oceans and increase sea levels. This will likely be exacerbated by water dilatation under the effect of rising temperatures.

According to the least-alarming forecasts of the *Intergovernmental Group on the Evolution of Climate* (GIEC), the ocean level may rise between 20 and 90 cm throughout the 21st Century, with an average increase of 50 cm (cf. 10 cm in the 20th Century). However, the larger international scientific scene asserts that a temperature elevation of 1 °C may lead to a sea level rise of 1 meter. This increase of 1 m would bring about a loss of land roughly equivalent to 0.05 % in Uruguay, 1 % in Egypt, 6 % in the Netherlands, 17.5 % in Bangladesh and up to 80 % in the Majuro atoll in Oceania (Marshall and Kiribati islands and, incrementally, the Maldives).

This first 1-metre increase will affect 50 million people in the developing world, but a second 1-metre rise in sea levels will bring about a more severe impact. Countries like Vietnam, Egypt, Bangladesh, Guyana or Bahamas will see their most inhabited places swamped at each flood event and their most fertile fields devastated by the invasion of salt water. New York, Bombay, Calcutta, Hô Chi Minh City, Shanghai, Miami, Lagos, Abidjan, Jakarta, Alexandria, are but a few of the cities that will produce more than 250 million climate refugees. Beyond this, an average of 9 % of GDP will be threatened globally without adequate planning and investment to build protections and alternative responses to threat of rising sea levels.

The threat of rising ocean levels makes it imperative that governments move from a mere strategy of reaction to one of adaptation and long-lasting anticipation. It is surprising that the governments of some islands that may well disappear have failed to take the necessary precautions to rethink about how their settlements will appear. More surprising is that people of the developed countries continue to build houses and buildings in these islands which are certain to be flooded in time.

3 Lilypad: A Prototype of a Self-Sufficient Amphibious City

The "Lilypad" design proposal for a floating high-density settlement presents itself as a solution to the loss of low-lying coastal land areas in response to the threat of rising ocean levels.

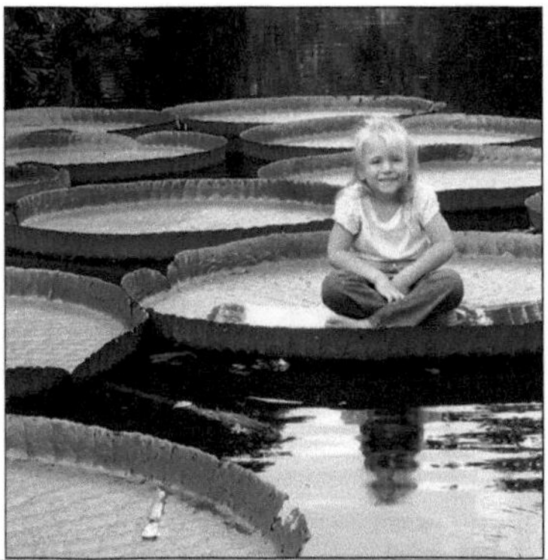

Type	Spontaneous proposal for the principality of Monaco
Client	Principality of Monaco
Programme	Mixed uses for smart cities
Surface area	750,000 m^2
Year	2008
Team	Vincent Callebaut, Philippe Steels and Maguy Delrieu

Whereas the Netherlands and the United Arab Emirates have spent the last decade on projects to "fatten" their beach with billions of euros to build short-lived polders and protective dams, the project "Lilypad" offers a tenable solution to the threat of rising sea levels by capitalizing on the buoyancy of the water, creating a structure that works with the surface of the water regardless of its level. The project serves as an example of how floating settlements such as this floating Ecopolis will not only increase, in a sustainable way, offshore territories of the most developed countries such as the Principality of Monaco , but will address the housing needs of the future such as those from Polynesian atolls.

Given its ability to be mobile, the settlement allows a new biotechnological prototype of ecological resilience dedicated to nomadism and the urban ecology in the sea. It is envisaged that some iterations of Lilypad could potentially travel on the currents of the oceans, from the equator to the poles following the marine streams such as the warm Gulf Stream or cold Labrador stream of current (as shown in Fig. 1).

The project operates as an amphibious half-aquatic and half-terrestrial city, and anticipates being able to accommodate 50,000 inhabitants along with the biodiversity to develop its fauna and flora around a central lagoon of soft water that collects and purifies rainwater falling on its superstructure. This artificial lagoon is entirely immersed, thus ballasting the city. It enables inhabitants to also live in the heart of the subaquatic depths.

The multifunctional programming is based on three marinas and three 'mountains' zoned for office space, retail, and entertainment. The entire superstructure is covered by a stratum of planted housing in suspended gardens and crossed by a network of streets and alleyways with an organic outline. The emphasis of this is to create a harmonious coexistence with the surrounding natural

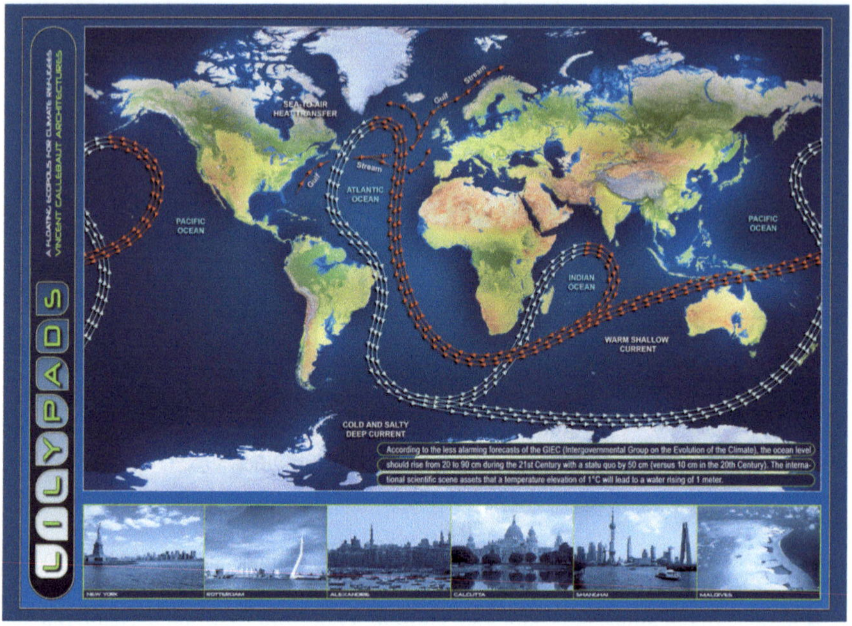

Fig. 1 Map of sea level rise

environment and to explore new modes of living beside and within the sea by building fluid and collective spaces in proximity. This will permit the creation of spaces of social inclusion suitable for all inhabitants—denizen or foreign-born, recent or old, young or aged people—to congregate and thrive.

The structural elements that 'branch' to create the surface and structure of the Ecopolis are directly inspired by the highly ribbed leave of the great lilypad (see Fig. 2) of the *Amazonia Victoria Regia*. Coming from the family of Nympheas, this aquatic plant with exceptional plasticity was discovered by the German botanist Thaddeaus Haenke and dedicated to the Queen Victoria of England in the 19th Century. The double skin is made of polyester fibres covered by a layer of titanium dioxide (TiO2) like an anatase, which reacts to the ultraviolet rays that allows the absorption of atmospheric pollution through photocatalytic action. Entirely sufficient, Lilypad takes up the four main challenges launched by the OECD in March 2008: climate, biodiversity, water and health. The floating Ecopolis will also be in a positive energy balance with zero carbon emission through the integration of all the renewable energies (solar, thermal and photovoltaic energies, wind energy, hydraulic, tidal power station, osmotic energies, phytopurification, biomass) allowing the production of more energy that it then consumes.

It is anticipated that this floating Ecopolis will adopt cutting-edge building services to allow the building to serve its surrounding oceanic ecosystems by producing and softening oxygen and electricity; recycling carbon dioxide and

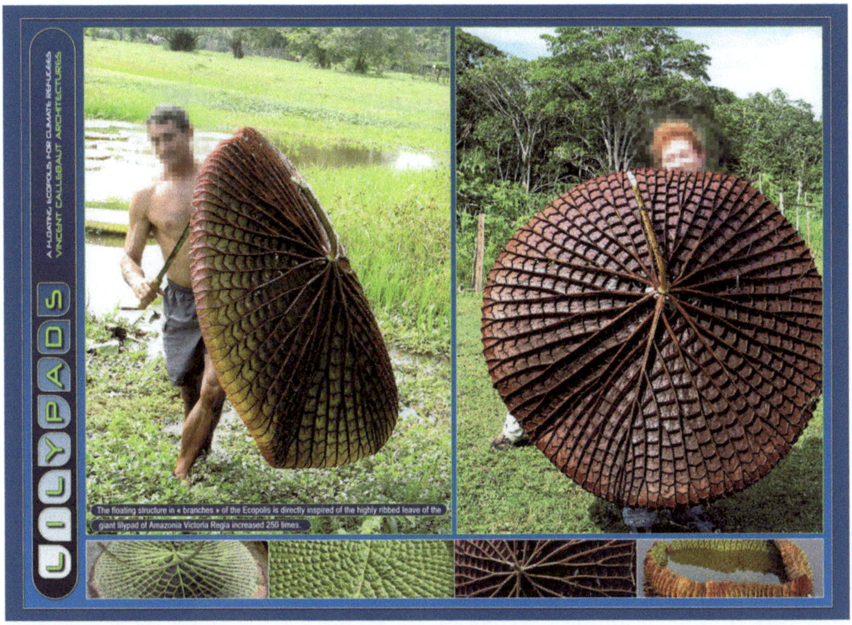

Fig. 2 Concept of Lilypad

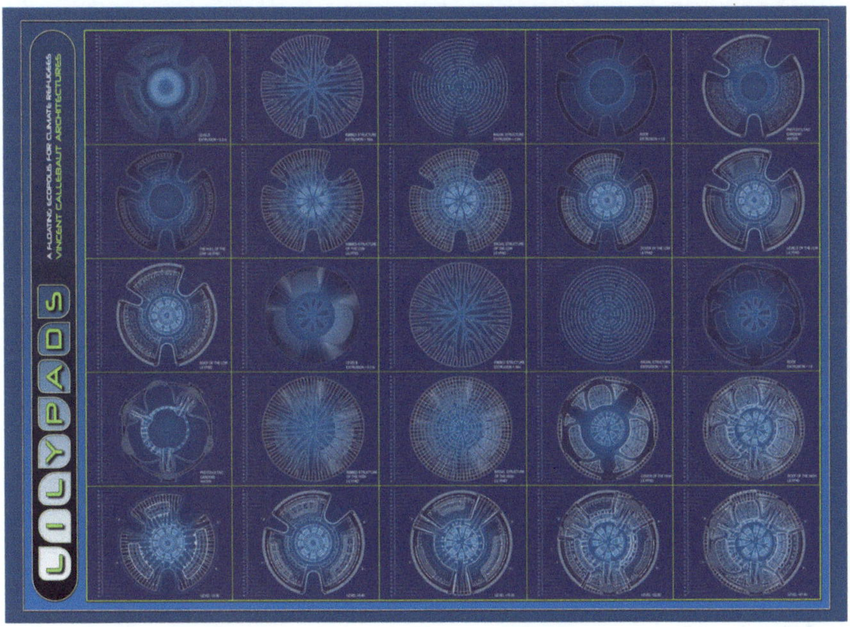

Fig. 3 Structural forms of Lilypad

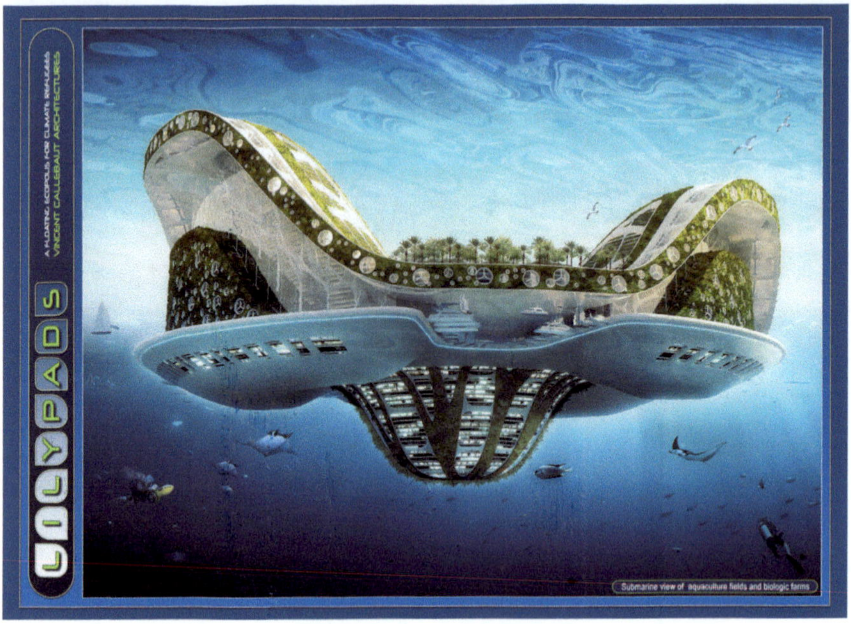

Fig. 4 Underwater view

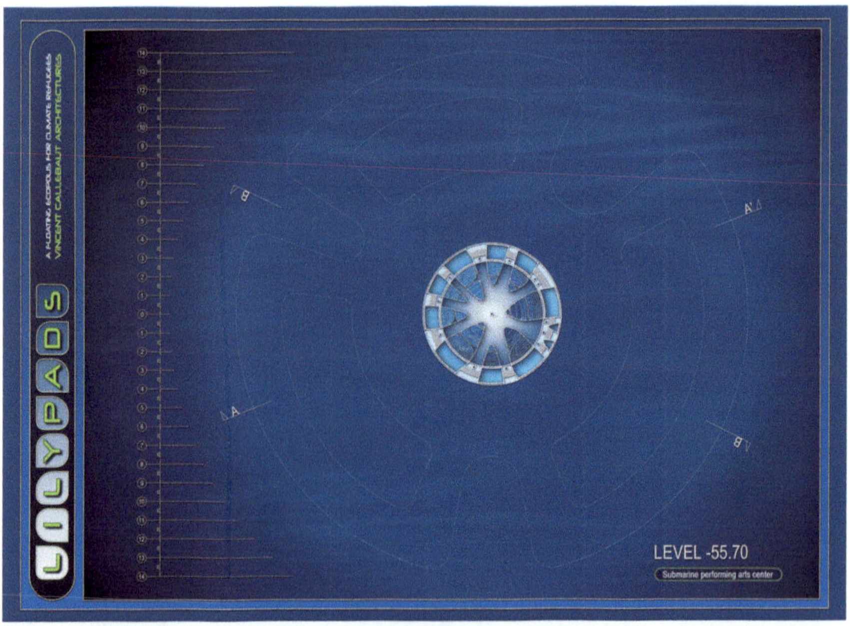

Fig. 5 Level –55.70

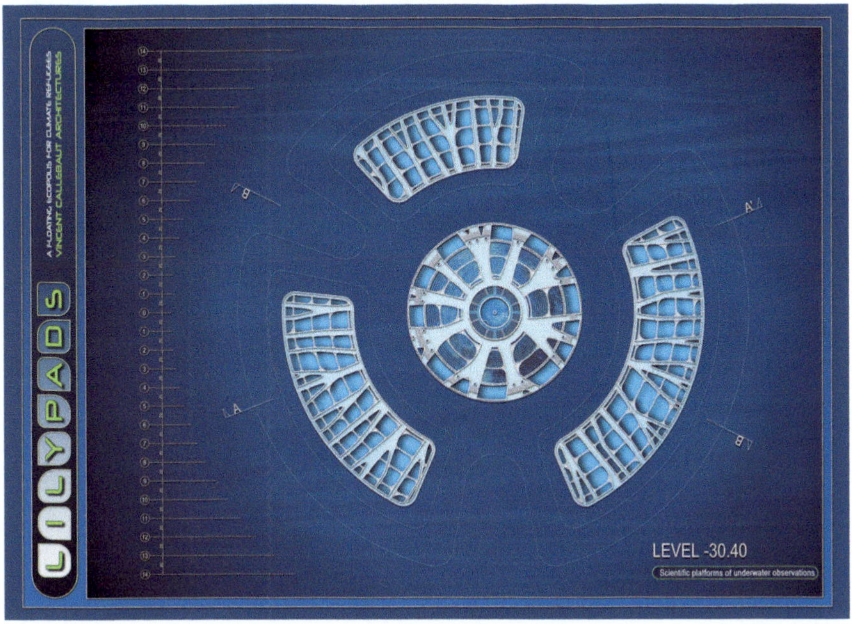

Fig. 6 Level –30.40

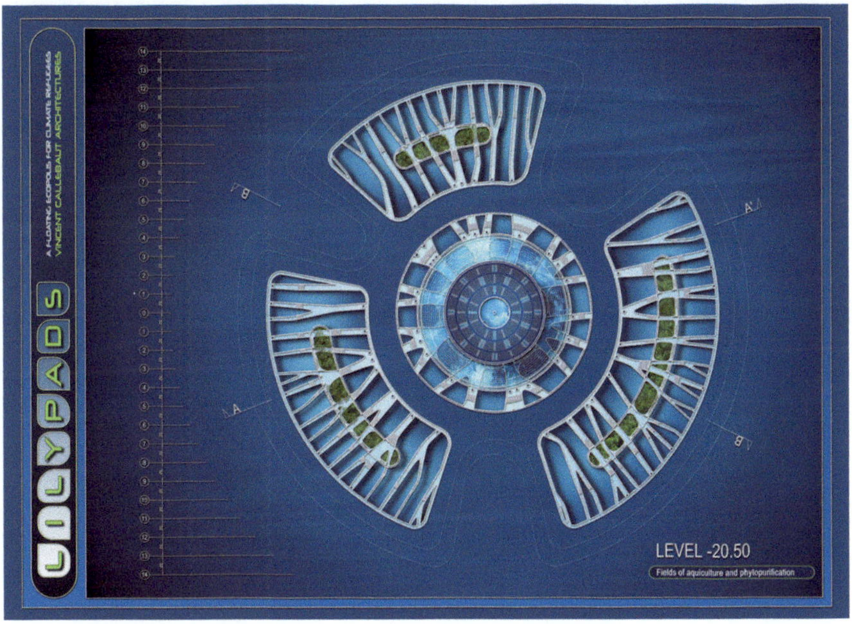

Fig. 7 Level –20.50

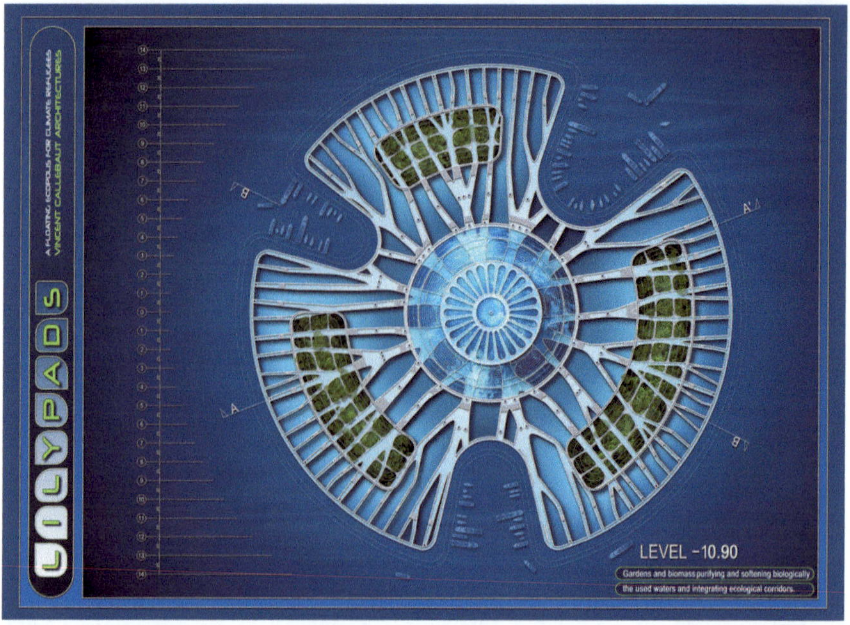

Fig. 8 Level –10.900

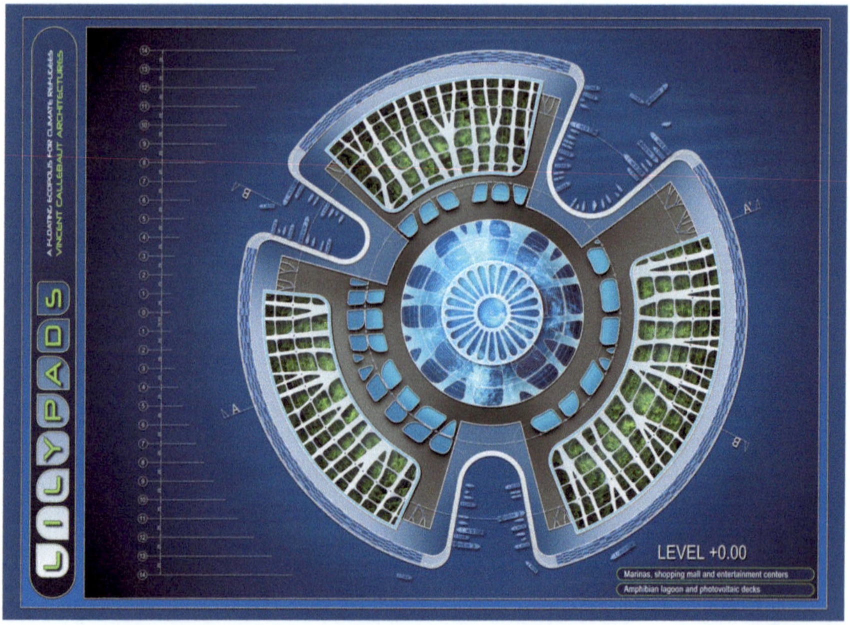

Fig. 9 Level 0.00

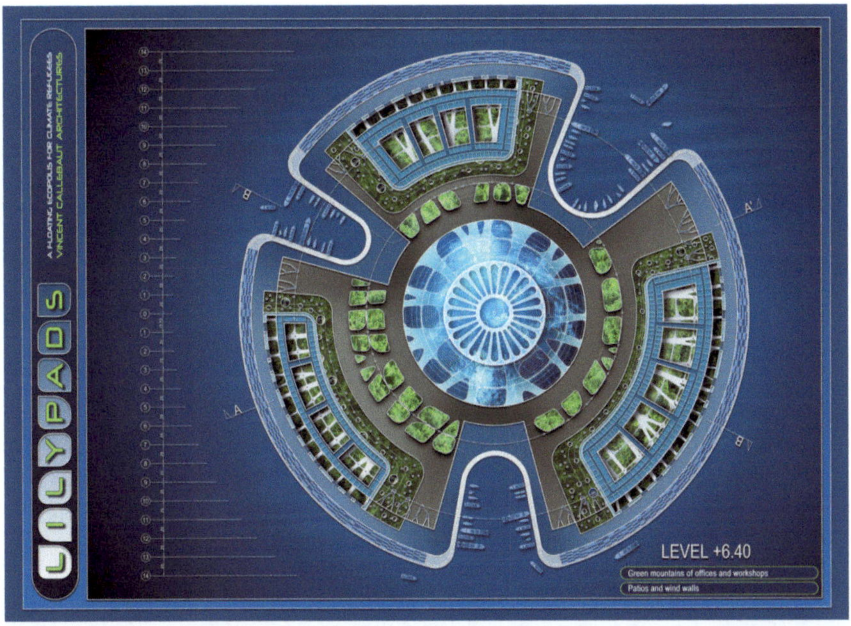

Fig. 10 Level +6.40

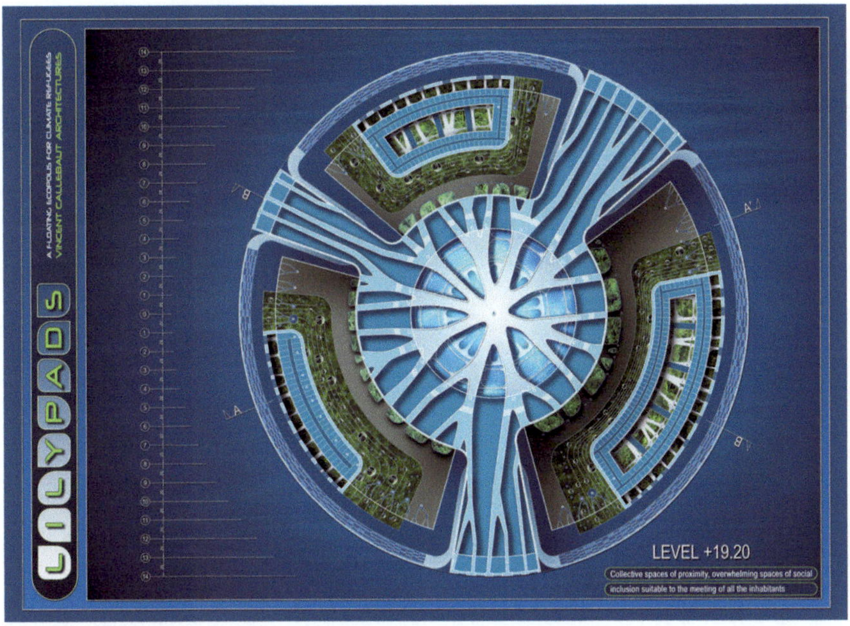

Fig. 11 Level +19.20

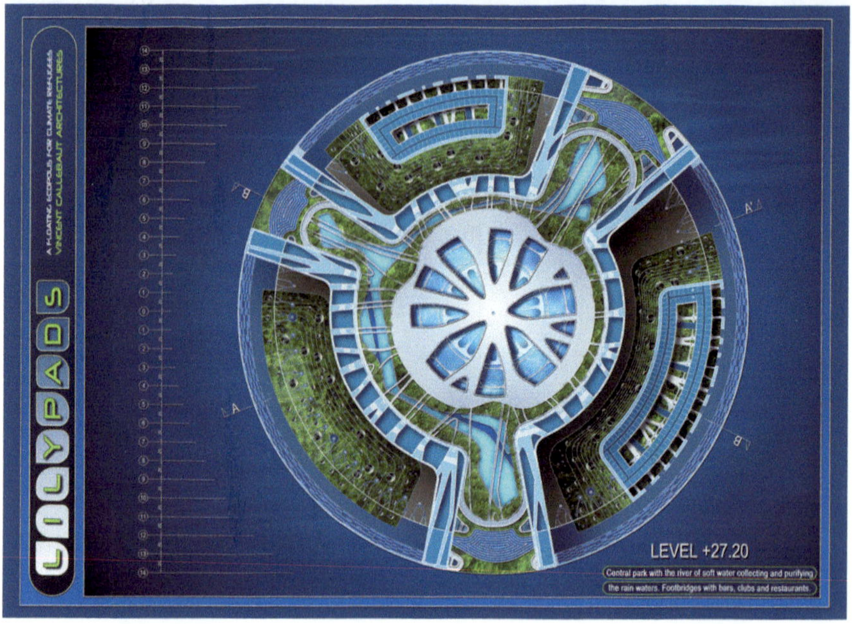

Fig. 12 Level +27.20

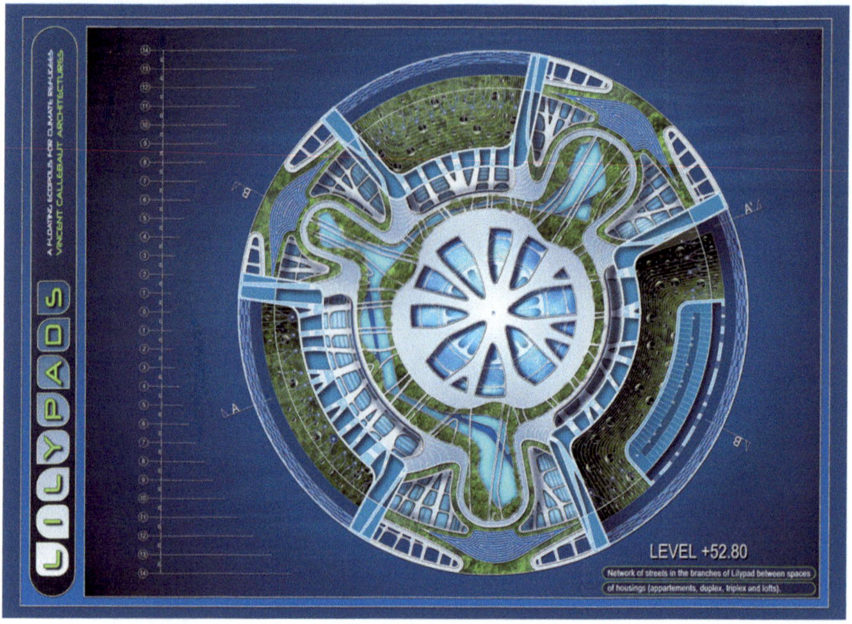

Fig. 13 Level +52.80

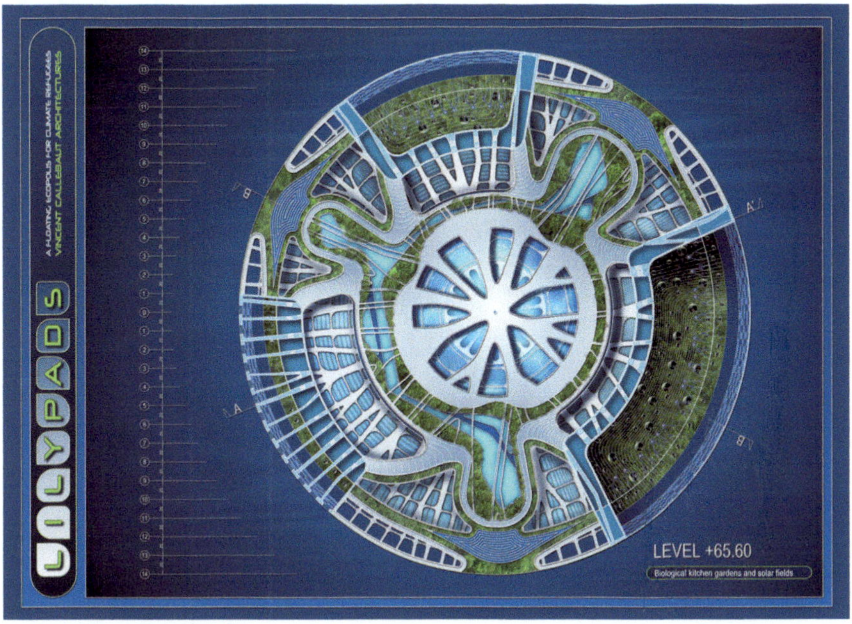

Fig. 14 Level +65.60

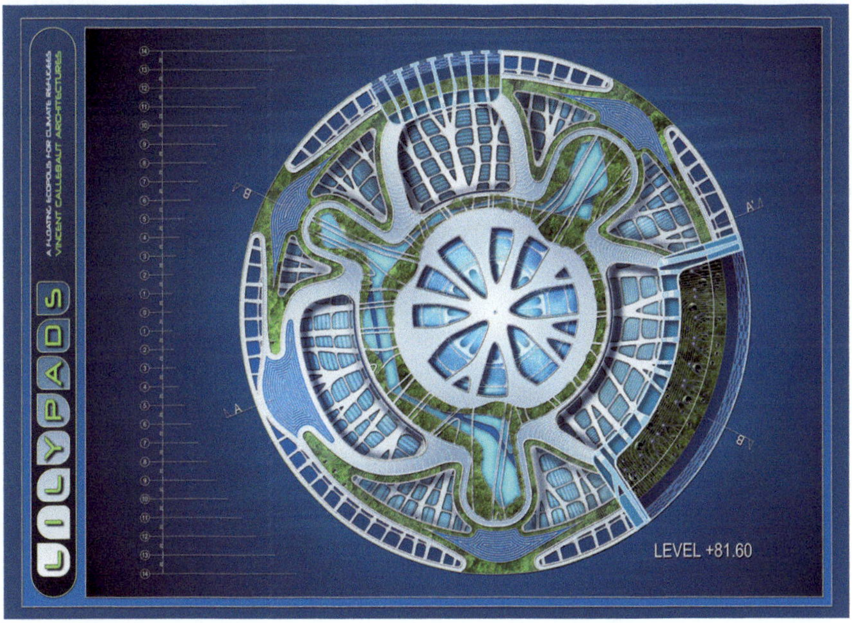

Fig. 15 Level +81.60

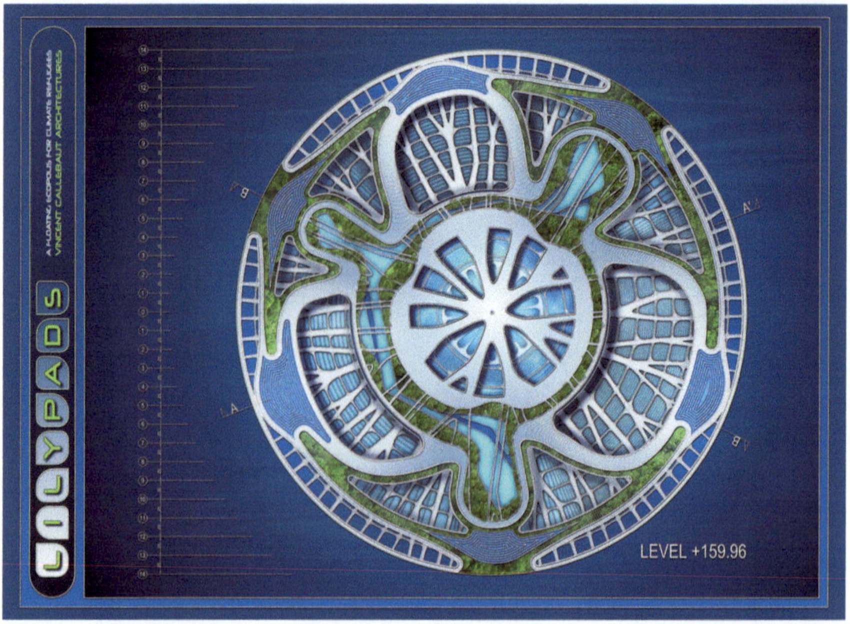

Fig. 16 Level +159.96

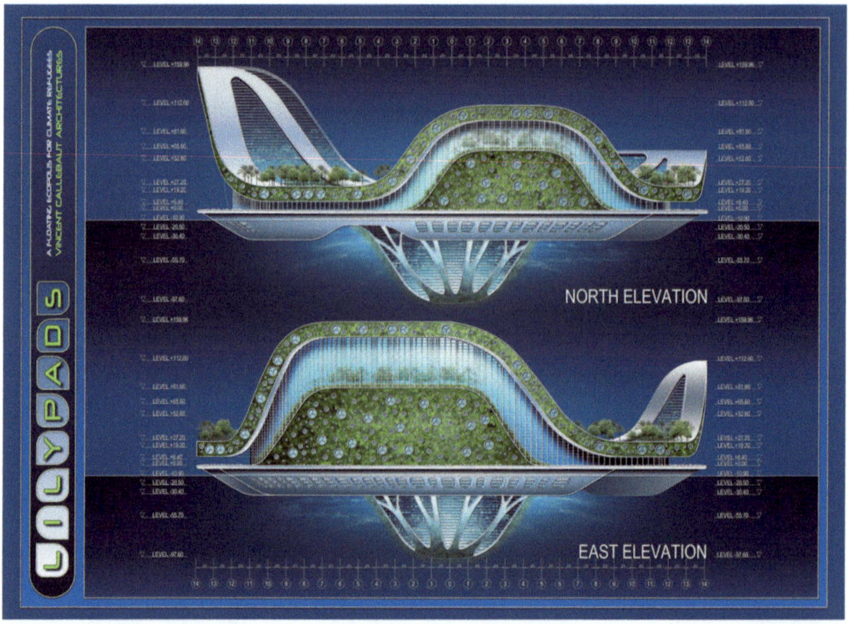

Fig. 17 North elevation and East elevation

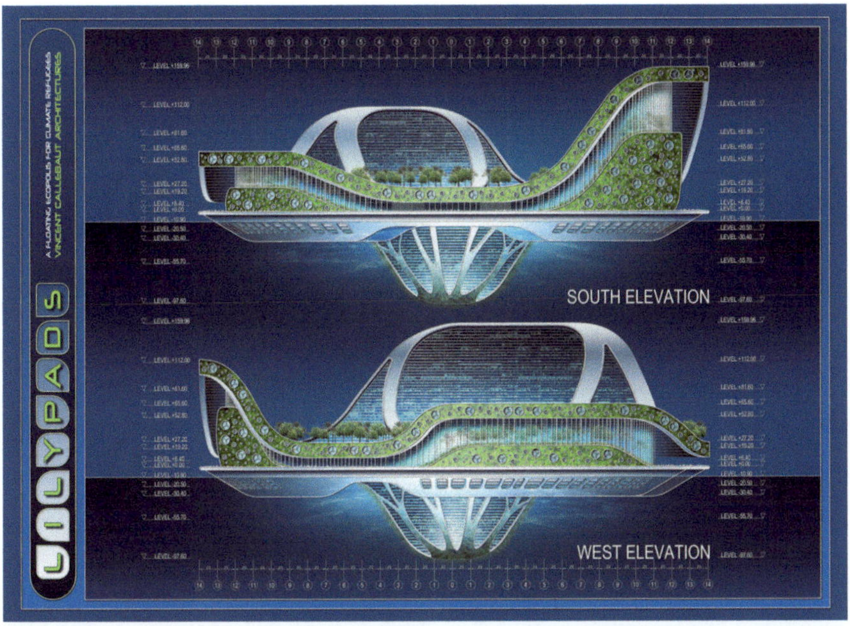

Fig. 18 South elevation and West elevation

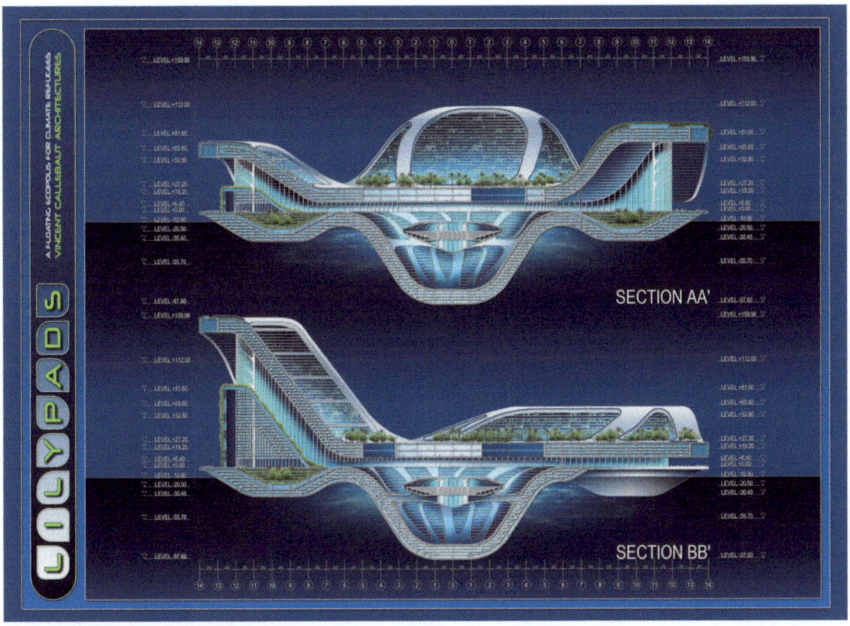

Fig. 19 Section AA' and section BB'

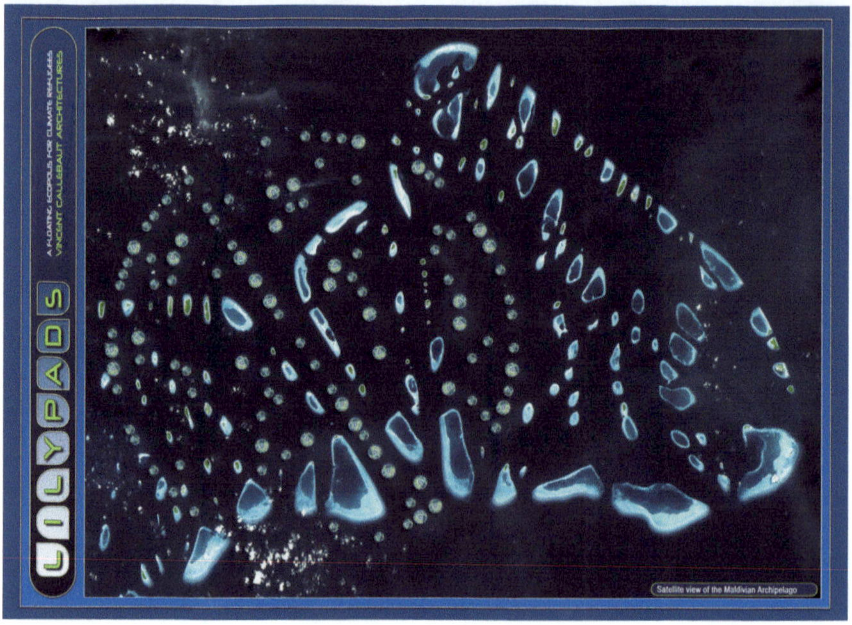

Fig. 20 Satellite view of the Maldivian Archipelago

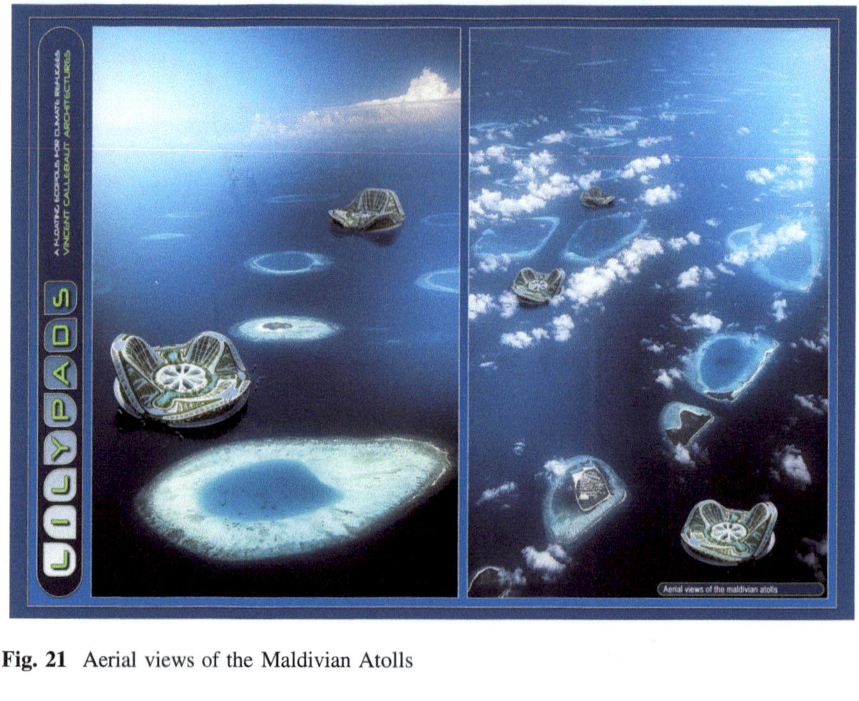

Fig. 21 Aerial views of the Maldivian Atolls

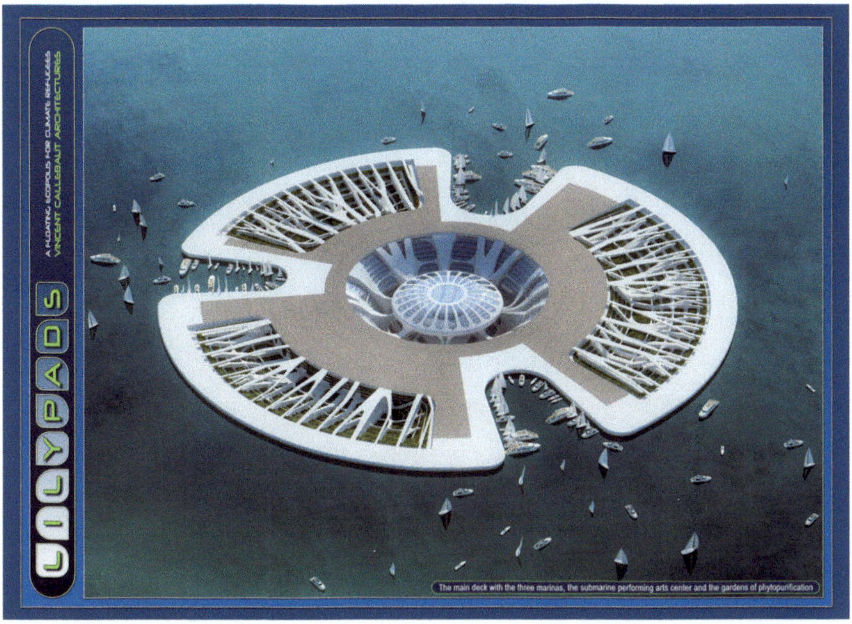

Fig. 22 Axonometry A

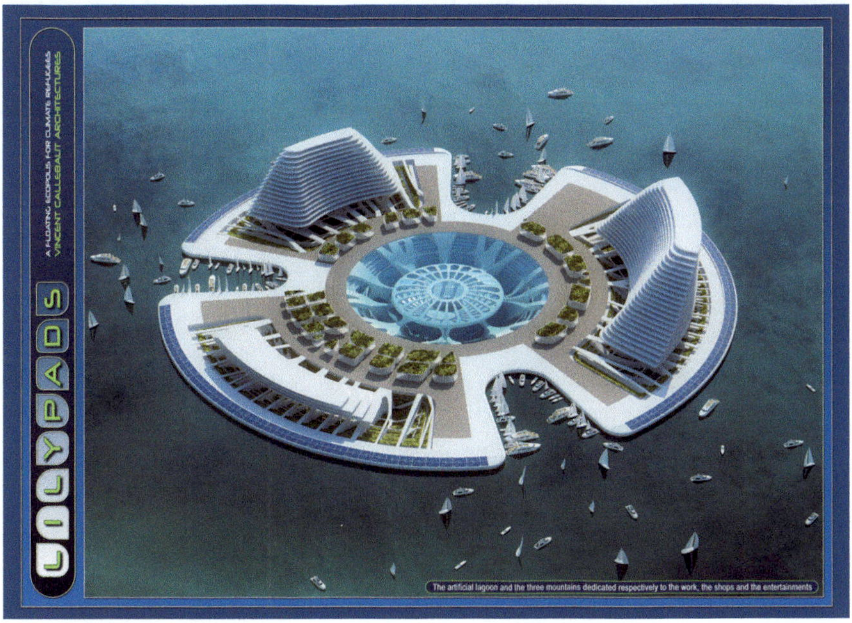

Fig. 23 Axonometry B

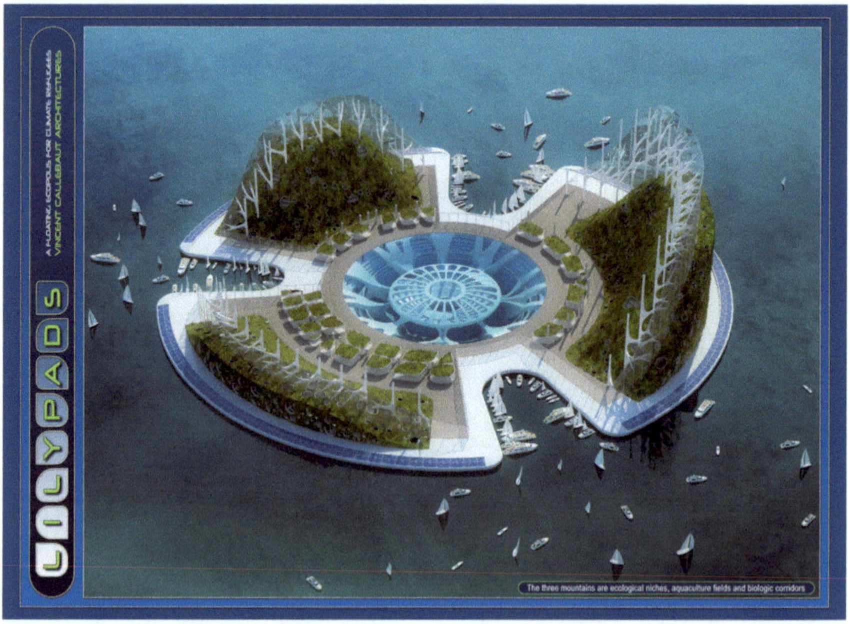

Fig. 24 Axonometry C

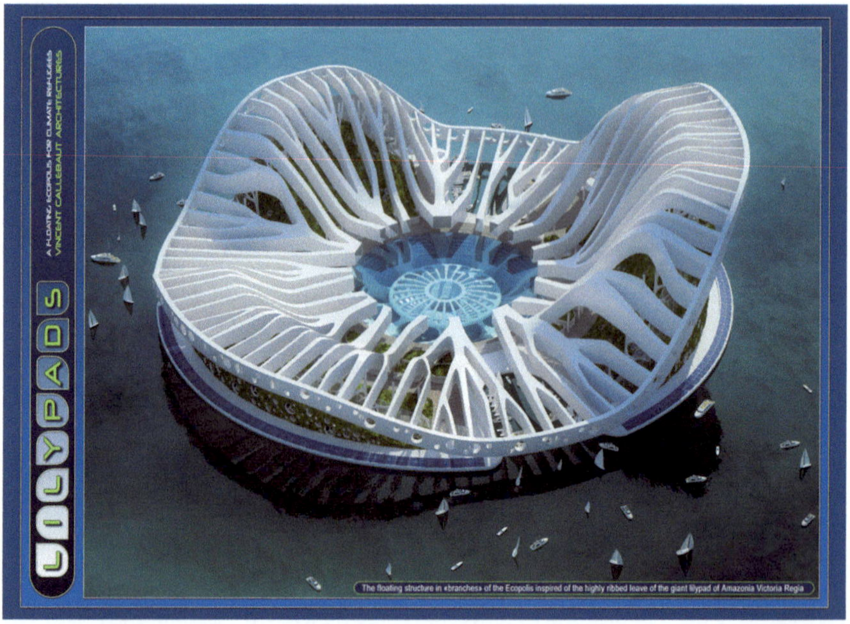

Fig. 25 Axonometry D

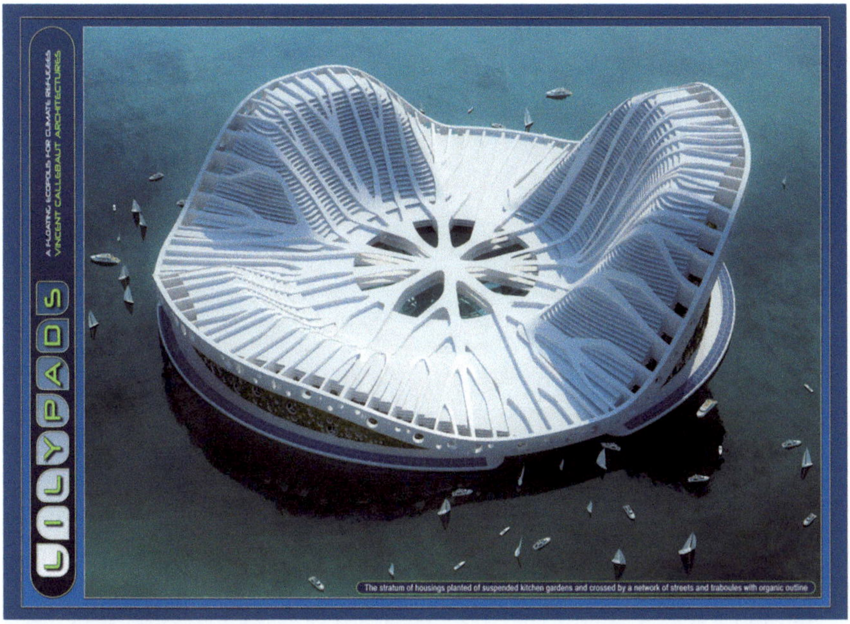

Fig. 26 Axonometry E

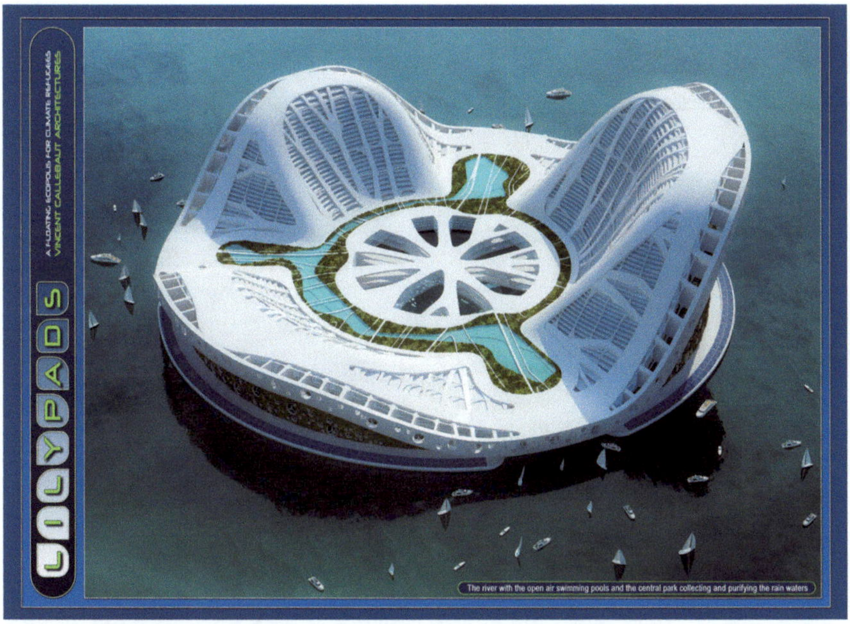

Fig. 27 Axonometry F

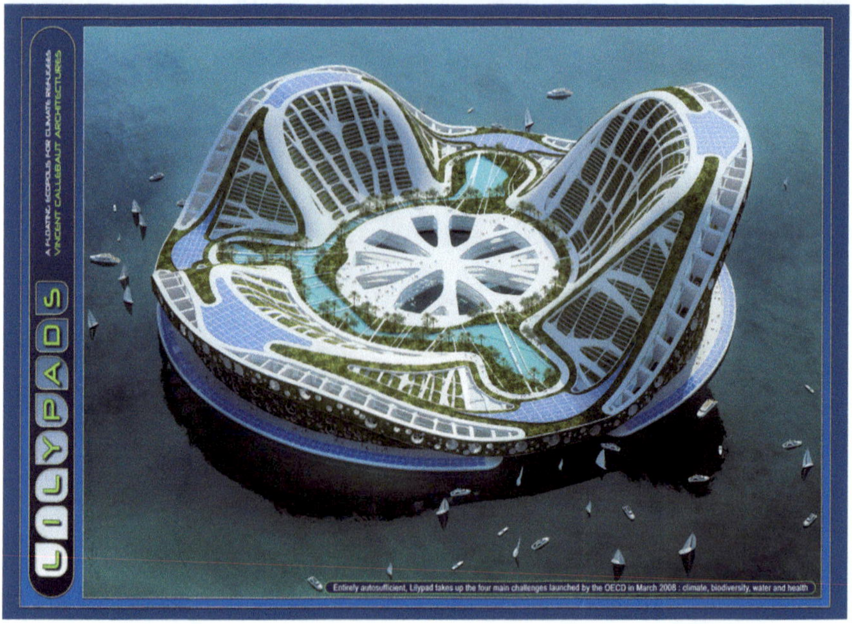

Fig. 28 Axonometry G

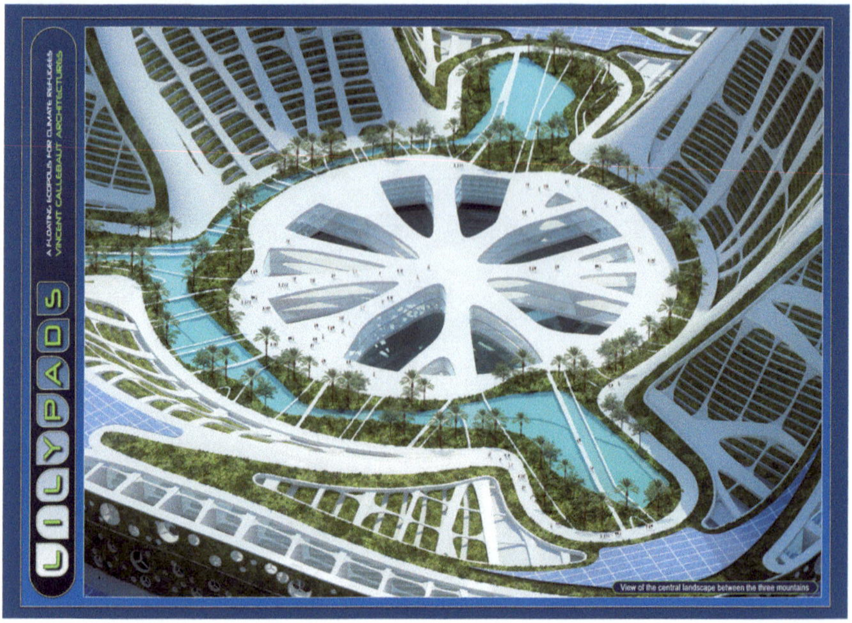

Fig. 29 Zoom on the central landscape

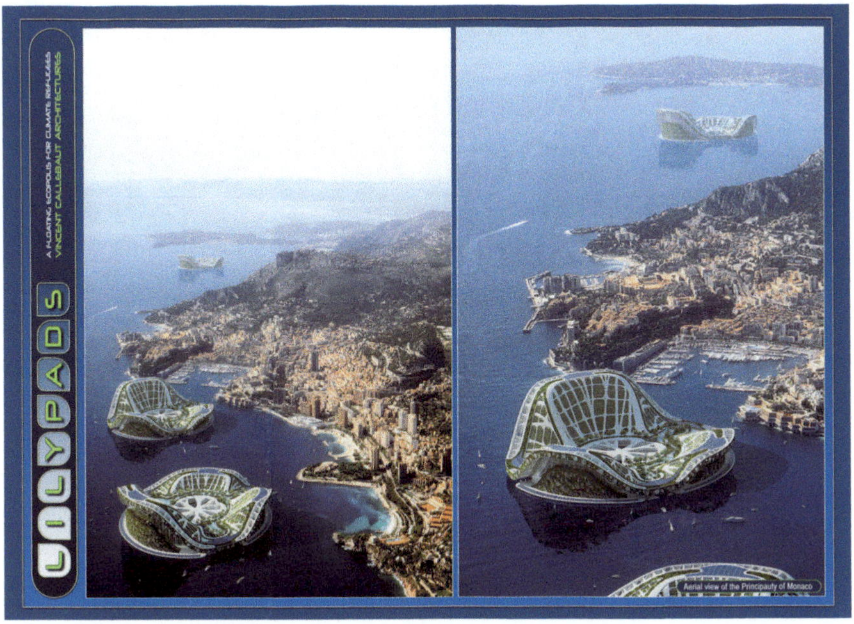

Fig. 30 Aerial view of Monaco Bay

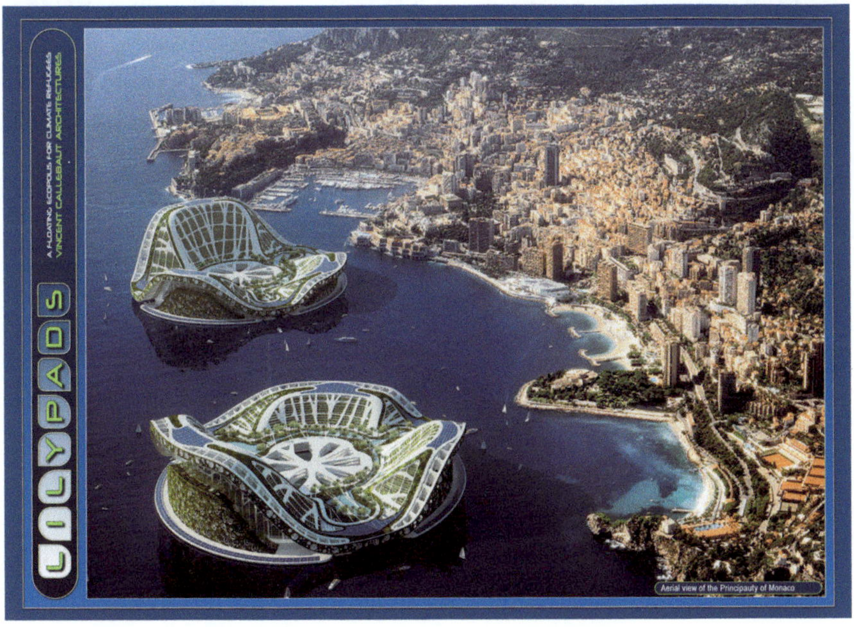

Fig. 31 Aerial view of the principality of Monaco

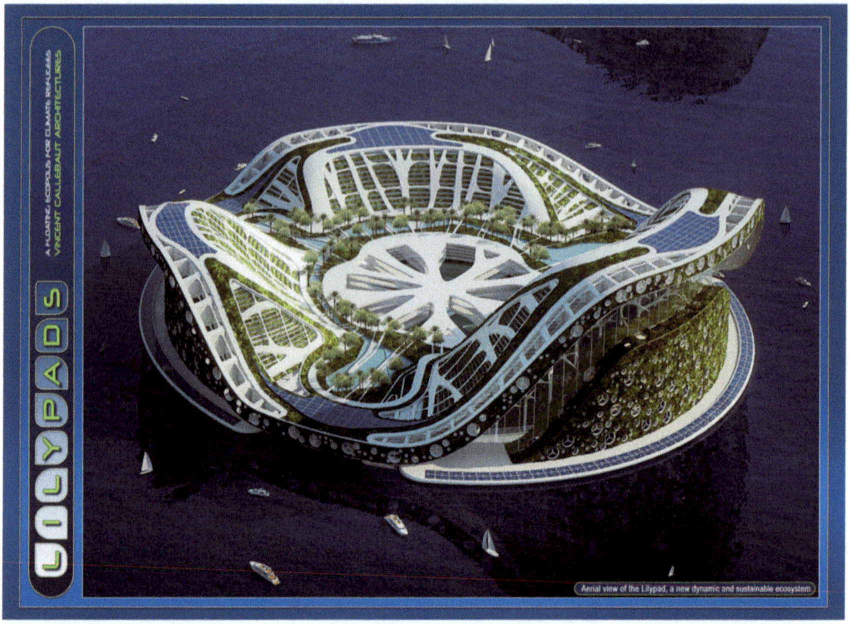

Fig. 32 A new ecosystem

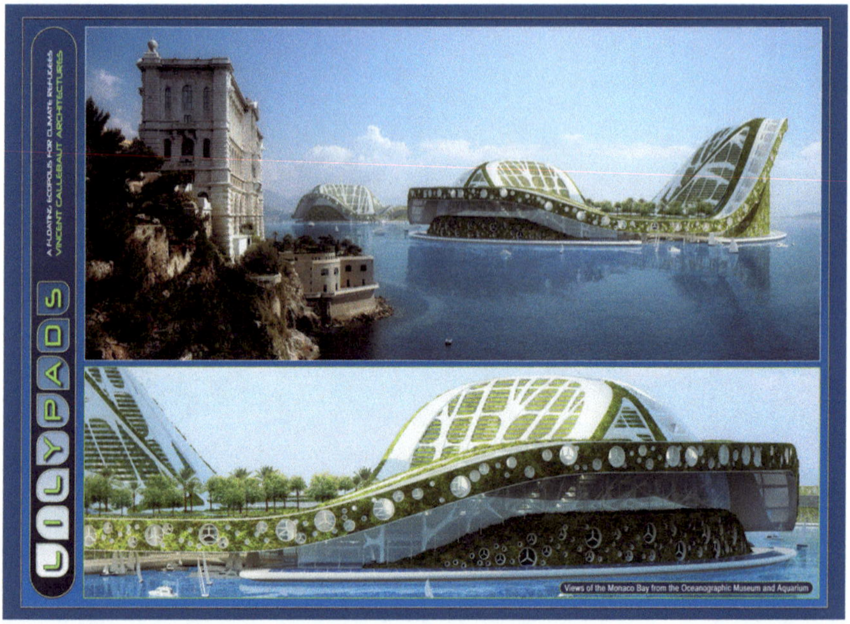

Fig. 33 A view of the Monaco Bay

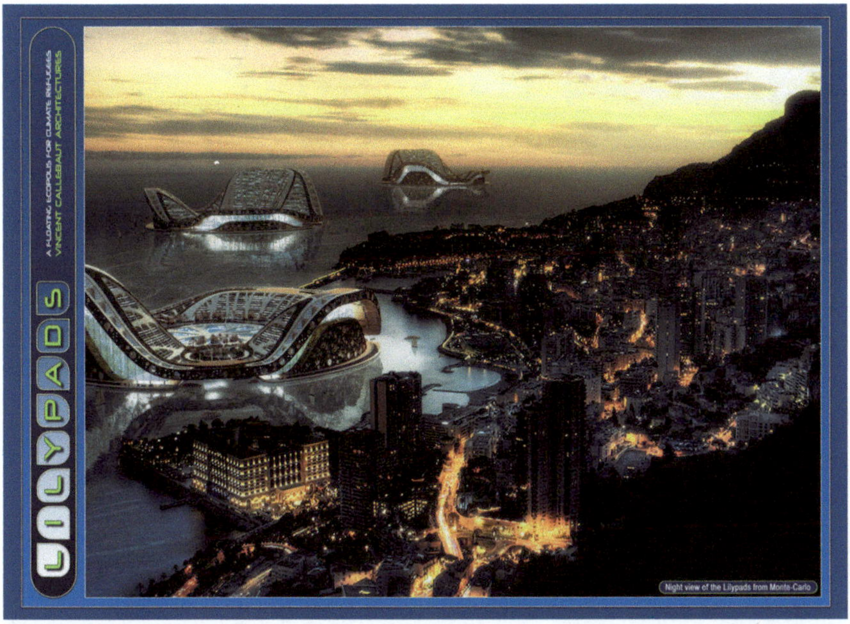

Fig. 34 Night view from Monte Carlo

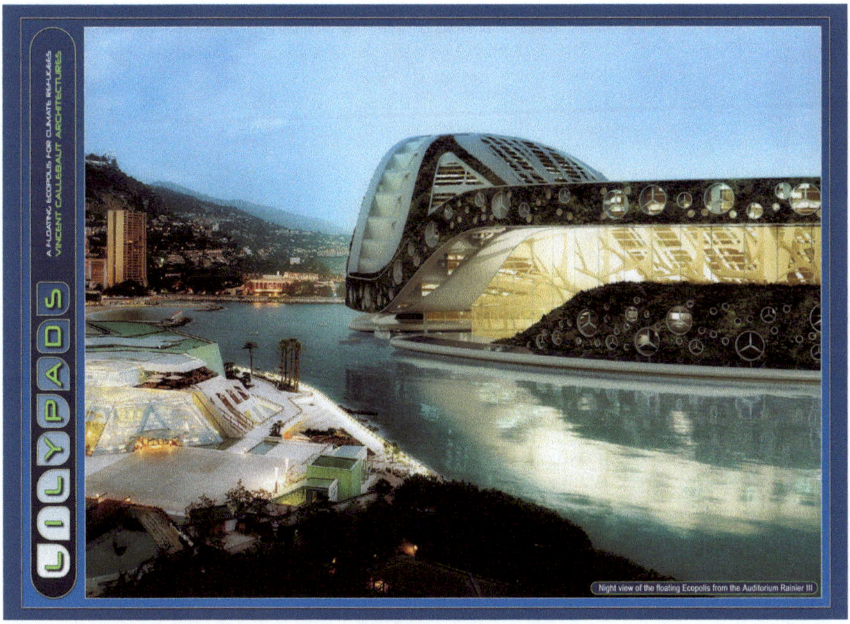

Fig. 35 Night view from the auditorium Rainner III

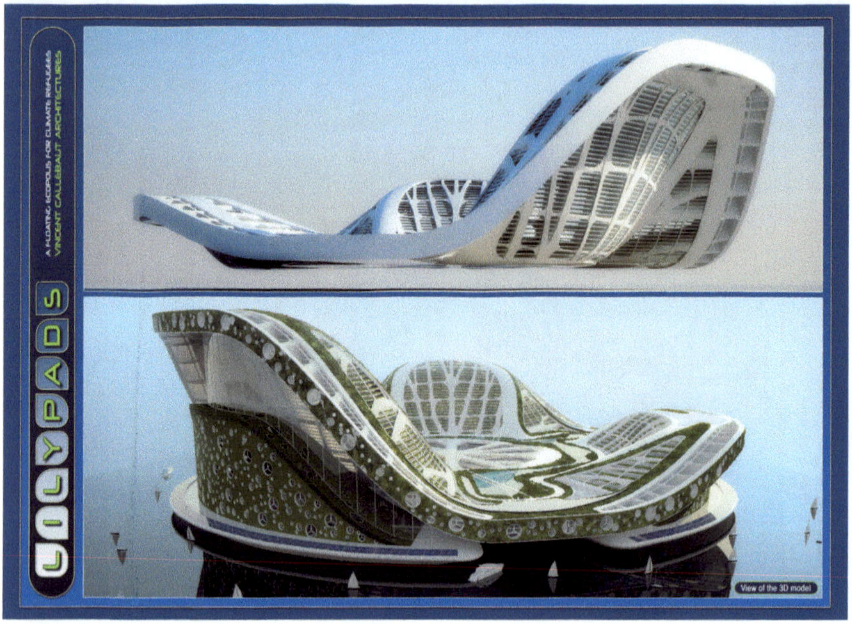

Fig. 36 Model studies

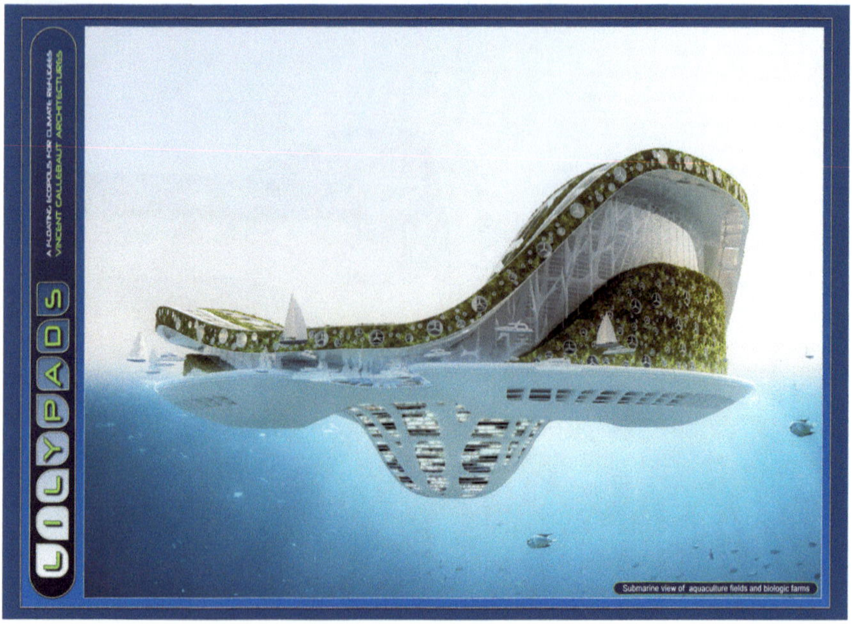

Fig. 37 Submarine view

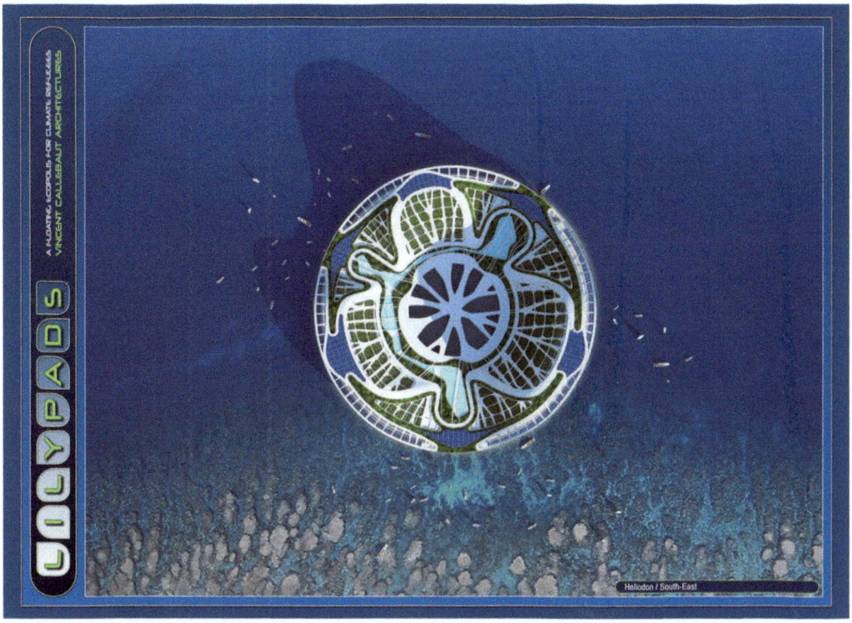

Fig. 38 Heliodon south east

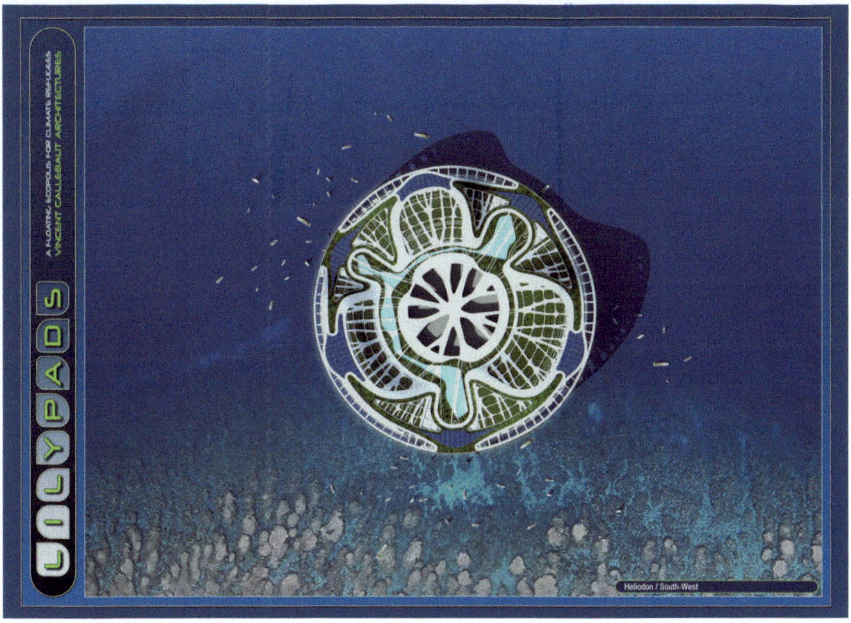

Fig. 39 Heliodon south west

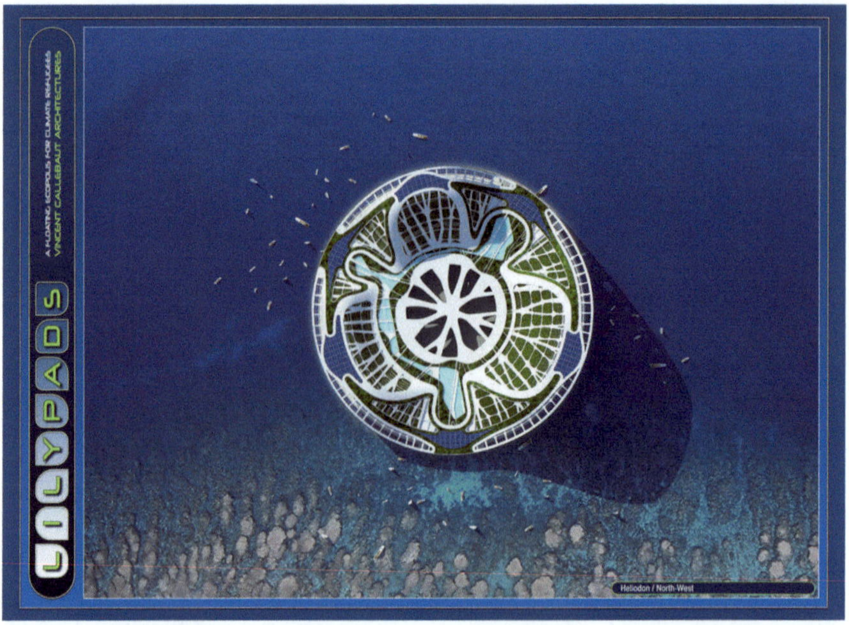

Fig. 40 Heliodon north west

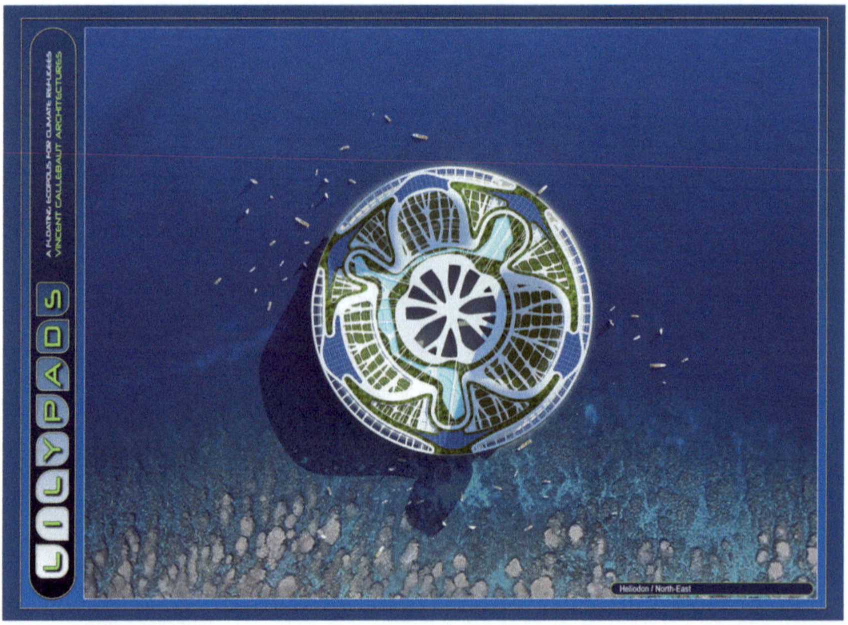

Fig. 41 Heliodon north east

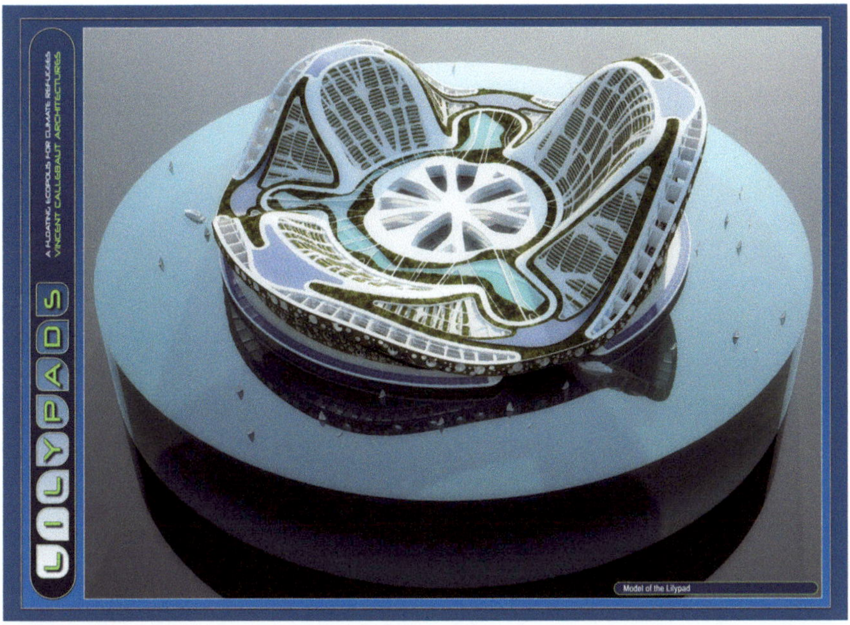

Fig. 42 Model

waste; purifying and softening biologically waste water; and integrating ecological niches, aquaculture fields and biotic corridors on and under its body to meet its own food needs.

Figures 3, 4, 5, 6, 7, 8, 9, 10, 11, 12, 13, 14, 15, 16, 17, 18, 19, 20, 21, 22, 23, 24, 25, 26, 27, 28, 29, 30, 31, 32, 33, 34, 35, 36, 37, 38, 39, 40, 41 and 42 show the level plans, elevation plans and the models of the Lilypad.

MIX
Papier aus verantwortungsvollen Quellen
Paper from responsible sources
FSC® C105338

If you have any concerns about our products,
you can contact us on
ProductSafety@springernature.com

In case Publisher is established outside the EU,
the EU authorized representative is:
Springer Nature Customer Service Center GmbH
Europaplatz 3, 69115 Heidelberg, Germany

Printed by Libri Plureos GmbH
in Hamburg, Germany